Water Science and Application 2

Land Use and Watersheds

Human Influence on Hydrology and Geomorphology in Urban and Forest Areas

---

Mark S. Wigmosta
Stephen J. Burges
Editors

American Geophysical Union
Washington, DC

Land Use and Watersheds: Human Influence on Hydrology and Geomorphology in Urban and Forest Areas
Water Science and Application 2

Library of Congress Cataloging-in-Publication Data
Land use and watersheds: human influence on hydrology and geomorphology in urban and forest areas / Mark S. Wigmosta, Stephen J. Burges, editors
p. cm.-- (Water science and application; 2)
Includes bibliographical references (p.).
ISBN 0-87590-351-7
1. Hydrology. 2. Urbanization. 3. Forest management. I. Wigmosta, Mark S. II. Burges, Stephen J. III. Series

GB661.2.L35 2000
551.48--dc21

2001022667

ISBN 0-87590-351-7
ISSN 1526-758X

Printed in the United States of America.

CONTENTS

PREFACE

Considerable basic and applied research has been and continues to be focused on understanding and evaluating the influence of land use changes, such as urbanization or forest management, on watershed hydrology and geomorphology. The motivation for the research is to develop land use policies that minimize adverse impacts and maintain the biodiversity and sustainability of human influenced ecosystems.

In this monograph we present recent data and interpretations from a variety of approaches on how forest management and urbanization have influenced the hydrologic and geomorphic responses of watersheds. Included are field studies of paired experimental/manipulated watersheds, plot studies, and spatially distributed models. While much of the material within the monograph concerns watersheds in a defined locale, the Pacific Northwest, the methods and management approaches also provide guidance for land use in other regions with similar hydro-climactic-geomorphic-ecological settings.

The papers in this monograph address three important issues—human use of forests, hydrological and geomorphic issues in urban areas, and wetland restoration—that are at the interface of science, societal needs, and public policy. They illustrate application of the current state of scientific understanding and show the significant progress being made in applying hydrologic and geomorphic theory, and in the use of new spatial data, particularly digital topography. These advances also demonstrate the need for carefully designed monitoring programs for hypothesis testing. The papers illustrate why the application of models for decision-making require intelligent interaction between the model and the operator/interpreter. The papers are indicative of the amount of effort that is required to apply planning and environmental decision-making models wisely and well. The societal need for such models is evident by the growing number of relatively inexperienced users who access the worldwide web in search of appropriate environmental decision-making tools.

The monograph is divided into two sections, each with distinct introductions. The first section covers the influence of urbanization on hydrologic and geomorphic processes. The four papers that comprise this section discuss wetland processes, channel disturbance, changes in hydrology, and susceptibility to landslides in the urban setting. The second section features discussion on the influence of forest management activities on hydrologic-geomorphic processes. More specifically, these seven papers in total consider the effects of timber harvesting and road construction on streamflow, sediment yield, and the occurrence of landslides.

The monograph derives from a special session held at the AGU 1997 Fall Meeting in San Francisco, California: "Impacts of Land Use on the Hydrologic-Geomorphic Responses of Watersheds I & II." In addition to seven papers selected from that special session, we invited other colleagues to broaden the scope of the subject matter with four additional papers.

We are indebted to the many reviewers who gave freely of their time. We also thank contributing authors for the time, effort, and expertise that they brought to the development and completion of their book chapters.

Mark S. Wigmosta and Stephen J. Burges

Section 1

Influence of Urbanization on Hydrology and Geomorphology

Introduction to Section 1—Land Use Changes in the Urban Setting

Stephen J. Burges

The four papers in this section discuss wetland processes, channel disturbance, changes in hydrology, and susceptibility to landslides in the urban setting. In many regions the urban setting is the most heavily human influenced ecosystem. Efforts to preserve biocomplexity in lands that are converted to high density urban use are often focused on small parts of the landscape including riparian corridors, relatively small natural areas, wetlands, and stream channels. In locations with few wetlands, stream channels occupy about one percent of the land form during non-flood conditions. In wetland regions, the fraction of water-covered landscape is larger. If the landscape was initially densely forested and the forest largely removed for urban land use, the entire terrestrial and aquatic ecosystems also change. The pre-development spatial and temporal partitioning of rainfall (or snow) to evaporation, transpiration, surface flow, subsurface flow and to recharge deeper groundwater changes substantially when the vegetation is removed, the soil disturbed, and the infrastructure to support the urban environment is put in place. It is the changes in the spatial patterns and timing of water movement that pose most of the problems of concern that are addressed in the four papers.

In addition to major changes in biogeochemical cycling and the form and distribution of vegetation in landscape engineered for dense urban use, other major changes occur: water is imported and redirected from place to place for human use and natural hillslope water movement and stream flow patterns are completely altered. Questions concerning flood damage mitigation, the temporal and spatial availability of fresh water for human use, and how much water should be available for ecology drive most aquatic environmental efforts. Changed hillslope flow paths and flow fluxes, and the altered distribution of vegetation, substantially influence wetland ecology, stream ecology, and fluvial and hillslope geomorphology. The latter is of particular concern for land stability and landslide hazard estimation and prediction. Most urban land use decisions are directed toward providing infrastructure subject to topographical and hazard mitigation considerations and minimizing adverse environmental influences. Most fresh water aquatic ecosystem environmental efforts in urban regions focus on preservation of, or construction of, wetlands and water bodies, and preservation of streams and stream riparian corridors. The four papers address practical aspects of these efforts as well as landslide hazard prediction.

Land Use and Watersheds: Human Influence on Hydrology and Geomorphology in Urban and Forest Areas
Water Science and Application Volume 2, pages 3-4

An ideal set of papers would cover a wide selection of urban configurations in a variety of hydro-climactic-geomorphic-ecological settings. It has been the collective experience of the authors of the four papers presented here that it takes a long time to reach partial understanding of a relatively small region. The first of the papers by Thom et al. draws general material about wetlands from numerous locations in North America, and provides some summary findings from wetland restoration efforts in the Portland Oregon area. A more extensive discussion concerns the Puget Sound Region of western Washington, where a major effort to characterize wetlands in urban settings has been made. The findings summarized by Thom et al. serve as a cautionary note to those who may view wetlands simply in terms of flood attenuation or for removal of pollutants. Mitigation wetlands in urban areas generally have different species and hydrologic characteristics than do naturally occurring wetlands. Naturally occurring wetlands in urban areas may experience changes in hydrologic characteristics, water quality, soil conditions, vegetation, invertebrates, amphibians, birds, and mammals.

The second paper by Booth and Henshaw describes efforts to track rates of erosion in small urban channels in the Puget Sound Lowlands, with findings important for urban land use planning. They define characteristics of channels with the greatest susceptibility to erosion that can be used to reduce the impacts from future urban development on natural stream systems. They also find that channel changes can occur "so rapidly after development begins that remediation, to be effective, must occur prior to development." The issue of data paucity is amply demonstrated in their work. The data needed to track channel erosion over time to elucidate possible causes and effects are indeed scarce. Their paper highlights and reinforces concerns raised by Leopold (1982) about the importance of relevant field data, and the relative lack of attention that the research community has given to such data.

The third paper by Jackson et al. addresses the complex social and technical tradeoffs in the development of regulatory practice for mitigating deleterious hydrologic influences of urbanization. They report on the development and implementation of urban hydrologic design standards for King County, Washington, the twelfth most populous county in the United States. This work required detailed assessment of precipitation records, development of continuous stream hydrograph threshold guidelines, education of the professional engineering and land development communities in newly required continuous simulation hydrology methods, and acceptance of the overall approach by the County Council. It took a decade from the first formulation of the approach to its implementation as a County Ordinance.

The fourth paper by Montgomery et al. focuses on landslide hazard mapping and landslide location prediction. The authors had an unusual opportunity to use records of street addresses kept by the City of Seattle where landslides have occurred since 1890 (these records did not discuss the extent of the landslide). Such spatial and temporal data permitted them to test a topographically based shallow landslide prediction model developed previously for non-urban landscapes. They concluded that their model could aid in identifying landslide hazard zones despite the complexity of the hydrology of the altered landscape.

The papers are by no means exhaustive. While they deal primarily with a limited region of North America, they address complex issues at the interface of science, societal needs, engineering, data collection, and public policy. They are indicative of the state of applications of scientific understanding to some problems of major concern to society. The papers are largely about environmental decision making under uncertainty where data limitations make it difficult to sharpen assessment of environmental risks or uncertainties. In all cases the need for improved data and sharpened quality assurance and quality control (QAQC) of data networks is evident. The difficulties that the Morgan Engineering Company encountered with rain data QAQC and the decisions they made in accepting or rejecting data when designing the dams and channel works of the Miami Conservancy District in Ohio circa 1914 (Morgan, 1951) have modern counterparts in this set of papers. Our data are precious.

REFERENCES:

Leopold, Luna, B. 1982 "Field Data: The Interface Between Hydrology and Geomorphology", pp. 105-108, in *Scientific Basis of Water-Resource Management,* National Academy Press.

Morgan, Arthur, E. 1951. *The Miami Conservancy District*, McGraw Hill, New York.

Influence of Urbanization on Ecological Processes in Wetlands

Ronald M. Thom and Amy B. Borde

Battelle Marine Sciences Laboratory, Sequim, Washington

Klaus O. Richter

Department of Natural Resources, King County, Washington

Lyle F. Hibler

Battelle Marine Sciences Laboratory, Sequim, Washington

Wetlands provide a wide variety of ecological functions and services critical to the overall operation of many ecosystems. These functions include water quality improvement, the attenuation of floodwater, groundwater recharge, cycling nutrients and processing carbon at high rates, providing habitat for fish and wildlife, and increasing the overall biodiversity because of the connection to both land and water. The ability of wetlands to provide these functions is highly dependent on the movement and distribution of water. Over half of the wetlands globally have been destroyed over the past 150 years or so. Much of this degradation has been due to alteration of hydrologic processes through a variety of impacts including the reduction of hydrologic inputs, construction of hard surfaces, channelization, damming of outflows, and the introduction of contaminants and exotic species. Studies in the Puget Sound trough and in the Portland, Oregon region, have intensively investigated hydrological alterations and their effects on wetlands. In general, increases of impervious surfaces and rerouting of natural hydrology have affected plant and animal species composition and altered the processes of primary production, nutrient cycling, groundwater recharge rates, and sediment dynamics. The close coupling of wetland functions and plant-and animal populations to hydrology is illustrated in these case studies. Case studies summarized here have provided guidance for the management of wetland systems in these areas.

INTRODUCTION

Urbanization, the development of metropolitan areas for business and residential purposes, has affected the distribution and functioning of wetland ecosystems. Wetlands and their processes support a variety of ecological services and functions. For example, wetlands commonly are points of groundwater recharge, absorb water and airborne pollutants, attenuate floodwater, control erosion, cycle minerals such as nitrogen, produce organic matter through carbon fixation, and provide feeding, refuge and reproductive habitats for a wide variety of fish and wildlife. The objective of this paper is to summarize how urbanization affects the ecological processes that generate hydrologic functions, which affect wetlands.

Land Use and Watersheds: Human Influence on Hydrology and Geomorphology in Urban and Forest Areas
Water Science and Application Volume 2, pages 5-16

Wetlands are "...lands transitional between terrestrial and aquatic systems where the water table is usually at or near the surface or the land is covered by shallow water..." [Cowardin et al. 1979]. They generally include marshes, swamps, bogs, wet meadows, prairie potholes, sloughs, and lowlands near rivers [Shaw and Fredine 1956]. Wetlands are estimated to cover 5.3 to 8.6 x 10^6 km^2 globally, and thus comprise approximately 6% of the land surface [Maltby and Turner 1983, Matthews and Fung 1987]. About 56% of the worlds wetlands occur in tropical and subtropical areas. However, vast peatlands (i.e., bogs and fens) occur in colder northern areas such as Canada, Alaska, and Siberia.

Because wetlands have historically been considered nuisance areas, they have been filled and modified to suit human needs for habitation, floodwater storage, and other types of development. The realization in the mid-1900's that wetlands were also critical to the maintenance of waterfowl populations and other services, led to the passage of legislation which curbed wetland losses. In the United States, 2.3 x 10^8 km^2 of the 4.94 x 10^8 km^2 pre-settlement wetland areas have been lost, with annual losses continuing at a rate of about 741 x 10^3 ha y^{-1} [Mitsch and Gosselink 1993]. Much of this loss has been due to urbanization.

Wetlands provide a variety of important ecologic, economic, and aesthetic functions. Although there are many ways to classify these functions, the National Research Council [1995] lists the following:

- maintenance of biodiversity
- wildlife and fisheries support
- water quality improvement
- groundwater recharge
- surface water retention
- nutrient processing
- sediment trapping
- carbon sequestration
- organic matter production

The level and dynamics of these functions is controlled, to a great extent, by the hydrologic processes regulating the distribution of water.

Wetland ecosystems are dominated by water-tolerant plants growing in hydric soils. The species, their size, and relative abundance are major factors in the development of animal communities and other ecological functions associated with these systems. For example, waterfowl breeding in the prairie pothole wetlands of the northern Central Plains is strongly correlated with plant size, plant composition, and the proportion of open water [Galatowitsch and van der Valk 1994]. The number and types of amphibian species associated with vernal pool wetlands in central California is dependent on the hydrology and plant community structure there [Zedler 1987]. Therefore, alterations in the structure of wetland systems will ultimately affect their ability to support animal species populations.

Impacts to wetland hydrology can be physical, chemical and biological (Figure 1). Physical changes include paving over wetlands or large portions of wetlands, channelizing surface water inflows or outflows, rerouting of natural water supplies to the system, and altering the soil structure with resulting changed soil permeability. Although not as well demonstrated or clearly linked, toxicological (chemical) impacts to wetland vegetation (e.g., through a spill of petroleum) may affect transpiration processes. Loss of emergent marsh species through the effects of chemical disturbances can result in a major shift in community structure to a system dominated by algae and pollution-tolerant taxa. Biological effects on wetland hydrology can result from invasion of non-native or undesirable species. For example, melaleuca invasion in Florida is thought to be capable of drying wetlands or reducing periods of high soil moisture through increased transpiration and rainfall interception, and through changes in topography that may occur as a melaleuca swamp succeeds to a more mesic "melaleuca hammock" [Shaw and Jacobs 1996].

IMPACTS TO HYDOLOGIC PROCESSES

Alterations in hydrology affect wetland structure and functions. Urbanization affects the three primary hydrological processes (Figure 2) in wetlands through modification of:

1. surface water exchange,
2. groundwater exchange, and
3. evapotranspiration

In the following sections, we provide examples of how each of these processes has been impacted by urbanization and how ecological functions are ultimately affected.

Modification of Surface Water Exchange

The inflow and outflow of surface water varies significantly among wetland systems. Systems can be highly open such as riverine wetlands where the water from the river flows through the system almost constantly. Examples of riverine systems are the delta marshes of the Mississippi River [Gosselink 1984], riparian wetlands of the Southwest [Faber et al., 1989], and bottomland hardwood swamps of the Southeastern United States

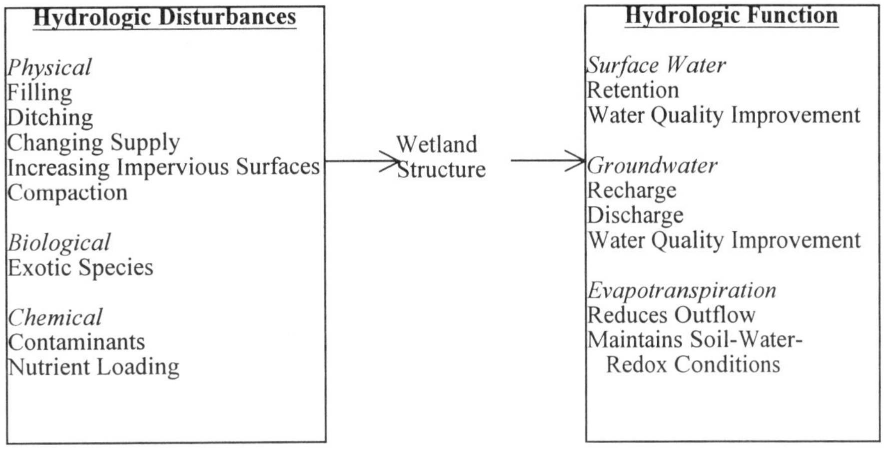

Figure 1. Conceptual model showing the relationship between hydrologic disturbances and the resulting hydrologic functions of wetlands.

[Wharton et al., 1982]. Alternatively, systems can be more isolated receiving only intermittent surface water input. These systems include vernal pools in California [Zedler 1987], prairie pothole wetlands in the Midwest [Kantrud et al., 1989], and playa lakes in the Southern High Plains. Surface water storage in these wetlands depends on the temporal and spatial dynamics of water exchange and the holding capacity of the individual system.

Urbanization affects surface water flows into and out of wetland systems through alteration of:

- wetland drainage,
- rates and location of water flowing into the system,
- holding capacity, and
- outflow rates.

Wetland losses reduce the potential for water storage and increase the velocity of water movement through the watershed. Wetlands affect surface water hydrology by reducing the height and velocity of flooding downstream because water stored in wetlands drains more slowly back into the system. Leschine et al. [1997] found that protection of the flood attenuation capacity afforded by wetlands in Washington today would cost less than flood protection measures in the future. In the Puget Sound region of Washington, urbanization has led to loss of 100 percent and 99 percent of the wetlands in the Puyallup and Duwamish River basins, respectively [Green 1998].

Although there are many examples of urbanization resulting in the complete cutoff of all surface water supplied to wetlands (resulting in system dry-out), perhaps the single greatest impact to surface water inflows has been through channelization of water flowing through urban areas. The net affect of this channelization is to increase water velocities and quantities entering a wetland system, resulting in unnatural wetland flooding [Booth 1991]. Because the soil processes in a wetland are affected by water level and duration of inundation, unnatural flooding will cause the soil system to become anaerobic. Although wetlands often have anaerobic zones, in severely reduced conditions substances such as methane, nitrogen gas, and hydrogen sulfide are produced. Severe flooding can kill vegetation and animals normally associated with the system that are not accustomed to prolonged or abnormal anaerobic conditions or water levels.

An increase in "effective impervious area" (EIA) (i.e. impervious surfaces with direct hydraulic linkage to the channel system) significantly altered rainfall runoff response in the basins. Booth [1991], showed that increases from 6% to 29% EIA through urban development resulted in not only magnification (up to two orders of magnitude increase) of peak discharges but also the creation of new peak runoff events. Smaller storm events, which produced no storm runoff before development, generated substantial flows following development. The capacity of the basins to store water in wetlands was changed enough by development to result in this increased frequency and magnitude of storm-generated flooding.

Booth and Jackson [1994] have also shown that loss of riparian wetland vegetation increases the bankfull channel width in small streams. They found that 0.6 m (17%), on average, of channel widening occurred wherever native bank vegetation had been altered or removed. In addition, storm flow increases due to increased urbanization will result in a decrease in bank stability. They found that with the conversion of more than 10% of the upstream area to impervious surface, channels became much more unstable.

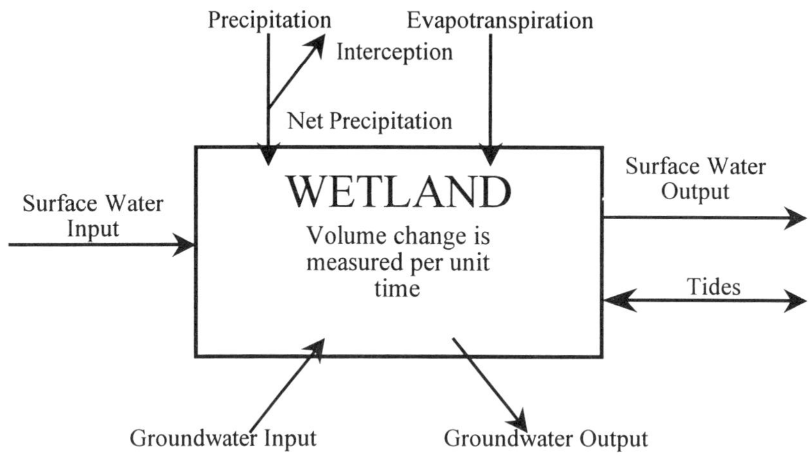

Figure 2. Diagram showing a generalized wetland water budget [from Mitch and Gosselink 1993].

Salmon in the Pacific Northwest spawn preferentially on clean gravels found in pool and riffle areas. Channel destabilization results in erosion and deposition of fine materials in the stream, which can potentially affect the quality of salmon spawning habitat.

Urbanization can lead to channelization of surface water outflows through road building, site preparation, or water diversion to avoid inundation of a particular area during high flows. Ditching wetlands affects surface water hydrology by lowering the water table and reducing the hydroperiod. Mortellaro et al. [1995] examined the effects of ditching on 27 wetlands in Sarasota County, Florida. During the period of study, the average dry season water levels were 0.3 to 0.6m lower in ditched wetlands than in unditched wetlands. The average hydroperiod of natural wetlands in the study was 313 days, while the ditched wetlands had a hydroperiod lasting only 173 days. Although the results of lowering the water table and shortening the hydroperiod were not examined in the study, the potential effects could be an increase in abundance of invasive species, a decrease in populations of obligate (i.e. those plants that require soil saturation) wetland species, or complete dry-out of the wetland.

Other alterations of surface outflows caused by urbanization include complete damming of the outflow, for example by a road bed, or a change in the outflow dynamics of the system due to culverts and other water control structures. Restrictions to outflow typically result in flooding of the wetland and uplands behind the structure, and may cause erosion of outflow pathways downstream of the structure. This type of condition has caused major problems with fish passage, especially for species that migrate downstream or upstream. At present, it is estimated by the U.S. Department of Transportation that there are thousands of problematic road culverts (both public and private) in the United States [J. Schafer, Washington State Department of Transportation, personal communication].

Wetland holding capacity is a function of soil and vegetation conditions, as well as size and morphology of the wetland. Excessive drying of wetland soils can lead to hardpan formation and loss of vegetation which will limit groundwater seepage and transpiration and thus decrease the amount of water that is held within the system. The net effect is that the stormwater retention service of the system is altered. Holding capacity can also be reduced by an increase in sedimentation that results in accelerated accretion in the wetland.

An example of alteration in holding capacity due to sedimentation is found in the prairie pothole region of the Midwest. Although increased sedimentation in this area is due to intensive agricultural practices, the effects on wetlands are the same as they are in urban situations where sedimentation may be caused by an increase in erosion and sloughing due to development. In the prairie pothole region, sedimentation has been shown to reduce wetland volume thereby reducing the water capacity and flood attenuation benefits of the wetlands [Gleason and Euliss 1998].

Increased sedimentation can also negatively impact the water quality function of wetlands by the effect it has on plants, organic substrates, and microbial populations [Gleason and Euliss 1998]. Plants and algae are important in the nutrient uptake, short-term storage, and cycling of nutrients in wetlands [Mitch and Gosselink 1993]. High turbidity, a result of sediment increase, reduces the light available for photosynthesis. In addition, sediment can bury seed banks prohibiting species reproduction. Organic

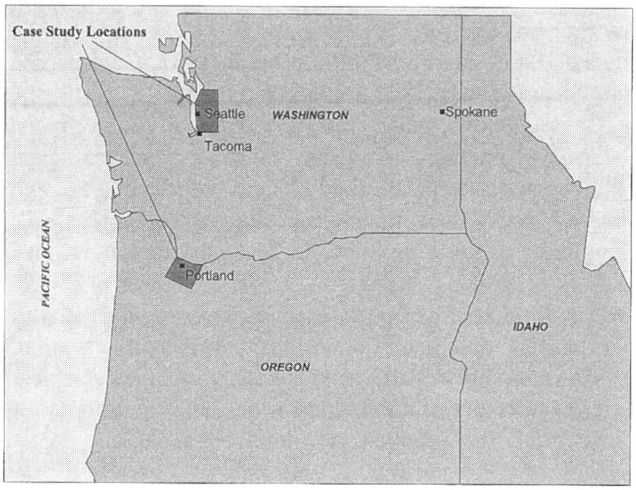

Figure 3. Location of the Portland, Oregon and Seattle, Washington case studies in the northwest United States.

substrates are important for the sorption of contaminants, especially at the water-soil interface; increased inorganic sediments would reduce the availability of these surfaces [Gleason and Euliss 1998]. The impact of sedimentation on microbes has not been widely studied. However, these microbes or the organic materials they rely upon could be adversely effected by an increase in sedimentation.

Modification of Groundwater Exchange

Wetlands can recharge groundwater or can be located in areas where groundwater is discharged to surface water. Movement of groundwater through a wetland is a function of the permeability of the soils, which is partially affected by vegetation and soil type. Alterations to permeability may be caused by drying of the wetland, the discharge of sediments from other areas (e.g., dredged material disposal), salt build up, and loss of organic matter.

In discharge wetlands, where groundwater emerges from the wetland, reductions in the source of groundwater will result in impacts to the wetland. Loss of sources of groundwater due to drawdown of the water table in urban settings has been documented in the forested wetlands of Florida [Mortellaro et al., 1995]. The effects of drawdown on wetlands include reduced hydroperiod, lower water levels, a shift to drier, non-wetland plant species, death of animal species, loss of fish and amphipods, reduced use by birds and wildlife, and increased fire damage.

Modification of Evapotranspirative Exchange

Evapotranspiration (ET) accounts for loss of water ranging from 20 to 83 % of the water budget for the wetland system [Mitch and Gosselink 1993]. Evapotranspiration plays a role in attenuation of floodwaters, as well as maintenance of the soil-water-redox conditions in the wetland. Changing the water supply, loss of plants, and changes in soil conditions will affect ET and in turn change the functioning of the system for a variety of the services listed above. For example, transpiration can exceed loss due to open-water evaporation in prairie pothole systems [LaBaugh et al., 1998]. Willow trees rimming pothole wetlands can strongly influence groundwater flow out of the wetland, and thereby control the water balance. In turn, significant alterations to the willow stands will change this water balance.

CASE STUDIES IN THE PORTLAND, OREGON, METROPOLITAN AREA

A comprehensive investigation of the effects of urbanization on wetlands and wetland hydrology was undertaken beginning in the mid 1980's by Mary Kentula and colleagues at the U.S. Environmental Protection Agency, National Health and Environmental Effects Research Laboratory, Corvallis, Oregon. In some of their initial work, Kentula's group attempted to locate and study 233 wetlands in the greater Portland, Oregon area (Figure 3) [Holland et al. 1995]. These wetlands were initially identified as part of the National Wetlands Inventory (NWI) conducted in 1981-1982 in the region. They found that of the 233 wetlands, 92 no longer existed, with human activities accounting for 73% of the loss and drought the remaining 27%. Urbanization, the combination of commercial, industrial and residential land loss, accounted for 63% of the losses attributed to human activity, with agriculture accounting for 31% of the disappearance. The cause of destruction could not be determined on the remaining 6% of the wetlands lost. Of the remaining142 wetlands, 25% were severely degraded. About half of those not found to be degraded were never-the-less affected by noise, grazing and litter.

Shaffer et al. [1999], studying water levels in 96 wetlands in the vicinity of Portland, found that there was no statistically significant relationship between hydrologic attributes (e.g., water level, period of inundation) and land use, soil association or wetland area. They did find significant differences in hydrologic attributes and the presence of water control structures. These structures resulted in more extensive inundation and less variability in water level in wetlands. The net effect was to alter the vegetative community as well as many of the functions associated with wetlands.

The group found that mitigation wetlands in urban areas resulted in large-scale shifts in wetland types from what they refer to as naturally occurring wetlands to mitigation wetlands [Gwinn et al., 1999]. The naturally occurring wetlands are largely (91%) members of wetland types typically found within the region such as riverine and slope wetlands. In contrast, mitigation wetlands were depressions-in-riverine-setting and in-stream-depressions, which are not typical of the region. Water control structures were often used to create the mitigation wetlands. These wetlands compensate for impacts of urban development projects. They noted that, with the increased use of water control structures (to create mitigation wetlands and for other purposes) in a watershed, the drier-end wetlands are being more rapidly lost from the landscape.

The effect of urbanization on wetland plant communities was pronounced in the Portland area [Shaffer et al., 1999]. Wetlands surrounded by agricultural and commercial/industrial land use had more introduced species than wetlands surrounded by undeveloped land. Wetlands showing the most hydrological changes (e.g., through water control structures) were highly degraded through (1) dilution of the native flora with non-native species, and (2) modification and replacement of naturally occurring wetlands with mitigation wetlands which have different species and hydrologic characteristics.

CASE STUDIES IN THE SEATTLE, WASHINGTON, METROPOLITAN AREA

One of the most comprehensive studies of the influences of urbanization on wetland ecological processes was the Puget Sound Stormwater and Wetland Research Program initiated in 1986 and extending through 1995 (Figure 3). In 1986 local state and federal agency scientists and managers with stormwater management and wetland protection responsibilities in King, Snohomish, Pierce and Thurston counties identified the need for research clarifying the relationship between stormwater runoff and wetlands. Managers and developers recognized that wetlands could be used to collect urban runoff to prevent flooding and stream channel erosion from high peak flow rates and to capture and retain pollutants in storm water. However, scientists and engineers cautioned that by providing for these select functions other wetland functions affecting humans and biological communities could be altered or damaged. For example, a number of studies raised concerns of wastewater treatment in wetlands including long-term toxic metal accumulations, biomagnification of toxins in food chains, nutrient toxicity and other detrimental impacts to wetland ecosystems and public health [Sloey et al., 1978, Benforado 1981, Guntspergen and Stearns 1981, USEPA 1985, Wentz 1987].

Consequently, under the guidance of a technical advisory committee of agency managers, environmental engineers, academicians, and citizens with knowledge and interest in this issue, scientists developed a comprehensive long-term research program. Collectively, they developed a work program spanning 10 years in which wetlands were selected, scientifically defensible surveys and studies undertaken, and interrelationships between urbanization, stormwater and wetland ecological communities assessed. A brief summary of the findings is described here.

In 1987, the research program conducted a synoptic survey of the water quality and soils of 73 wetlands, 46 (63%) of which received urban runoff and 27 (37%) of which did not [Horner 1989]. A multiyear study then followed in which 19 paired control/treatment wetlands were studied to determine differences attributable to urbanization. Control wetlands being in pristine areas and presumed to remain unaffected by urban runoff versus treatment wetlands identified to be effected by runoff.

Hydrological

Increased stormwater caused higher wetland water levels for longer periods of time than for the natural state. Greater surface runoff increased baseline volumes [Hopkinson and Day 1980, Booth 1991], current velocities, and water level fluctuations. Diminished infiltration from urbanization also reduced stream baseflows and ground water supplies, drying the wetlands. Hence, permanently flooded wetlands become seasonally flooded wetlands which in turn dry out earlier in summer.

Urbanization directly impacted wetland hydrology by altering the depth, duration and frequency of flooding. These changes have a far more serious effect on wetland ecology than other impacts. In a comprehensive study of the water balance of two wetlands within watersheds containing different levels of urbanization Reinelt and Horner [1993] found that in the urban wetland a greater percentage of precipitation became inflow, and was delivered faster and in greater short term volumes, than in the rural wetland. Moreover, wetland watershed characterization, outlet description and water level monitoring at all 19 wetlands indicated that the largest range of water levels, as well as the largest mean and maximum water level fluctuations were found in wetlands within basins having the greatest percent of impervious area while simultaneously exhibiting restricted outlets [Taylor 1993]. Two thresholds were identified at which

Table 1. Summary of water quality characteristics in wetlands within watersheds of moderate and high urbanization

Moderately Urbanized Watersheds	Highly Urbanized Watersheds
Highly likely (≥ 71 % of cases) to have median conductivity > 100μS/cm but median TSS in the range of 2-5 mg/L.	Most likely have median conductivity > 100 μS/cm.
Highly likely (≥ 71 % of cases) to have NH3-N <50 μg/L.	Very likely to have median NH3-N <50 μg/L.
Highly likely (≥ 71 % of cases) to have total Zn < 10 μg/L.	Very likely to have total Zn > 10 μg/L but in most cases not exceed the chronic criterion of 59 μg/L for relatively soft waters.
Highly likely (≥ 71 % of cases) median fecal coliforms <50 CFU/100 mL, but also have many individual measurements above 200.	Likely to have median fecal coliforms > 50 CFU/100 mL.
Highly likely (100% of cases) to have TP ≥ 20μg/L and likely (57% of cases) to have TP >50μg/L.	Very likely to have median TP > 50μg/L. These wetlands are likely, but somewhat less than moderately urbanized wetlands to have autotrophic growth limited by phosphorus.
Likely (57% of cases) to have NO3+NO2-N < 100 μg/L. Although it is highly likely (86 % of cases) < than 500 μg/L.	Very likely to have median NO3+NO2-N > 100 μg/L.
Unpredictable pH and DO from consideration of urbanization status alone, being dependent on other factors	Tendency for pH to be closest to neutral and DO and is similar to moderately urbanized watershed.

water level fluctuations became significantly higher in the urban wetlands. The first at 3.5% imperviousness (usually 15% low density residential within the watershed), the second at 20 % imperviousness (24% low density residential within watersheds). Both represented points at which flow volumes and increased water level fluctuations (flashiness) dominated other factors influencing a wetland's hydroperiod.

Water Quality

Several surveys and analyses were carried out to establish water quality relationships to urbanization. In 46 urban and 27 nonurban wetlands, fecal coliforms and enterococcus were significantly higher in urban wetlands [Horner et al. 1988] and although exhibiting mean counts within water quality standards, exceeded standards in high density human use areas.

A second study compared water quality in 10 wetlands within moderately urbanized watersheds of 4 - 20% imperviousness and 7 - 40% forest cover with the water quality in 2 wetlands within highly urbanized watersheds of ≥ 20 % imperviousness and ≤ 7% forest cover. Horner et al. [2000a] found that wetlands in highly developed watersheds most likely have higher conductivity, Zn, fecal coliforms and NO3+NO2-N (Table 1).

The third effort relating water quality conditions to watershed and wetland morphological conditions indicated conductivity, pH and total suspended solids (TSS) increased most directly and dramatically with urbanization [Taylor et al., 1995, Ludwa 1994]. Percent forest cover was the best predictor of water quality, followed by percent total impervious area, then forest to wetland area ratio within watersheds, and finally wetland morphology. All water quality variables except NO3+NO2-N were greater at wetlands in urbanized watersheds of less than 14.7% forest cover [Taylor et al., 1995]. Furthermore, conductivity, TP and fecal coliform (FC) rose significantly when the percentage of impervious surface in watersheds exceeded 3.5% and 20%, respectively, of total watershed area. Finally, the forest-to-wetland area ratio strongly influenced conductivity, TSS, NO3+NO2-N, Total Phosphorus (TP), soluble reactive phosphate (SRP) and FC when ratios fell below 7.2.

Collectively these findings reveal that a decrease in watershed forest cover in combination with increasing imperviousness facilitates the movement of stormwater runoff containing inorganic particulate and dissolved matter.

Soils

Wetland soils moderately affected by urbanization in the Puget Sound Basin were characterized by a somewhat higher pH of 6.5 as opposed to a base pH that ranges between 5.1-6.1. Overall, highly urbanized sites had the highest median pH, followed by moderately urbanized and non-urbanized wetlands [Horner et al., 2000b]. The synoptic study of urban and non-urban wetland soils suggested that concentrations of soil Pb, Cd and Zn were significantly higher at both the inlets and within the emergent zone of urban than non-urban wetlands [Horner et al., 2000b].

Using regression relationships between widely distributed crustal metals (Al and Li) and toxic metals in relatively unimpacted wetlands to establish enriched concentrations in wetland soils, Valentine [1994] found that the most urbanized wetlands exhibited higher rates of Pb and As enrichment than moderately urbanized wetlands. Cu and Ni enrichment was also identified at urban wetlands but only for regressions with Al and Li, therefore again suggesting soils of palustrine wetlands in the Puget Sound Basin are impacted by urbanization.

Microtox results of urban and nonurban soils differed depending on their location. Soils of the open water zone in urban areas exhibited greater toxicity (i.e., Microtox test with significantly lower effective concentrations in both 5-minute and 15-minute exposures) than soils in the open water zone of nonurban wetlands. However, no significant Microtox differences in soils at the inlet and within scrub-shrub zones of urban and nonurban wetlands were found [Horner et al., 2000b]. Repeat soil surveys in 1993 failed to confirm either of these previous findings [Horner et al., 2000b] suggesting that urban wetland soils may be no more toxic those of rural wetlands.

Vegetation

Unique vegetation classes were identified within distinct hydrological regimes. The aquatic bed class, with its representative plant community, was identified at median water depth. The emergent zones were in shallower water and peaked at 96 cm median depth at flooding in some wetlands. The shrub-scrub communities were found at median flooding conditions of 18 cm water depths, although communities were identified at 0 to 40 cm depths [Cooke and Azous 2000].

Moreover, the aquatic bed community was found in areas where annual water levels fluctuated an average of 14 cm, shrub communities were identified where water levels fluctuated an average of 21 cm and forested communities were typically found above inundation elevations.

Within these vegetation communities plant species differed in their tolerance to flooding during the early growing season. Black cottonwood (*Populus trichocarpa*), for example, was mostly found at locations with little or no surface water during the growing season. Hard hack spirea (*Spirea douglasii*), however, was found under a wide range of flooding conditions. It was found at sites that were flooded only during the spring growing season and at sites that remained flooded throughout the year [Cooke and Azous 2000]. Several emergents (*Carex exsiccata*, *Scirpus atrocinitus*) were more likely found in hydrologically undisturbed (few flood pulses) wetlands, whereas other species tolerant of water level changes were most often found in hydrologically disturbed wetlands (e.g., *Scirpus microcarpus*). Several invasive species, including the aggressive reed canarygrass (*Phalaris aurundinaceae*), were found under conditions of large seasonal water level fluctuations.

From the surveys it appeared that water depths, and particularly water level fluctuations, during the early growing season may be a key factor in the development of plant associations in wetlands [Azous and Richter 1995, Cooke and Azous 2000, Azous et al., 2000]. However, few significant changes were found in the vegetation communities along transects, despite some hydrological changes suggesting that vegetation responses may require greater hydrological changes or take a longer time period to respond to hydrological changes. Regardless, it is predicted that given more time, vegetation communities in wetlands of highly urbanized watersheds will shift towards monocultures of aggressive native or introduced species.

Invertebrates

Emergence trap findings represented the first comprehensive description of the distribution and abundance of emergent macroinvertebrates within the Puget Sound region. A total of 115 aquatic and semi aquatic taxa, including 17 of 35 North American dipteran families including 76 chironomid taxa were found in 1986 alone [Richter 2000]. The geographic range was extended for several taxa and the study encountered taxa not previously reported in wetlands. Despite some taxa associations with habitats, there was considerable variation in invertebrate taxa and abundances between and within wetlands, and between years. Nevertheless, greater species diversity values were recorded for total insects and in just the chironomids at permanently flooded wetlands.

Although unique insect assemblages associated with watershed urbanization were not found, effects of land use and wetland morphology were indicated [Ludwa 1994, Ludwa and Richter 2000]. Specifically, multiple regression indicated that the diversity index decreased with increasing watershed urbanization (i.e., impervious area), wetland channelization, and incidence of drying. Correspondingly, a second analysis revealed that hydrology and water quality parameters explained as much as 73 percent of the variation in index scores [Ludwa and Richter 2000]. Threshold analyses additionally confirmed that index scores were significantly higher with increasing watershed forest cover and lower with increasing impervious area. Finally, the analysis of the insect community revealed that urbanization effects the emergent macroinvertebrate community. Urbanization decreased overall taxa richness, eliminating or reducing taxa belonging to scraper and shredder functional feeding groups, reducing Ephemeroptera, Plecoptera, Odonata and Trichoptera richness and relative abundance, and eliminates or reduces specific Chironomidae taxa.

Amphibians

Amphibian species richness declined with increasing watershed urbanization and water level fluctuations exceeding 20 cm [Richter and Azous 1995, Azous and Richter 1995]. Furthermore, species richness and the size of a breeding population were established to be a reflection of the availability of preferred micro-habitats within a wetland and the nature of the wetland buffer and adjacent habitats [Richter and Azous 2000a]. Most importantly, pond-breeding amphibians spawned in low or zero velocity water and prefer to spawn at specific depths below the water surface depending on the species [Richter 1997]. Consequently, flashiness, higher peak flows, and longer summer drying from decreasing infiltration and increasing impervious areas resulting from increasing urbanization [Booth 1991, Schueler 1994] was shown to account for increasing egg and larvae mortality of salamander and frog at some wetlands [Richter personal observations). Finally, controlled field studies revealed that amphibians preferentially bred along shorelines with high exposure to sunlight, in partially vegetated sites of thin-stemmed emergents such as aquatic grasses, rushes, sedges or rootlets, twigs and small branches. These are all wetland characteristics that are altered by urbanization.

Birds

Wetlands and their buffers were used by a wider diversity of birds than terrestrial systems in the region. Consequently, they are probably the most important habitat for total bird richness in the Puget Sound Basin [Richter and Azous 2000c]. Bird species richness reflected wetland area, wetland complexity (e.g., interspersion of the size and number of vegetation classes), and adjacent landscape cover type (characteristic of buffer and upland development) within which wetlands were situated [Martin-Yanny 1992, Richter and Azous 2000c]. Open water wetlands additionally included waterfowl and shorebirds, species not found in wetlands dominated scrub-shrub or forested vegetation classes.

Bird richness decreased, although abundances remained the same, in wetlands within developed or developing watersheds, but richness was unchanged in wetlands within rural, relatively pristine watersheds that remained undeveloped during the study. The diversity of a wetland was correlated to development within the entire watershed and within 500 m - 1,000 m from the wetland but not to development within 500 m of the wetland. Species of birds known to avoid human development further decreased within wetlands of highly urbanized watersheds and wetlands undergoing additional decreases in watershed habitat. Simultaneously, species exploiting urbanization (e.g., American crow, European starling, and house sparrow) significantly increased in numbers at wetlands in non-urbanized watersheds.

Based on these findings it was predicted that the species and abundance of birds will continue to change in wetlands as urbanization continues and intensifies. Decreasing diversity and abundance among wetland obligate species (i.e. species requiring wetland conditions for survival) is predicted, along with simultaneous increases of aggressive species, feeding generalists and nest predators.

Small Mammals

Similar to bird richness, small mammal communities at wetlands were among the most diverse communities in the Puget Sound Basin [Richter and Azous 2000b]. Twenty-one species were found, which is significantly more than that regularly captured in second-growth forests of the Puget Sound Basin. Deer mice (*Peromyscus maniculatus, P. oreas*) were numerically the most abundant species, captured at all but one wetland suggesting they most likely play an important role in the trophic dynamics of wetland ecosystems. Other species observed in many, but decreasing numbers, of wetlands included Trowbridge shrew (*Sorex trowbridgei*), Vagrant Shrew (*S. vagrans*), and Creeping Vole (*Microtus oregoni*).

Strikingly, no small mammals were captured in the wetland with Norway rats *(Rattus norvegicus*), suggesting that this aggressive species may displace native species. In other wetlands where Norway rats are commonly captured, native small mammals are similarly absent [Richter personal observation]. Correspondingly some, but not all, wetlands at which black rats (*Rattus rattus*) were captured had lower richness of native species. Generally, wetlands severely altered by urbanization harbored minimal populations of native species. These were visited by cats, dogs, opossums and raccoons whose activities most likely accounted for the fewer captures of native small mammals than that captured at more rural wetlands.

Small mammal richness was associated with the combined factors of wetlands size, adjacent land use and the relative quantity of large woody debris within the wetland buffer [Richter and Azous 2000b]. The presence of development and it's associated human and animal impacts (i.e., rats, cats, and dogs) did not show the strong relationship that forest land and the distribution of large woody debris exhibited. This suggests that certain densities of development and pets may be tolerated if enough intact forest remains available for food, cover and shelter.

CONCLUSION

There is a strong basis to conclude that wetland hydrology is critical to the survival of wetlands and their inhabitants, as well as the ecological functions performed by wetland systems. These systems are part of a landscape mosaic that provides overall watershed functions. Removal or alteration of the wetland system, through alteration of hydrology, significantly affects the health and functioning of the broader landscape. Even impacts to the hydrological cycle in a watershed far from a wetland can effect the wetland. There is a growing awareness that, to preserve these critical components of the watershed, the entire watershed must be considered, not just the immediate area surrounding the wetland.

Urbanization affects surface water flows into and out of wetland systems through alteration of wetland drainage, rates, and location of water flowing into the system, holding capacity, and outflow rates. Channelization of flow is perhaps the single greatest impact to surface water inflows in urban areas. Channelization increases water velocities and quantities entering a wetland system, resulting in wetland and upland flooding. Another factor that affects surface water flow is the conversion of more than 10% of the upstream area to impervious surface which can destabilize channels. Roads can block or severely alter outflow dynamics of the system. Increased flow rates through culverts can be a major impediment to wetland functioning as well as fish migration.

Urbanization can also affect wetland function, through changes in groundwater turbidity, and evapotranspiration. Reductions in the source of groundwater will result in impacts to the wetland. For example, the effects of drawdown on wetlands include reduced hydroperiod, lower water levels, a shift to drier, non-wetland plant and animals species, and increased fire damage. Turbidity in streams, rivers, and estuaries results from enhanced erosion due to altered hydrology. Increased turbidity affects vegetation production, and the quality of the water for fish. Finally, changing the water supply, loss of plants, and changes in soil conditions affects evapotranspiration rates and alters the functioning of the wetland system.

The conclusions are based largely on assessments of findings of wetland investigations from Florida and the Portland, Oregon, and Puget Sound Basin regions. Regulatory groups in King County, Washington, in Florida, and Portland, Oregon now manage water sources and distribution partially through assessment of the effects of development on wetlands. The investigations into wetland features have led to changes in the way development is permitted in these areas. The experience in these three locations provides guidance for land use in other regions.

Acknowledgements. We sincerely thank Mark Wigmosta for handling this manuscript. In addition, we thank Mary Kentula for providing us with unpublished information. Eileen Stoppani assisted in the revision of the manuscript. Finally, comments by three anonymous reviewers greatly improved the manuscript.

REFERENCES

Azous, A.L. and K.O. Richter, Amphibian and plant community responses to changing hydrology in urban wetlands, pp. 156-162, in *Proceedings Puget Sound '95*, January 12-14, 1995 Seattle, W.A., edited by E. Robichaud, pp. 156-162, Puget Sound Water Quality Authority, Olympia, W.A., 1995.

Azous, A.L., L.E. Reinelt, and J. Burkey, Managing wetland hydroperiod: Issues and Concerns, in *Wetlands and Urbanization: Implications for the Future*, edited by A.L. Azous and R.R. Horner, Chapter 13, Lewis Publishers, Boca Raton, F.L., in press, 2000.

Benforado, J., Ecological considerations in wetland treatment of wastewater, in *Selected Proceedings of the Midwest Conference on Wetland Values and Management*, St. Paul, M.N., June 17-19, 1981, edited by B. Richardson, pp. 307-323, Freshwater Society, Navarre, M.N., 1981.

Booth, D. B., Urbanization and the natural drainage system - Impacts, solutions, and prognoses, *Northwest Environmental Journal*, 7, 93-118, 1991.

Booth, D.B and C.R. Jackson, Urbanization of aquatic ecosystems—degradation thresholds and the limits of mitigation, in *Effects of human induced changes on hydrologic systems*, edited by R.A. Marston and V.R. Hasfurther, pp. 425-434, Proceedings of annual summer symposium of the

American Water Resources Association, June 26-29, 1994, Jackson Hole, W.Y., 1994.

Cooke, S. S. and A. L. Azous, The hydrological requirements of common Pacific Northwest wetland plant species, in *Wetlands and Urbanization: Implications for the Future*, edited by A.L. Azous and R.R. Horner, Chapter 10, Lewis Publishers, Boca Raton, F.L., in press, 2000.

Cowardin, L.M., V. Carter, F.C. Golet, and E.T. LaRoe, Classification of wetland and deepwater habitats of the United States, U.S. Fish and Wildlife Service Publ. FWS/OBS-79/31, Washington, D.C., 1979.

Faber, P.M., E. Keller, A. Sands, and B.M. Massey, The ecology of riparian habitats of the southern California coastal region: A community profile, U.S. Fish and Wildlife Service, Biological report 85(7.27), Washington, D.C., 1989.

Galatowitsch, S.M. and A.G. van der Valk, Restoring prairie wetlands, Iowa State Univ. Press, Ames, I.A., 1994.

Gleason, R.A. and N.H. Euliss, Sedimentation of prairie wetlands, *Great Plains Research*, 8, 97-112, 1998.

Gosselink, J.G., The ecology of delta marshes of coastal Louisiana: A community profile, U.S. Fish and Wildlife Service, Biological Services FWS/OBS-84/09, Washington, D.C., 1984.

Green, T. H., Towards an economic valuation of natural resources: Estimating the value of flood control provided by wetlands in western Washington, M.S. thesis, University of Washington, Seattle, 1998.

Guntspergen, G., and F. Stearns, Ecological limitations on wetland use for wastewater treatment, in *Selected Proceedings of the Midwest Conference on Wetland Values and Management*, St. Paul, MN, June 17-19, 1981, edited by B. Richardson, pp. 272-284, Freshwater Society, Navarre, M.N., 1981.

Gwinn, S.E., M.E. Kentula, and P.W. Shaffer, Evaluating the effects of wetland regulation through hydrogeomorphic classification and landscape profiles, *Wetlands*, 19, 477-489, 1999.

Holland. C.C., J. Honea, S.E. Gwinn, and M.E. Kentula, Wetland degradation and loss in the rapidly urbanizing area of Portland, Oregon, *Wetlands*, 15, 336-345, 1995.

Hopkinson, C. S. and J. W. J. Day, Modeling the relationship between development and stormwater and nutrient runoff, *Environmental Management*, 4, 315-324, 1980.

Horner, R.R., Long-term effects of urban storm water on wetlands, in *Design of Urban Runoff Quality Controls, Proceedings of an Engineering Foundation Conference on Current Practice and Design Criteria for Urban Quality Control*, Potosi, Missouri, July 1988, edited by L.A. Roesner, B. Urbonas and M.B. Sonnen, pp 451-465, American Society of Civil Engineers, New York, N.Y., 1989.

Horner, R.R., F.B. Gutermuth, L.L. Conquest, and A.W. Johnson, Urban stormwater and Puget Trough wetlands, in *First Annual Meeting of Puget Sound Research: Volume 2*, Seattle, Washington, March 1988, Pp. 723-746, Puget Sound Water Quality Authority, Seattle, W.A., 1988.

Horner, R.R., S. Cooke, L. Reinelt, K.A. Ludwa, N. Chin, and M. Valentine, The effects of watershed development on water quality and soils, in *Wetlands and Urbanization: Implications for the Future*, edited by A.L. Azous and R.R. Horner, Chapter 9, Lewis Publishers, Boca Raton, F.L., in press, 2000.

Horner, R.R., S.S. Cooke, L.E. Reinelt, K.A. Ludwa, and N.T. Chin, Water Quality and Soils, in *Wetlands and Urbanization: Implications for the Future*, edited by A.L. Azous and R.R. Horner, Chapter 2, Lewis Publishers, Boca Raton, F.L., in press, 2000.

Kantrud, H.A., J.B. Millar, and A.G. van der Valk, Vegetation of wetlands of the prairie pothole region, in *Northern Prairie Wetlands*, edited by A.G. van der Valk, pp. 132-187, Iowa State Univ. Press, Ames, I.A., 1989.

LaBaugh, J.W., T.C. Winter, and D.O. Rosenberry, Hydrological functions of prairie wetlands, *Great Plains Research*, 8, 17-37, 1998.

Leschine, T.M., K.F. Wellman, and T. Green, The economic value of wetlands: Wetlands' role in flood protection in Western Washington, GMA Working Paper 97-7, June 1997, School of Marine Affairs, University of Washington, Seattle, W.A., 1997.

Ludwa, K. A., Urbanization effects on palustrine wetlands: Empirical water quality models and development of macroinvertebrate community-based biological index, M.S. thesis, University of Washington, Seattle, 1994.

Ludwa, K.A. and K.O. Richter, Emergent macroinvertebrate communities in relation to watershed development, in *Wetlands and Urbanization: Implications for the Future*, edited by A.L. Azous and R.R. Horner, Chapter 11, Lewis Publishers, Boca Raton, F.L., in press, 2000.

Maltby, E. and R.E. Turner, Wetlands of the world, *Geog. Mag.*, 55, 12-17, 1983.

Martin-Yanny, E., The impacts of urbanization on wetland bird communities, M.S. thesis, University of Washington, Seattle, 1992.

Matthews, E. and I. Fung, Methane emissions from natural wetlands: global distribution, area, and environmental characteristics of sources, *Global Biogeochemical Cycles*, 1, 61-86, 1987.

Mitsch, W.J. and J.G. Gosselink, Wetlands, Second edition, Van Nostrand and Reinhold, New York, N.Y., 1993.

Mortellaro, S., S. Krupa, L. Fink, and J. VanArman, Literature review of the effects of groundwater drawdowns on isolated wetlands, Technical Publ., 96-01, South Florida Water Management District. West Palm Beach, F.L., 1995.

National Research Council, Wetlands characteristics and boundaries, National Academy Press, Washington, D.C., 1995.

Reinelt, L.E. and R.R. Horner, Pollutant removal from stormwater runoff by palustrine wetlands based on comprehensive budgets, *Ecological Engineering*, 4, 77-97, 1993.

Richter, K.O., Criteria for the restoration and creation of wetland habitats of lentic-breeding amphibians of the Pacific Northwest, in *Wetlands and Riparian Restoration: Taking a Broader View, Contributed Papers and Selected Abstracts*, Society for Ecological Restoration International Conference,

September 14-16, 1995, edited by K.B. Macdonald and F. Weinmann, pp. 72-94, University of Washington, Seattle, W.A., Publication EPA 910-R-97007, USEPA, Region 10, Seattle, W.A., 1997.

Richter, K.O., Macroinvertebrate distribution, abundance and habitat use, in *Wetlands and Urbanization: Implications for the Future*, edited by A.L. Azous and R.R. Horner, Chapter 4, Lewis Publishers, Boca Raton, F.L., in press, 2000.

Richter, K.O. and A.L. Azous, Amphibian occurrence and wetland characteristics in the Puget Sound Basin, *Wetlands*, 15, 305-312, 1995.

Richter, K. O. and A. L. Azous, Bird distribution, abundance and habitat use, in *Wetlands and Urbanization: Implications for the Future*, edited by A.L. Azous and R.R. Horner, Chapter 6, Lewis Publishers, Boca Raton, FL., in press, 2000a.

Richter, K. O. and A. L. Azous, Terrestrial small mammal distribution, abundance, and habitat use, in *Wetlands and Urbanization: Implications for the Future*, edited by A.L. Azous and R.R. Horner, Chapter 7, Lewis Publishers, Boca Raton, FL., in press, 2000b.

Richter, K.O. and A.L. Azous, Bird communities in relation to watershed development, in *Wetlands and Urbanization: Implications for the Future*, edited by A.L. Azous and R.R. Horner, Chapter 12, Lewis Publishers, Boca Raton, F.L., in press, 2000c.

Schueler, T., The importance of imperviousness, *Watershed Protection Techniques*, 1, 100-111, 1994.

Shaffer, P. W., M. E. Kentula, and S. E. Gwinn, Characterization of wetland hydrology using hydrogeomorphic classification, *Wetlands*, 19, 490-504, 1999.

Shaw, D.T. and K.J. Jacobs, The functionality of Melaleuca-dominated wetlands in south Florida: A critical review of scientific literature, Resource and assessment division, South Florida Water Management District, West Palm Beach, F.L., 1996.

Shaw, S.P. and C.G. Fredine, Wetlands of the United States, their extent, and their value for waterfowl and other wildlife, U.S. Department of the Interior, Fish and Wildlife Service, Circular 39, Washington, D.C., 1956.

Sloey, W. E., F. L. Spangler and C. W. Fetter, Jr., Management of freshwater wetlands for nutrient assimilation, in *Freshwater Wetlands: Ecological Processes and Management Potential*, edited by R.E. Good et al., pp. 321-340, Academic Press, New York, N.Y., 1978.

Taylor, B. L., The influence of wetland and watershed morphological characteristics on wetland hydrology and relationships to wetland vegetation communities, M.S. thesis, University of Washington, Seattle, 1993.

Taylor, B. L., K. A. Ludwa, and R. R. Horner, Urbanization effects on wetland hydrology and water quality, in *Proceedings Puget Sound '95*, January 12-14, Seattle, W.A., edited by E. Robichaud, pp. 146-154, Puget Sound Water Quality Authority, Olympia, W.A., 1995.

U.S. Environmental Protection Agency, Freshwater Wetlands for Wastewater Management: Environmental Assessment Handbook, EPA 904/9-85-135, U.S. Environmental Protection Agency, Region 4, Atlanta, G.A., 1985.

Valentine, M., Assessing trace metal enrichment in Puget Lowland wetlands, M.S.E. thesis, University of Washington, Seattle, 1994.

Wentz, W. A., Ecological/environmental perspectives on the use of wetlands in water treatment, in *Aquatic Plants for Water Treatment and Resource Recovery*, edited by K.R. Reddy and W.H. Smith, pp. 17-25, Magnolia Publishing, Orlando, F.L., 1987.

Wharton, C. H., W. M. Kitchens, E. C. Pendelton, and T. W. Sipe, The ecology of bottomland hardwood swamps of the southeast: A community profile, U.S. Fish and Wildlife Service, Biological Services Program, FWS/OBS-81/37, Washington, D.C., 1982.

Zedler, P. H., The ecology of southern California vernal pools: A community profile, Biological Report 85(7.11), U.S. Fish and Wildlife Service, Washington, D.C., 1987.

Ronald M. Thom, Amy B. Borde, and Lyle F. Hibler, Battelle Marine Sciences Laboratory, 1529 West Sequim Bay Road, Sequim, Washington 98382

Klaus O. Richter, Resource Lands and Open Space Section, King County Department of Natural Resources, 201 South Jackson Street, Suite 600, Seattle, WA 98104-3855

Rates of Channel Erosion in Small Urban Streams

Derek B. Booth and Patricia C. Henshaw[1]

Center for Urban Water Resources Management, University of Washington, Seattle, Washington

We address four objectives, focused on urban and urbanizing watersheds that drain forested (or once-forested) landscapes in humid regions: to document rates of channel change, to evaluate the relationship between development intensity and the rate of channel change, to evaluate geologic and topographic controls on channel change, and to determine if predevelopment conditions can be used to predict erosion-susceptible reaches. We used an 11-year data set covering 21 urban and suburban channels in western Washington, draining watersheds from 0.1 to 20 km^2, a range that covers both seasonal and perennial channels that generally respond readily and rapidly to watershed disturbance. The results indicate:

1. Rates of vertical channel change vary from below the range of measurement error (<20 mm vertical change between visits) to about 1 m (width-averaged) per year. The median rate for this data set was 60 mm per year.
2. Within these lightly to moderately urbanized watersheds, rates of channel change did *not* correlate with development intensity.
3. The nature of the geologic substrate strongly influenced whether or not significant channel change occurred, but no unconsolidated substrate appears immune to change given sufficiently severe watershed disturbance. Erosion rates are independent of channel gradient within the range investigated (0.013-0.52).
4. Channels with the greatest susceptibility to rapid vertical change share the following characteristics, which can be used to predict and so reduce the consequences of future urban development on natural stream systems:
   - Erosion-susceptible geologic substrate;
   - Moderate to high gradient;
   - Absence of natural or artificial grade controls;
   - Predevelopment runoff inputs predominantly via subsurface discharge, likely to be converted to surface (point) discharge in the post-development condition.

[1]Now at Northwest Hydraulic Consultants, 16300 Christensen Road, #350, Seattle, WA 98188

Land Use and Watersheds: Human Influence on Hydrology and Geomorphology in Urban and Forest Areas
Water Science and Application Volume 2, pages 17-38

INTRODUCTION

Changes to channel morphology are among the most common and readily visible effects of urban development on natural stream systems in humid environments. The actions of deforestation, channelization, and paving of the uplands can produce tremendous changes in the delivery of water and sediment into the channel network. In channel reaches that are alluvial, subsequent responses can be rapid, dramatic, and readily documented. Channels widen, deepen, and in extreme cases may incise many meters below the original level of their beds. Alternatively, they may fill with sediment derived from farther upstream and braid into multiple rivulets threading between gravel bars. In either case, they become aesthetic eyesores and biological invalids; natural populations of benthic invertebrates and fish are decimated, to be replaced by limited numbers and taxa of disturbance-tolerant species.

We review here the current understanding of the behavior and physical changes reported from stream channels in urban and urbanizing watersheds. Our focus is on those draining forested, or once-forested, landscapes in humid regions because the subsequent channel changes there appear to be most dramatic and detrimental to physical, biological, and aesthetic attributes. The historic data on channel changes is supplemented by our own 11-year data set on urban and suburban channels in western Washington, draining watersheds from 0.1 to 20 km^2, a range that covers both seasonal and perennial channels that generally respond readily and rapidly to watershed disturbance.

Previous Studies

Although urban-induced channel changes are widely recognized, their magnitudes, rates, and controls are largely matters of sparse data and anecdotal information. As a result, we have only limited understanding of the physical determinants of channel change and even less predictive ability of the likely consequences of urban development on downstream channels. Although a variety of papers have explored the phenomenon of channel change in urban environments, they do not establish a consistent framework because they have analyzed the process from multiple perspectives:

- *Watershed hydrology*, where the most detailed analysis is conducted on the hydrologic changes brought by urbanization, and any channel response is shown or presumed to follow those changes directly;
- *Sediment loading*, where the delivery of sediment into the channel from an urban or urbanizing watershed is analyzed most completely, and the observed channel response is explained in whole or part through the trends of that sediment delivery over time; or
- *Conceptual models of channel erosion and sedimentation*, where a sequence of channel responses, initiated by watershed change, follows a predictable path somewhat independent of the details of the initiating upstream activity.

Watershed Hydrology. Changes in the hydrologic response of an urban watershed, notably the increase in stream-flow discharges, are demonstrably the clearest single determinant of urban channel change. Even where the channel is physically isolated from both physical disturbance and new inputs of coarse sediment, the occurrence and magnitude of increased discharges generally are mirrored by observed increases in channel dimensions. Previous studies that present such relationships include Hammer [1972], Hollis and Luckett [1976], Morisawa and LaFlure [1982], Neller [1988], Whitlow and Gregory [1989], Moscrip and Montgomery [1997], and Booth and Jackson [1997]. Yet this relationship, although common and intuitive, is not universal. A few studies note a *reduction* in channel width or depth with increases in watershed urbanization and, presumably, the discharge that accompanies it [e.g., Leopold, 1973; Nanson and Young, 1981; Ebisemiju, 1989a; Odemerho, 1992].

Sediment Loading. Delivery of sediment into the channel network is a common consequence of urban development with potentially significant expression in the channel form. The broad relationship between stages of watershed development and resulting sediment loads have been presented in studies such as Wolman [1967], Graf [1975], and Douglas [1985; Table 1, below]. Increased sediment loads, generated at particular stages in the forest-agriculture-urban sequence of North American land development, exert an opposing tendency on the channel to that of increasing discharge and probably explain much of the channel narrowing or shallowing that is sometimes measured.

Conceptual Models of Channel Erosion and Sedimentation. Previous efforts to integrate the generally similar, but locally disparate, observations of channel change [see Park 1997] into a unified model generally articulate a *sequence* of anticipated changes over time. For example, Douglas [1985] suggested a specific pattern of watershed development and channel response (Table 1).

Simon [1989] evaluated the consequences of channelization and described a predictable evolutionary sequence of undercutting, bank failure, channel widening, and restabilization that closely resembles that of urbanization. Arnold *et al.* [1982] also recognized the interplay of spatial factors, notably upstream stream erosion and downstream deposition, that can result in multiple "responses" along the same channel, a theme of complex spatial and temporal response echoed by Gregory *et al.* [1992] and Park [1997].

TABLE 1. Conceptual relationship between stages of development, sediment yield, and channel stability [from Douglas, 1985].

"Stage"	Land Use	Lag time	Sediment Yield	Channel Stability
A	Natural forest or grassland	100 years	low	Relatively stable with some bank erosion
B	Heavily grazed	80	low to moderate	Somewhat less stable
C	Cropping	75	moderate to heavy	Some aggradation and increased bank erosion
D	Abandonment of cropping; permanent grass	85	low to moderate	Increased stability
E	Urban construction	40	very heavy	Rapid aggradation and some bank erosion (can braid)
F	Stabilization	25	moderate	Degradation and severe bank erosion
G	Stable urban	15	low to moderate	Relatively stable

Lessons from Previous Work

In any given locality, observed correlations between channel size, rate of channel change, and watershed characteristics are likely to be fortuitous or at least non-universal. In general, the dimensions of channels in an urban stream network will tend to follow the overall pattern of discharge across that network—larger flows beget larger channels. Consequently, a naïve prediction of *channel* change based on the magnitude of anticipated *hydrologic* change [Booth, 1991] is also probably justified as a first-order estimate (*e.g.*, Figure 1 as an example from the Pacific Northwest). Yet details of the channel, the watershed, and the timing and location of the measurement itself may overwhelm this presumption of channel-discharge equilibrium for the following reasons:

- *Location of the measurement station in the channel network:* Is the measurement located in a "transport" reach, where water and sediment are passed downstream with little channel adjustment, or a "response" reach, where channel form readily adjusts to changing conditions? Not every channel responds to increasing sediment load or water discharge in the same way [*e.g.*, Montgomery and Buffington, 1997]. Local channel gradient and the pattern of gradient changes across a channel network are particularly important factors, but they are rarely reported or incorporated into case-study analyses.
- *Location of urban development relative to the channel network:* This includes the obvious factors that headwater development will affect more of the channel network than one that drains into the stream farther downstream, and that the influence of a particular area of disturbance will be proportionally greater on progressively smaller catchments. Similarly, developments that concentrate urban effects in only a few areas tend to have less impact on the channel network as a whole than equivalent development spread across the watershed [Ebisemiju, 1989b]. In addition, flow increases introduced at one point in the channel network may be far more effective at eroding sediment than at another, because of the spatial variability of watershed soils and the distribution of alluvial and bedrock (or other non-alluvial) reaches.
- *Interplay of the timing of watershed development, large storms, and stream-channel observations:* Many of the "relationships" advocated in the literature between channel form and the magnitude and age of the watershed development are probably artifacts of a particular combination of unique temporal or geomorphic factors [Henshaw, 1999]. "Stable" stream channels may simply reflect a lack of recent rainfall [*e.g.*, Bird, 1982]. They are expected in mature systems where fluvial equilibrium has truly been reestablished (as anticipated for example by Hammer 1972, Neller 1988, Ebisemiju 1989b]; but alternatively, they may simply be the product of flushing all mobile sediment from the system to produce a relatively static, non-alluvial channel, where change may still occur [*e.g.* Tinkler and Parish, 1998] but at rates generally slower than reported elsewhere. If equilibrium can be achieved in a disturbed fluvial system, it will depend not only on the at-a-station fluvial processes but also on factors outside of, and perhaps wholly unaccounted by, fluvial conditions in the immediate channel reach. These factors include adjacent hillslope stability, which may have a

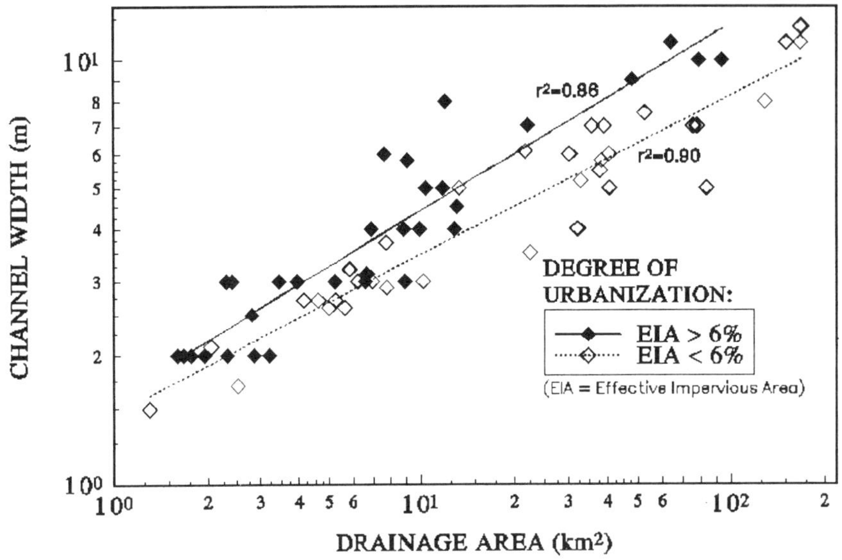

Figure 1. Bankfull channel widths, segregated by percent effective impervious area (EIA; see Dinicola, 1990) contributing to the measurement point. A discrimination at 6 percent EIA was chosen because it approximates the rural-to-suburban transitional land use in this region. From Booth and Jackson [1997].

dramatically longer time scale for stabilization than the fluvial system; and channel stability farther upstream, particularly the absence of large upstream sediment sources.

Most previous studies have reported examples of particularly dramatic channel changes. These sites are commonly erosional, because this response is generally more rapid and more localized than deposition, and because the occurrence of channel erosion (particularly downcutting, commonly the first such fluvial response) can initiate adjacent hillslope failures that mobilize substantially more sediment over a wider area than the original fluvial process. Thus they call attention to themselves from researchers and the public alike. That attention is entirely appropriate—such changes are among the most serious environmental disruptions for both human and biological use of streams in the urban environment. However, such a level of attention introduces a bias into our assessment of what constitutes "urban channel changes."

Channel Types and Classification

Principles and Limitations. Geomorphologists and biologists have been organizing and categorizing the myriad array of stream channels for about a century. The purpose of such an organization is fundamental: if a channel of interest can be placed in a group, and the properties of that group are already known, then the properties of the new channel will also be known with little additional work [Kondolf, 1995]. Those "properties" depend on the organizational scheme, but they include such attributes as the channel's response to environmental change (such as increased sediment load or placement of an artificial habitat-enhancement structure) or its importance in supporting stream biota [Mosley, 1987]. Intrinsic differences between channels will strongly influence channel response to urban development.

Yet the influence of a classification scheme can be detrimental, by suggesting an overly simplistic range of channel conditions that obscures critical differences between channels that are ostensibly "the same." It may also impart a false understanding if the classification method is taken outside of where it was developed to where the dominant landscape processes, or range of landscape conditions, are significantly different: channels may be "classified" but the predictive power of that classification will be low or misleading.

Two examples, both relevant to urban stream channels of the Pacific Northwest, illustrate this problem. The classification method of Rosgen (*e.g.*, 1994 and prior informal publications), applied widely throughout the United States, does not include the influence of large logs and other woody debris on channel processes, reflecting the non-forested environment in which this method was first developed. A forested stream may be "classified" by this method but the nature of its response to human disturbance may be poorly predicted. In contrast, the classification of Montgomery and Buffington [1997] was established explicitly to address the channels found in forested watersheds of the Pacific Northwest, where such large woody debris (LWD) is ubiquitous and its influence can be dominant. Yet this method was developed in mountain drainage basins sharing a typical downstream progression from steep headwater catchments underlain by bedrock to gentler, larger watershed areas in broad alluvial valleys. This orderly sequence may not be matched in a lowland setting—the smallest watersheds of urbanizing Puget Sound can be quite flat, with steeper reaches located some distance farther downstream. Sediment-delivery processes and sources of channel roughness are very different in lowland urban channels than in nearby mountainous channels, and so this classification system also may not fully predict the response of a particular urban stream.

Criteria. Despite these caveats, different channel "types" display different intrinsic channel behaviors and have different responses to watershed disturbance. No framework has been fully developed for our environment of interest, but we are using the conceptual approach of Montgomery and Buffington [1997] because of its orientation on channel-forming processes, its development in the same climatic region as the present study, and its explicit recognition of the influence of LWD and other such obstructions on channel morphology. By their terminology, most of our channels are either "plane-bed" or "forced pool-riffle" channels—relatively flat-bottomed channels lacking well-defined bedforms and instead displaying long, and commonly channel-wide, reaches of uniform riffles or glides which can aggrade or degrade rapidly in response to changing water and sediment fluxes. Development of a more heterogeneous morphology depends on the presence of immobile material, most commonly LWD. By restricting our evaluation to such channels, we may be limiting the potential utility of our work. Yet the vast majority of the small, responsive, urban streams in our region fit these categories, and this selectivity helps avoid the risk of transferring results inappropriately.

A Conceptual Framework to Assess Channel Change in Urban Watersheds

Past studies and repeated observation suggest a "typical" scenario for channel change in an urbanizing lowland watershed. Recognizing that this scenario does not encompass the full range of potential watershed conditions or stream-channel responses, it nonetheless characterizes the most common "problems" of urban channel change and highlights those settings where an *unexpected* response suggests the presence of atypical channel or watershed conditions.

Consider a watershed of some tens of hectares up to several square kilometers, where development has blanketed the upper watershed and so the first-order channel(s) are the most fully affected of any in the channel network. In most cases, channel expansion of at least several times the original cross sectional area accompanies the progression from rural to suburban to urban land uses. Whether or not the response of the channel to these flow increases is "orderly" (*i.e.* channel-size increases in approximate proportion to discharge increases in the sense of Booth, 1990) or "catastrophic" (*i.e.* rapid incision) is largely independent of the magnitude of the watershed disturbance (see below). Even low levels of land-cover changes, if accompanied by an efficient collection system (*e.g.*, road ditches) can produce significant increases in headwater channel discharges, which in turn will initiate increased in-channel erosion and sediment transport.

Because such land-use changes typically occur over a period of many years or decades, they tend to produce continuous changes in the downstream channel subject only to the variability of seasonal runoff. Any tendency towards "equilibrium," either dynamic or static, is completely obscured during this period. Sparse long-term data suggest that true equilibrium may be possible in watersheds with constant land use, over a years-to-decades time lag [Henshaw, 1999], but actually *observing* such a condition depends on achieving stable hillslope conditions as well, which may take many times longer. With these complications, it is not surprising that "reequilibration" may be more useful as a theoretical construct than as a widely observed condition.

The sediment released by this scenario of headwater flow increases may or may not accumulate as it passes through the downstream channel network. The potential input of additional urban-flow-induced sediment from other lateral tributaries will combine to influence whether sediment, eroded from upstream reaches, can remain in active transport or will accumulate in noteworthy volumes. Curiously, the vagaries of human infrastructure, particularly small roadway culverts that were sized and installed during an earlier pre-headwater-development era when only lower discharges of water (and tremendously lower discharges of sediment) occurred, appear to be one of the strongest single determinants of whether the urban channel change is perceived to be mainly a problem of "erosion" or one of "deposition" (Figure 2).

Figure 2. Deposition of stream-channel sediment, eroded from upslope reaches of tributary 0143G (station C8).

FIELD INVESTIGATIONS

No single study can cover all settings in which urban-induced channel change is observed. Yet even a geographically limited set of new data can increase our understanding and prediction of this threat to aquatic-system integrity. This study was initiated to provide some of that new data, focused on a part of western Washington state where (and beginning at a time when) the rate of new urban development was accelerating to historically unprecedented rates. It also began when the social and political desire to alleviate the worst environmental consequences of that development far exceeded the concrete knowledge necessary to achieve that goal.

Starting in 1986, 35 stations along an equal number of independent streams were established to monitor long-term channel changes in urbanizing watersheds. The purpose of this effort was four-fold:

1. To document erosion and deposition rates in a variety of physiographic settings;
2. To test the hypothesis that urban development consistently increases the rate of channel change, and that higher levels of urban development are correlated with faster rates of channel change;
3. To test the hypothesis that certain geologic and/or topographic settings are particularly susceptible to urban-induced channel changes; and
4. To improve identification of the most susceptible sites *before* development, and thus before degradation, has begun.

Methods

Study sites. The choice of channel reaches for monitoring began in early 1986, following a particularly large storm in January that resulted in many instances of channel modification and property damage from high discharges. These first sites were chosen because of known stream-channel erosion, reported downstream problems, or knowledge of impending development that might prove problematic. Over the next several years, a number of additional sites were identified and some unsuitable sites were relocated or abandoned, mainly due to unrepresentative channel conditions but also because of subsequent obliteration by development activity. A range of channel conditions, particularly slope, degree of upstream development, and geographic location, were covered by the final set of 21 selected sites (see Figures 3 and 4 and Table 2). Previous observations had suggested that channel changes were particularly rapid downstream

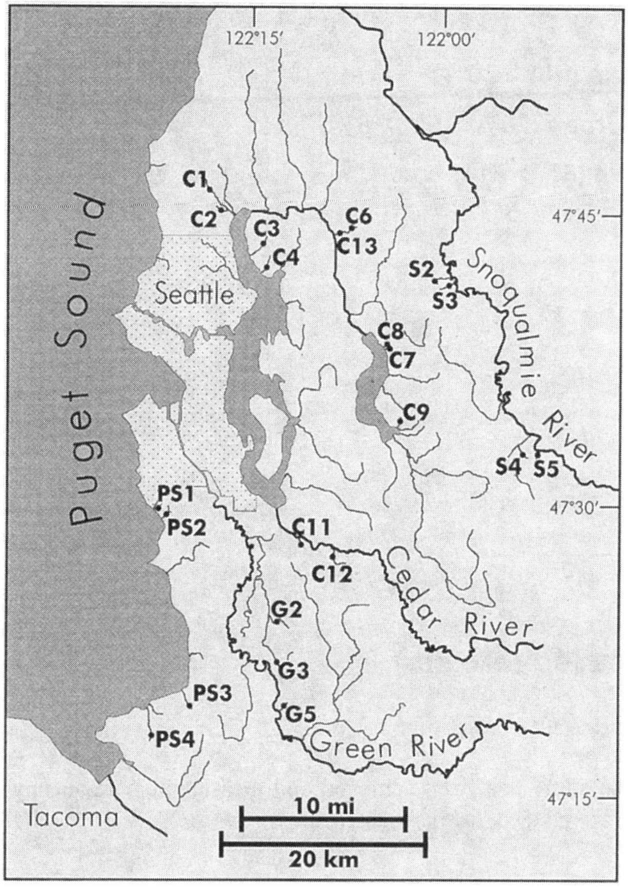

Figure 3. Location of stream measurement stations listed in Table 2.

of recent urban development in small headwater catchments and in channels traversing hillslope deposits of a specific regionally common geologic deposit, so these characteristics were emphasized in the initial site selection.

Most of the sites were, broadly speaking, *alluvial channels* [Leopold *et al.*, 1964]: carved by running water into the very sediment carried by that flow in the past, and that presumably could be carried by that flow in the future. These "self-formed" channels are free to adjust their shape in response to subsequent changes in flow and thus were anticipated to respond most sensitively to future development. However, as the channel changes in response to increased flows (and particularly if it begins to incise) the underlying hillslope deposit becomes more dominant as the channel-bounding sediment and the alluvial "character" of the channel can be reduced. In contrast, a channel formed in alluvial sediment but also choked with immovable roughness elements, such as logs, is not strictly "alluvial." Yet if those logs are removed, or if progressive bed erosion strands those logs above the elevation of the flow, the channel will become *more* characteristically alluvial over time.

The sample population was chosen to explore the influence of the underlying geology by emphasizing sites located on a particularly erodible substrate. Most of the stations have as their underlying substrate a thick and widespread sandy deposit with local concentrations of pebble to cobble gravel, laid down by glacial outwash streams during the last advance of the continental ice sheet (regionally named the "Vashon" by Armstrong and others [1965] and spanning an interval of about 17,000-13,000 years ago). This emphasis was established to quantify rates in what previously had been observed locally to be the most erosion-susceptible deposit. A moderate number of sites with other substrates were also included to test this hypothesis more precisely.

The work was initially sponsored by the jurisdiction of King County, Washington, so all sites were located within its boundaries. By virtue of its location and size, however, this 5000-square-kilometer area cuts a remarkably diverse and representative swath across the Puget Lowland and Cascade Range of western Washington; it also spans a range of land uses from forested wilderness through agriculture, suburbia, and intensely urban. In the rapidly developing suburban fringe targeted by this study, annual precipitation averages about 1000 mm, falling primarily from October to April as large frontal storms of several days' duration. Unlike much of the rest of the continent, short but intense storms are rare. The 100-year 6-hour rainfall intensity in this region is only about 5 mm/hr [Miller *et al.*, 1973], whereas the largest stream discharges are generally associated with moderate-intensity rainfall following a period of extended wintertime precipitation or snowmelt.

Measurement Techniques. Cross sections were measured using two procedures, modified from that suggested in Dunne and Leopold [1978]. A specific location along a relatively straight and uniform part of the channel, qualitatively judged to be "representative" of the reach in question, was selected. Two endpoints were established to define a line approximately perpendicular to the channel. Where available, streamside trees were used and the precise endpoint was marked with a 12d galvanized nail driven nearly flush with the trunk. Where trees were unavailable, 0.6-meter-long pieces of reinforcing bar ("1/2-inch rebar") were driven into the ground to provide a suitable monument. We rarely had much difficulty in

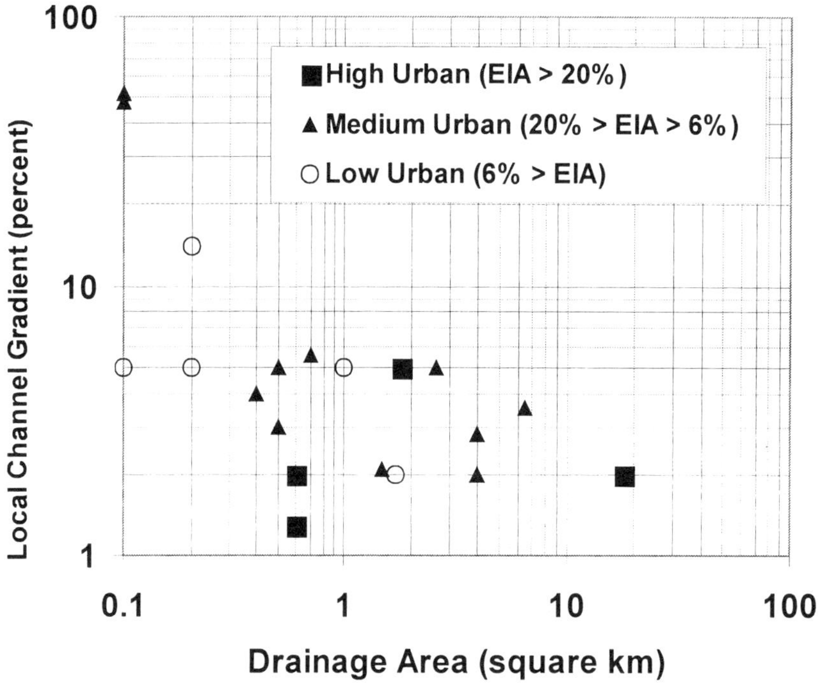

Figure 4. Summary attributes of measurement stations listed in Table 2.

relocating such markers, even after four years' absence, with sufficiently detailed notes.

In the early years of this study, a 50-m steel tape was stretched and held at constant tension between endpoints to provide both horizontal and vertical reference. The vertical distance between the tape and the ground surface was recorded at 0.3-m intervals, together with additional intermediate measurements at marked breaks in slope. At each station, the vertical angle and the catenary sag of the tape were measured and used to correct the raw data. This method had the advantage of speed and minimal field equipment but had some inherent inaccuracies. In the last year of the study we used an automatic level and surveyor's rod to determine channel depths, relying on the stretched tape only to specify horizontal distance. By direct comparison of these two measurement methods, we could determine the precision of the early method; its error was consistently less than 0.02 m, at or below the degree of inescapable measurement imprecision imposed by ground irregularities and sediment clasts on the channel bed.

In the early years of the project, measurements at most sites were made annually in the summer, the season when qualitative observations suggest that little or no channel changes occur from about June until October. After collecting data in the summer of 1990 from changes during the large storms of January 1990, the primary objectives of the study had been achieved and measurement frequency was reduced, with visits to most sites only in 1993 and 1997.

Results

Overview. Rates of erosion and deposition vary by over two orders of magnitude (see Figure 5). In this population, the minimum amount of the annual width-averaged vertical channel change was below the level of measurement error (about 20 mm); the maximum was about 1 m per year. Over the 11-year period, 80 percent of all measurements show an annual width-averaged vertical change (erosion *or* deposition) of less than 0.2 meters, with the median of all measurements at 60 mm/year.

The most consistent pattern is the correlation of rainfall with channel change. This outcome is qualitatively intuitive, although the nature of this relationship is more complex than might be first anticipated. For example, 1990 channel changes (*i.e.* occurring between the 1989 and 1990 measurements) are the largest, by a significant degree, at nearly all sites (Figure 6). Although the 1990 rainfall intensities are also the largest in the period as well (Figure 7), they do not exceed other "large" years (1991 and 1996) by nearly as much as the erosion/deposition measurements would suggest.

TABLE 2. Stream erosion station characteristics.

Station	Station Name	Drainage Area (km^2)	Local Slope %	% EIA[1]	Channel Width (m)	Substrate Type	Length of Record (yrs)
C1	McAleer Creek	18	2	22	9.8	silt-clay	7
C2	McAleer tributary	0.6	2	20	4.1	silt-clay	11
C3	Holmes Point trib.	0.5	5	16	2.4	sand	4
C4	Juanita Point trib.	0.4	4	12	1.7	sand	11
C6	Skookum tributary	1.7	2	5	2.5	sand	9
C7	Timberline trib. 0143F	0.2	14	1	2.3	sand	7
C8	Timberline trib. 0143G	0.1	48	15	2.3	sand	11
C9	upper 0164A	0.5	3	16	4.9	sand	7
C11	Ginger Creek	1.8	5	22	7.9	sand	5
C12	mid-Madsen Creek	6.5	3.5	10	6.6	sand	10
C13	Hollyw'd Hills trib.	4	2	7	4.6	sand	7
G2	Garrison Creek	2.6	5	10	1.8	sand	2
G3	Mill Creek	4	2.8	19	4.3	sand	3
G5	Cobble Creek	0.7	5.5	11	4.8	silt-clay	11
PS3	Easter Lake outlet	0.6	1.3	40	3.7	sand	11
PS4	Olympic View Park	1.5	2.1	17	2.4	sand	11
PS7	Boeing Creek	5	2	20	8	sand	2
S2	Chasm Creek mainstem	0.2	5	3	2.4	sand	4
S3	Pepper Creek	1	5	5	6.6	sand	7
S4	Lk. Alice Estates trib.	0.1	52	10	2.6	sand	11
S5	Joule short plat trib.	0.1	5	5	2.4	sand	4

[1]EIA = Effective impervious area

Most noteworthy of this data set, however, is the overall *absence* of general relationships between measured channel changes and simple, physical parameters of the stream or of the watershed, such as slope (Figure 8) or imperviousness (Figure 9). This condition bodes poorly for the kinds of simple predictive methods favored by local governmental jurisdictions in the prediction and avoidance of environmental impacts. Only the role of geologic materials shows any consistency, with cohesive silt-clay substrates generally permitting only low rates of channel adjustment.

The poor correlation between effective impervious area (EIA) and channel change is quite robust. It is displayed by both the station averages for the period of record (Figure 9) and the single-year (1989-1990) data (Figure 10). We therefore *reject* the first of our initial hypotheses, that urban development consistently increases the rate of channel change, and that higher levels of urban

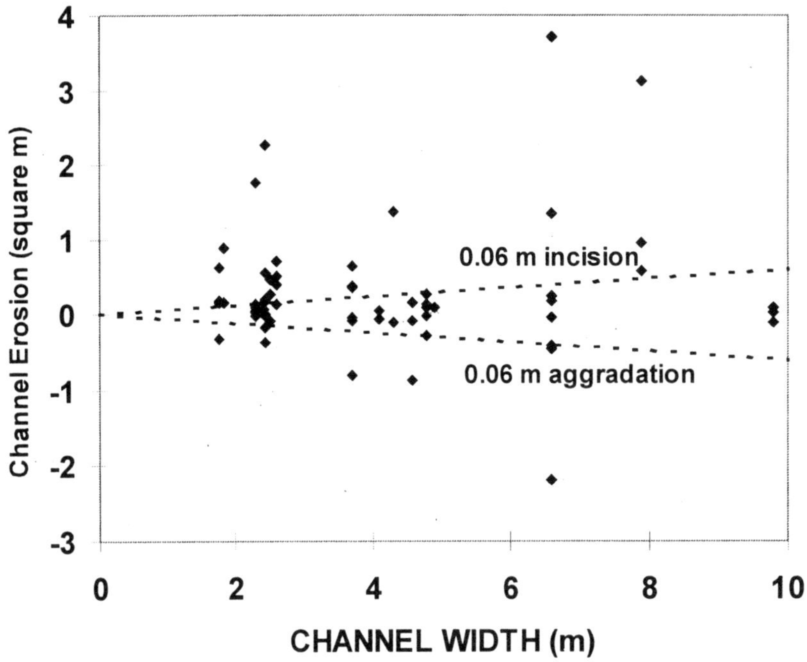

Figure 5. Results of all measurements, expressed as the average change in cross-section area per year between visits. Dotted lines plot the median vertical change of 0.06 m.

development are correlated with faster rates of channel change.

Evidence against this hypothesis is particularly clear at the following stations, discussed below in greater detail:

- Moderate to high development, moderate to minimal changes:
  - Easter Lake outlet (40% EIA, moderate change)
  - Olympic View Park (17% EIA, very little change)
  - McAleer Tributary (20% EIA, very little change)
- Little development, large changes:
  - Pepper Creek (3% EIA, very large change)

Although hydrologic processes may impose a general tendency for increased urbanization to yield greater channel change, the expression of that change is completely swamped by the vagaries of local conditions.

Our other initial hypothesis, the association of particular topographic or geologic conditions with rate of change, finds much more consistent support from the data. Granular hillslope deposits, normally mantled by alluvium but accessible to streamflow in an incising environment, were anticipated to display the greatest changes for a given degree of upstream urbanization; indeed, a majority of sites were chosen on the basis of this attribute. The type of deposit appears to exert a significant influence on channel-change rates; the most common alternative, cohesive silt-clay deposits, consistently showed low or very low rates of change.

Specific Site Conditions (see Figure 10)

Olympic View Park. This channel is located in a lightly developed parkland (Figure 11), established around the long-protected riparian corridor of the stream. The surrounding watershed has been almost fully developed for several decades, primarily with single-family residences. The ravine that contains the channel and associated park is excavated into deposits of the sandy Vashon advance outwash. Incision has clearly been part of the channel's past history; several hundred meters downstream of the measured cross section, large gabion baskets stabilize what must have been a major knickpoint in the 1970's. Yet the current decade of measurements is noteworthy in its near-negligible change from one year to the next (Figure 12), although the channel morphology is distinctly unappealing from either a biological or an aesthetic standpoint. It is relatively uniform, slightly sinuous, with virtually no heterogeneity or variability in size, shape, or roughness. Much of this uniformity is surely the result of close human contact—foot traffic up and down the channel (commonly dry in the summertime) is frequent, and any sticks or twigs

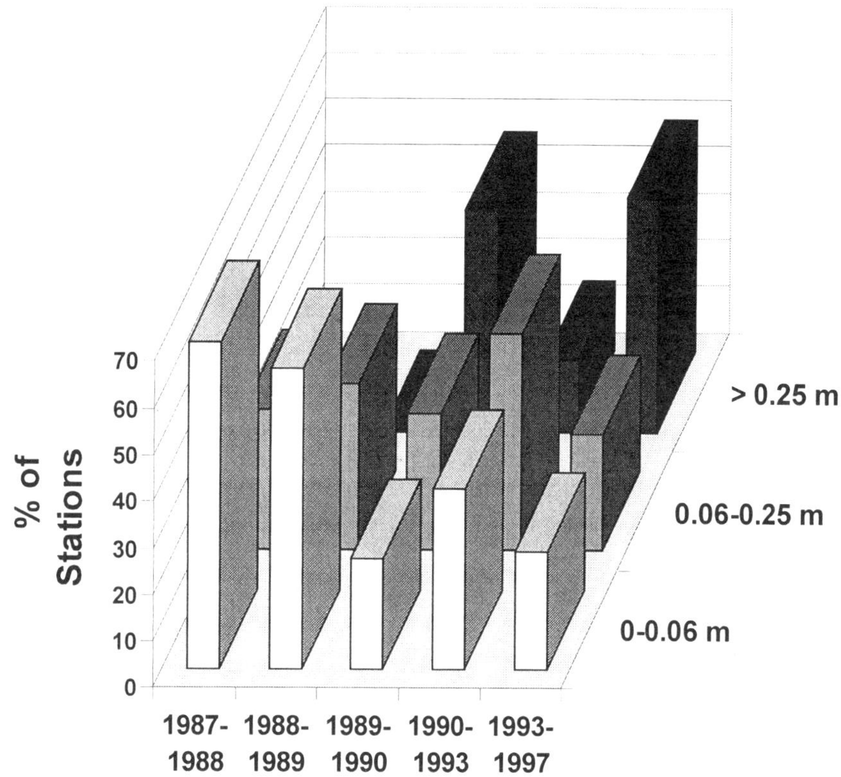

Figure 6. Distribution of stations below, above, and greatly above the median vertical change of 0.06 m/year. Channel changes are calculated across the full period between measurements and are *not* annual averages. Note that the distribution of stations in the 1989-1990 interval is nearly identical to that of 1993-1997, even thought the earlier period is only one-quarter as long.

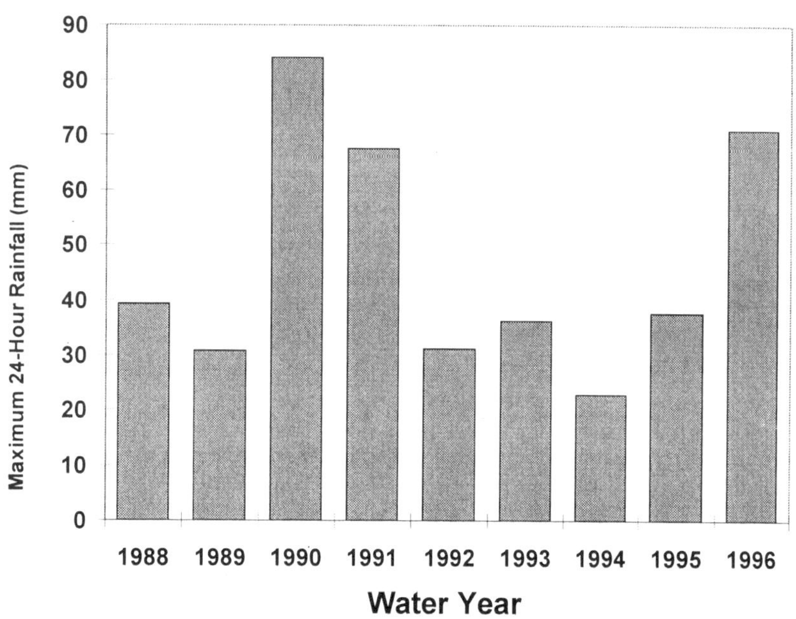

Figure 7. Maximum 24-hour rainfall for each year of the interval 1988-1996. The chosen gauge ("Maplewood") was centrally located for the study sites (near station C11 on Figure 3).

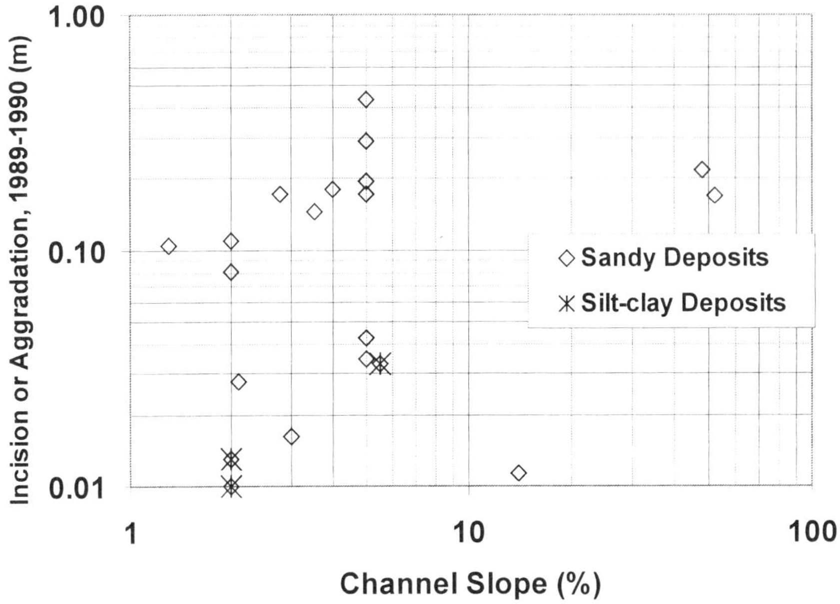

Figure 8. Demonstration of the poor correlation between local channel slope and the magnitude of one year's width-averaged vertical change. A more consistent pattern is suggested by the relatively low change shown at each of the three stations underlain by cohesive (silt and clay) hillslope deposits.

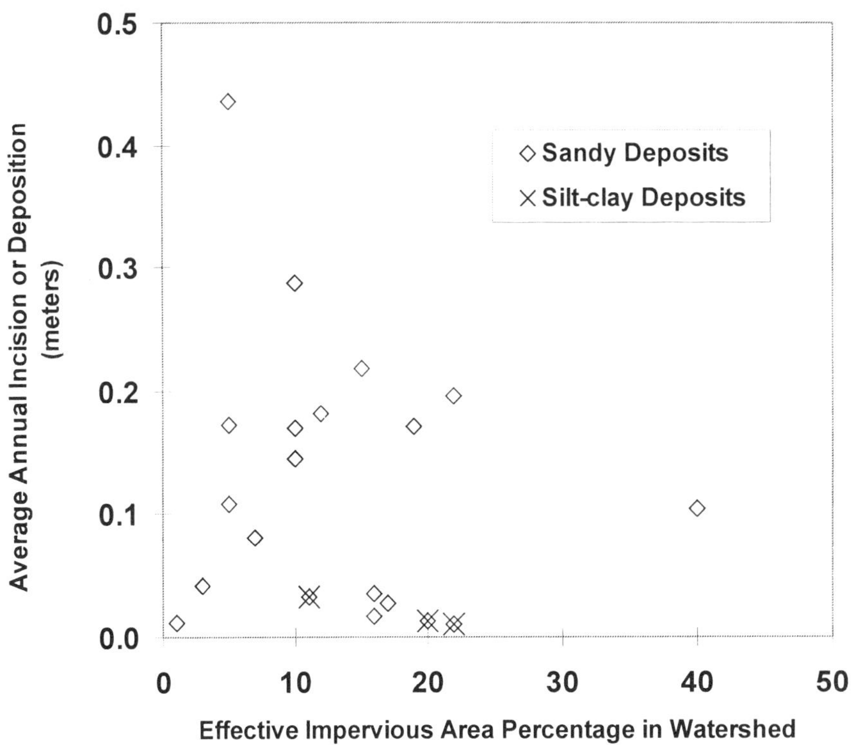

Figure 9. Demonstration of the poor correlation between contributing impervious area and the magnitude of the annual width-averaged vertical change, averaged over the full duration of each station's measurement history.

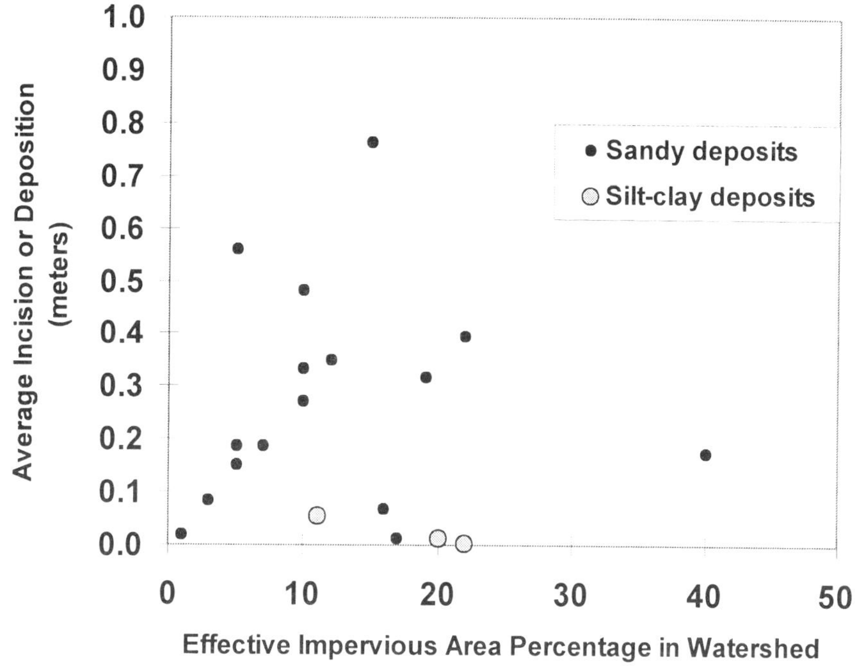

Figure 10. The equivalent parameters as for Figure 9 but considering *only* the measurement interval with the greatest change, 1989-1990. There are no appreciable differences between the pattern expressed by either the single-year or the time-averaged results. Labeled points are discussed in the following section.

Figure 11. View of the channel of Olympic View Park (Station PS4) in 1986, looking upstream. Line across the channel (below the scale bar) is the tape measure used to determine horizontal distance at the cross section location.

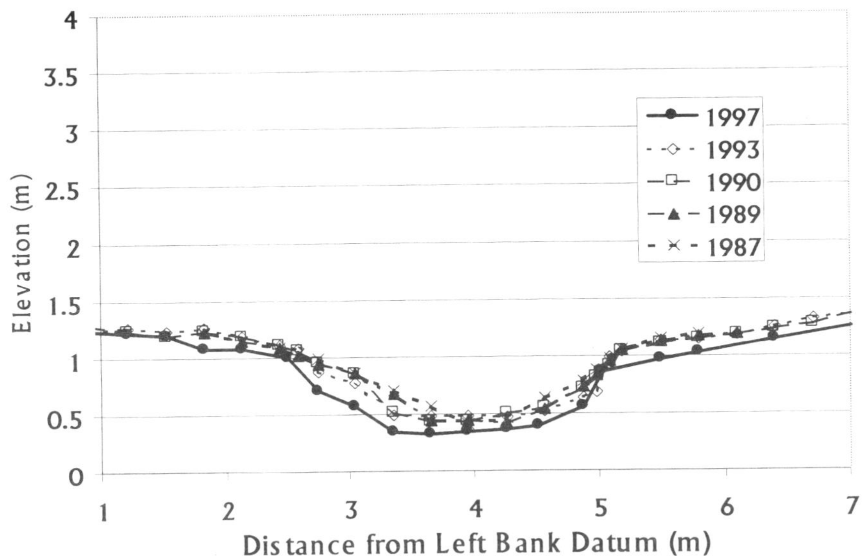

Figure 12. All cross section measurement of Station PS4, showing only minor changes from 1987 to 1993 (the total width-averaged erosion during this interval is 8 cm) and from 1993 to 1997 (total average erosion 7 cm).

would be promptly "cleaned up." Yet even where encroaching riparian shrubs limit the immediate access of people, conditions are essentially unchanged.

Easter Lake Outlet. This channel is also in a well-established part of King County, with a high level of urban development in the contributing watershed but with most of it predating 1970. About 80 percent of the runoff from that watershed drains through Easter Lake, which occupies about 10 percent of the contributing surface area of the watershed and which provides significant hydraulic control of discharges. Continuous hydrologic modeling of this lake [King County, 1990] displays a marked reduction in the unit-area discharges here relative to other catchments without lakes in the immediate region.

The channel-measurement station lies within a reasonably well-protected riparian buffer, located entirely on private property and generally not accessed by nearby residents because of topography (Figure 13). The channel is incised about 1.5 m into a narrow upland terrace, probably the old floodplain, set within a broader valley. A complex of logs and a large tree root, about 20 m downstream of the section, have formed the lip of a 1-m knickpoint and probably have inhibited more dramatic degradation.

This site displays the interplay of fluvial and hillslope processes and it also demonstrates why cross-sectional measurements can be a very incomplete characterization of stream-channel conditions. Net change at this site has not been dramatic, but bed scour has clearly oscillated with bank failure on a multi-year scale (Figure 14). For example, the steep high right bank remaining after the 1986 storms degraded over the next several years, with progressive bank collapse and channel widening contributing sediment to an aggrading channel bed. Renewed scour during the large flows of January 1990 lowered the bed by nearly 0.5 meters in the active part of the channel, and additional sediment was flushed out through at least 1993. The 1993-1997 interval was a period of substantial channel *widening*; but in contrast to many other stations during this time, channel *deepening* did not occur, and in fact aggradation was substantial.

Changes in channel cross section were accompanied by a progressive simplification of in-stream morphological features. Over the 11 years of observations this stream has become more like a drainage ditch, with a marked lack of heterogeneity in either sediment or bedforms. This has not required any direct human intervention; simplification of the channel, with attendant loss of aesthetic and biological benefits, has occurred only through the indirect effects of watershed disturbance.

McAleer Tributary. The third example of a relatively highly impervious, minimally changing channel is a small stream which, for a 200-m-reach, is surrounded by a surprisingly intact riparian buffer and wetland system. It is isolated from most human traffic by private property and distance from the adjacent county road. This part of the valley has been eroded into resistant silt and clay deposits, which are nowhere visible in the stream bed itself but do

Figure 13. View of the channel of the Easter Lake outlet (Station PS3) in 1986, looking upstream.

lie within a few decimeters of the ground surface and which contribute to the generally silty cohesive banks.

Systematic change at this station over the past decade has been minimal (Figure 15). The channel has aggraded slightly, and the channel banks have become more rounded and have retreated a few tenths of a meter. The clearest interval of change was between 1990 and 1993, when two threads of the flow evolved into a more broadly flowing single channel. Some textural changes in the bed sediment have been noticed over the years, with areas of gravel riffle in one visit becoming patches of silty sand in the next, but no long-term trends in these changes are evident.

Pepper Creek. In contrast to the modest changes observed at the previous stations, this site has shown tremendously variable conditions (Figure 16). It collects runoff from a watershed in the very earliest stages of urbanization; the major hydrologic changes have been related to channelization and road-ditch interception of shallow subsurface flow [Burges *et al.*, 1989], whereas the total fraction of contributing imperviousness is still quite low. The channel is extremely well protected from direct human intrusion, lying several hundred (very brushy) meters from the nearest structure or public road and entirely on private property. It is eroded into sandy valley-bottom deposits, delivered by episodic landslides from the surrounding hillsides and locally reworked by past fluvial action.

The likely magnitude of channel changes was first suggested by extensive deposition on the downstream alluvial fan of the stream, beginning in about 1980 and coincident with the first extensive road construction and forest removal in recent history. Following first measurements in 1986, two episodes of significant erosion were evident: 1986-1987 and 1989-1990. Channel readjustment, but little net erosion/deposition, occurred in 1987-1988. Near-static conditions persisted during the low-rainfall year of 1989. Substantial erosion continued following 1993, but landsliding off the hillside above the right bank, probably in 1996, completely obliterated the measurement station.

Timberline Tributaries. In a topographic and geologic setting remarkably similar to that of Pepper Creek, these smaller tributaries drain adjacent areas generally subjected to intensive urban development. A primary difference between them, however, is that the upland storm-drain system bypasses one (0143F) and discharges into the other (0143G). The stability of 0143F (Figure 17) throughout the measurement period demonstrates that there is nothing inherently unstable about these channels, even where gradients are steep and deposits are erodible. Curiously,

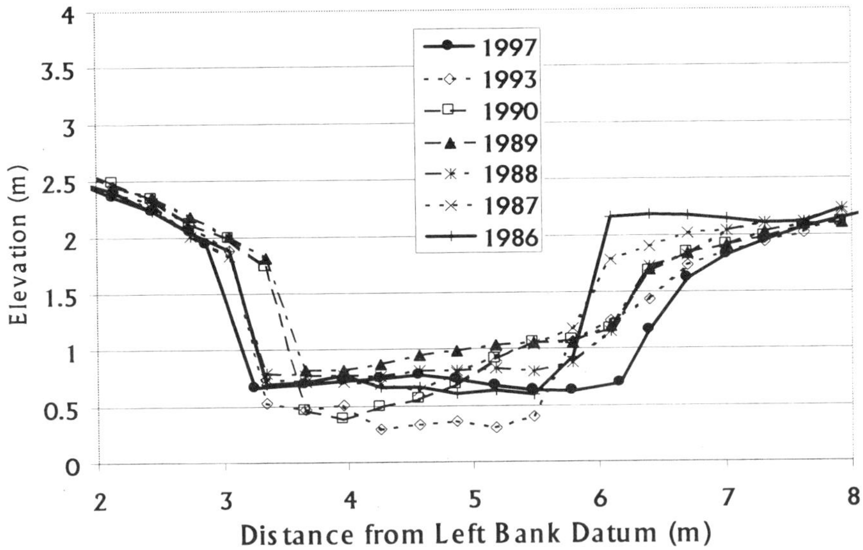

Figure 14. All cross section measurement of Station PS3, showing episodic deepening in 1990 and 1993 but aggradation both before and after. Widening has been progressive but particularly during the interval 1986-1988.

the discharge into 0143G was constructed with some advance awareness of the susceptibility of the downstream channel to erosion: rather than simply releasing the runoff from the end of a pipe, the discharges are first attenuated in a detention pond and then released at the head of a swale through a 20-m-long level dispersion pipe. Despite these efforts, first documented in Booth [1989], channel incision of more than one meter occurred in 1989-1990 (Figure 18). In contrast, the station without such flows remained virtually unchanged throughout the period of its measurement.

The respective (in)stabilities of these two channels emphasize a fundamental point about streams in urbanizing environments. Prior to watershed disturbance, "stability" (whether static or dynamic) is the norm. Although gradients are steep and substrate is easily erodible, woody debris and other in-channel roughness elements maintain an overall balance with the tractive

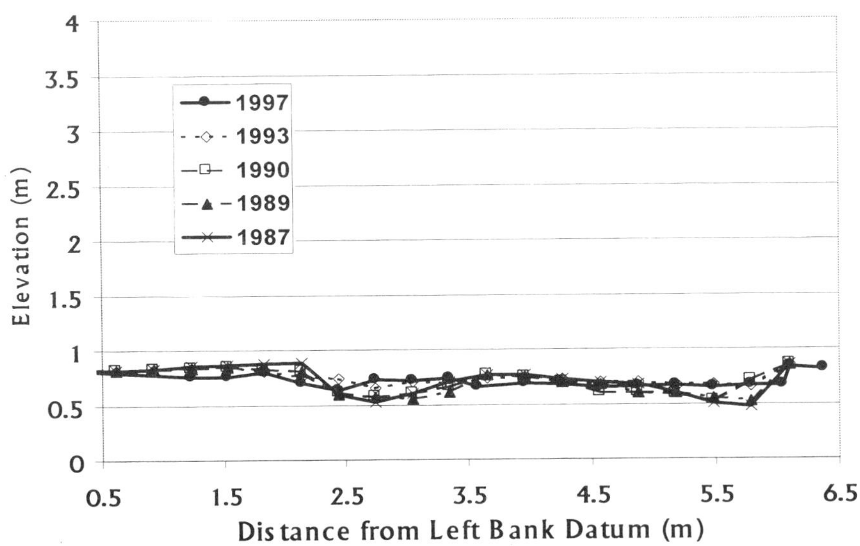

Figure 15. All cross section measurement of Station C2, one of the three sites underlain by cohesive deposits and showing almost no change 1987-1997.

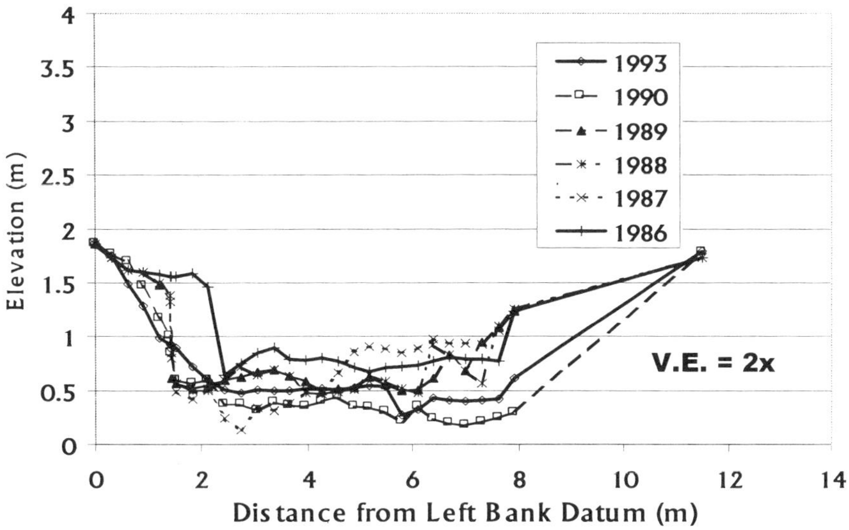

Figure 16. All cross section measurement of Station S3, showing active change in every measurement interval. Vertical exaggeration 2:1 (note expanded horizontal scale).

stress of the steeply flowing water and the delivery of sediment from farther up in the watershed. If urban development alters *any* element in this balance, however, the relative stability of the entire system can be lost with rapid and sometimes catastrophic results.

Rates of Channel Restabilization

In the final days of 1996, a major storm in the Seattle metropolitan area resulted in the failures of a road embankment and adjacent berm of a regional stormwater detention facility. Massive quantities of water and sediment were flushed down the North Fork and main stem of Boeing Creek, a stream draining several hundred hectares of primarily residential land use, and filled the valley of that channel as deep as 2 m with deposited sediment. Following this deposition, the channel of Boeing Creek began to reincise immediately, presenting an unexpected opportunity to document the establishment of a channel where watershed land cover was essentially stable

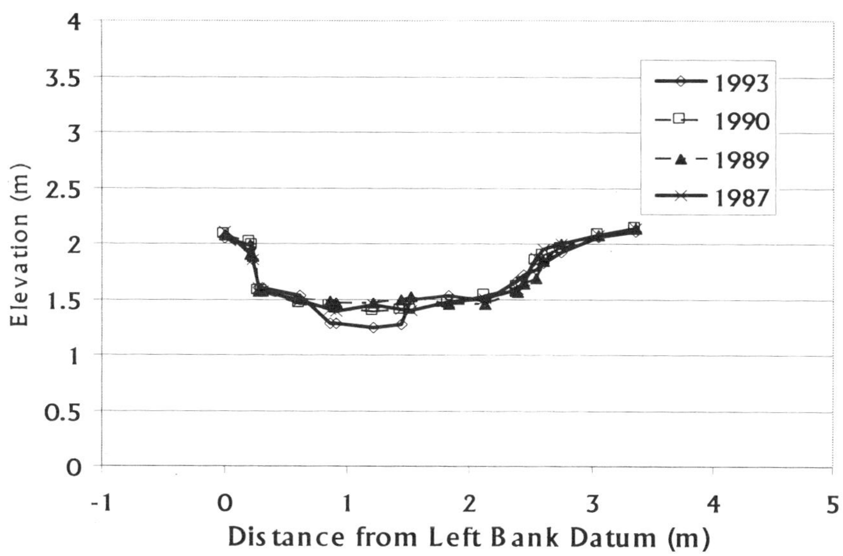

Figure 17. All cross section measurement of Station C9, the Timberline tributary that does *not* receive any appreciable runoff from the stormwater system of the upslope development.

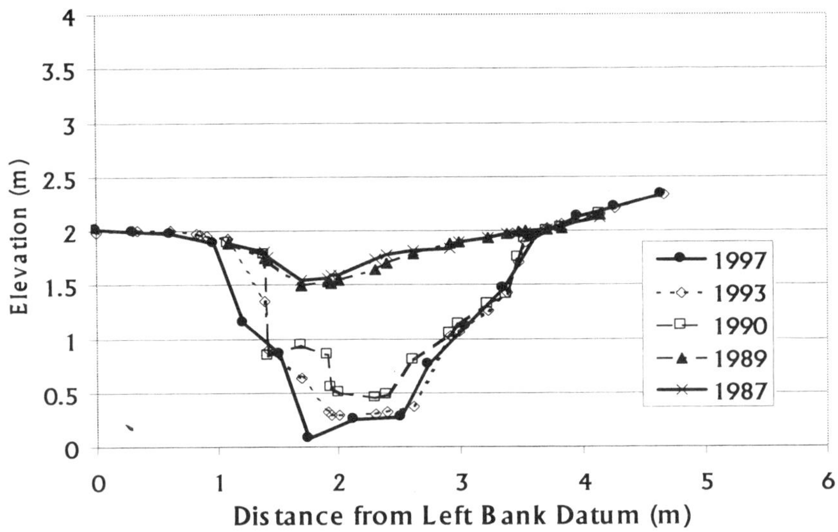

Figure 18. All cross section measurement of Station C8, the Timberline tributary that *does* receive significant stormwater runoff from the upslope development and which responded with abrupt downcutting in 1989-1990, and further incision by 1997,.

(and highly developed) and the sediment to be eroded was very easily transportable by even modest stream discharges. Seven cross sections were established in the 300-m reach between the failed detention pond and Hidden Lake, an artificial pond that marks the downstream end of the alluvial channel on the main stem of Boeing Creek. The normal protocol of annual measurements was replaced by monthly, and in some instances weekly, field visits.

The recovery of the channel of Boeing Creek was anticipated to take a period of some years and to be controlled in part by the reestablishment of floodplain vegetation to help bind and stabilize the very sandy fill that was deposited. Instead, reestablishment of an apparently stable, "equilibrium" channel was very rapid and occurred at most cross sections in a matter of a few months (Figures 19 and 20). The subsequent year's measurements demonstrated that this stability was not an artifact of the termination of rainfall in spring 1997; the 1998 channel, after another winter's high discharges, remained largely unchanged.

Discussion

The factors anticipated to influence the annual rate of channel change are generally well represented by the results. They include:

- Abundant rainfall,
- Easily erodible substrate, and
- Presence or absence of watershed urbanization.

These factors resist simple quantification, however, because of the tremendous variability imposed by the multiplicity of local geologic conditions, channel type, downstream grade controls (natural or artificial), the location of a chosen site in the context of the upstream channel network, and the variety of development ages and styles (*e.g.* residential density, or sewered vs. unsewered). Hammer [1972] recognized the last of these complications, and he developed a complex regression equation to express the observed relationship between different development types and channel dimensions. Yet these results are not readily transferred anywhere else, and the ever-changing patterns and styles of development render only the most general conclusions of lasting value.

The set of sample sites was neither varied enough nor large enough to allow a systematic evaluation of every relevant condition. However, several useful observations can be drawn:

1. The average annual rate of change can increase in a single channel by as much as 2 orders of magnitude between dry and wet years (e.g., the winter of 1989-1990, in this sample population); more typically, the greatest interannual change is about 5-fold. Because the study years include some of the largest lowland storms in recent memory as well as several quite unexceptional years, this variability is probably representative of most long-term conditions. In virtually all cases, the rate of channel change returns to nearly equivalent pre-event levels within one measurement interval (typically one or two years).
2. The previously recognized characteristics of erosion-susceptible channels are broadly correct—moderate to steep slopes, susceptible geologic materials, and

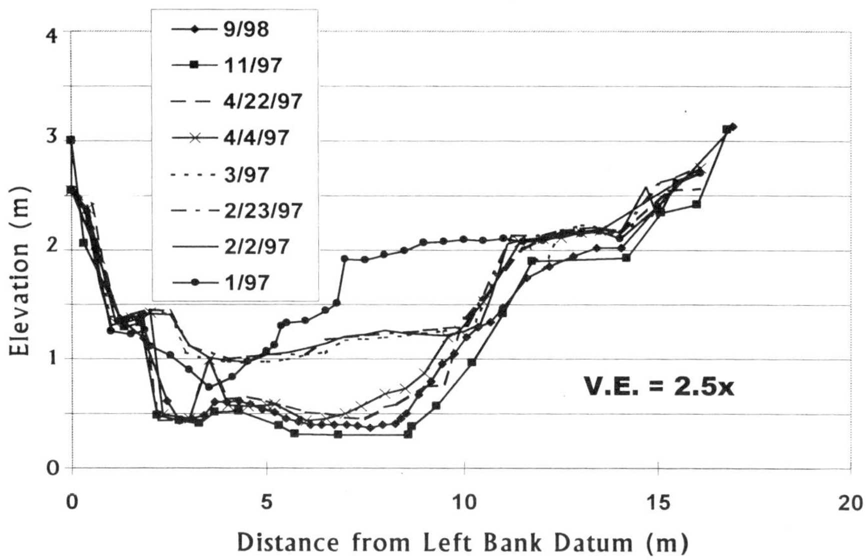

Figure 19. All measurements at Cross Section 4 of Boeing Creek, spanning 20 months since shortly after the site was obliterated by failure of the upstream detention pond embankment. Within weeks the channel achieved a "metastable" form that persisted for about two months (2/97-3/97); following several rainstorms in late March and early April 1997, channel form has remained nearly stable. Vertical exaggeration 2.5:1.

significant (and recent) upstream development. The unique factors of any given catchment, however, can greatly influence these predictions. No unconsolidated substrate appears immune from change, given sufficiently severe watershed disturbance. The streams draining large basins are more resistant than those draining small ones. Steep slopes in and of themselves are not critical, but they may increase the magnitude of the response to disturbance.

3. The age of the upstream development appears to be significant (as first recognized by Hammer, 1972) but the reason for this influence is enigmatic [see Henshaw, 1999]. In general, channels draining established neighborhoods [C1, C2, PS1, and PS4] show low rates of change. Possible explanations include (1) reequilibration of channel dimensions and sediment size with the increased (but now stable) flow regime; (2) removal of all erodible sediment from the channel perimeter, leaving non-erosive bed and banks; (3) cementation of channel sediments, a ubiquitous condition at these sites; or (4) reestablishment of bank vegetation following initial disruption of the channel by increased flows. Each of these explanations applies to certain sites, although (1) and (2) appear to be the most significant in a majority of cases. The reestablishment of equilibrium, however, does not necessarily coincide with a reestablishment of overall stream function or habitat quality: the channel capable of resisting the frequent, flashy urban discharges is generally inhospitable to most aquatic organisms.
4. Results are most unpredictable in the smallest basins (those of a few tens of hectares). In these basins, even a relatively small amount of development can have significant downstream effects if flow concentration occurs as a result of ditches or road crossings [S4, S5]. These effects, however, are not well represented by traditional methods of characterizing urbanization, such as impervious-area or disturbed-area percentages.
5. Any potential influence of channel slope is not well displayed by this data set (see Figure 8). All channels here have slopes of at least 1.3 percent, so true low-slope channels (also correlating, typically, with larger channels) are not represented. Within this data set, the only apparent relationship is that the very largest changes (>0.3 m/yr) appear to require a steep slope (4 percent or greater).

The experiences from this study also suggest several cautionary notes for future long-term channel monitoring. Most importantly, single-site measurements do not reliably characterize the overall status of the channel. Conditions both upstream and downstream of the selected site can be very different from those at the measured station. For example, delivery of sediment from an unstable upstream source can completely obscure the local behavior of the measured section. The headward migration of a downstream knickpoint, one obvious (and common)

Figure 20. View of the channel of Boeing Creek (Station PS7) at Cross Section 4 on September 9, 1998 (the last measurement plotted in Figure 19), looking downstream.

manifestation of an unstable stream reach [Booth, 1990], can cause dramatic changes in the bed elevation after a long period of apparent channel stability as measured at a single monitoring site.

Furthermore, the "stability" of a channel, as measured by the absence of change at single cross sections, does not necessarily equate with other desirable conditions, such as high-quality aquatic habitat. The converse statement, however, is generally correct: instability *does* correlate well with low habitat quality. Therefore, evaluating only channel stability does not provide unequivocal information on habitat conditions; if that information is needed, additional measurements are required.

SUMMARY

This study was begun with four objectives: to document rates of channel change, to evaluate the relationship between development intensity and the rate of channel change, to evaluate geologic and topographic controls on channel change, and to determine if predevelopment conditions could be used to predict susceptible reaches. Our results indicate the following conclusions:

1. Rates of vertical channel change vary from below the range of measurement error (<20 mm vertical change between visits) to about 1 m (width-averaged) per year. The median rate for this sample population was 60 mm per year.
2. Within these lightly to moderately urbanized watersheds, rates of channel change did not correlate with development intensity.
3. The nature of the geologic substrate strongly influenced whether or not significant channel change occurred. Other likely influences included local downstream grade control, riparian vegetation, and the age of the upstream development. Gradient was not a significant factor across the range of channels measured (0.013-0.52), but the importance of local grade controls suggest that low-gradient channels may show lower rates of change for a given level of disturbance and geologic susceptibility.
4. Channels with the greatest susceptibility share the following characteristics, which could be used to reduce the consequences of future urban development on natural stream systems:
   - Erosion-susceptible geologic substrate
   - Moderate to high gradient
   - Absence of natural or artificial grade controls
   - Water inputs via predominantly subsurface discharge, likely to be converted to surface (point) discharge in the post-development condition.

MANAGEMENT IMPLICATIONS

These results imply several consequences for watershed management. First, urban development is an obvious force in channel change, yet not all channels respond equivalently. The locations of potential susceptibility can be determined well in advance, at least in the Puget Lowlands of Washington, based on geologic conditions. Finally, channel changes, if and when they occur, can happen so rapidly after development begins that remediation, to be effective, must occur *prior* to development.

The results of this study also suggest that channel changes are very responsive to varying rainfall. This source of variability is obviously beyond the ability of surface-water or land-development agencies to control, yet its effects can be as significant as those of urban development. The most extreme effects of high rainfall are felt in the urbanized channels, and so one result of large storms is to amplify the differential response of developed and undeveloped watersheds. This imposes a challenging task for watershed managers: during low-rainfall years, any "warning" of impending channel-erosion disaster is muted, along with the public's concern for such issues. When a large storm arrives, however, the magnitude of channel change in urbanizing watersheds can cause significant damage, and its consequences invariably surprise almost everyone.

Acknowledgements. We are indebted to the associates and assistants who have helped with project design, site selection, and field assistance over the years, particularly Laura Casey, Kelin Whipple, Adelaide Johnson, John Buffington, Fred Bentler, Arny Stonkus, Karen Comings, and Marit Larson. Joan Blainey was instrumental in establishing the initial sites and making the first-year measurements on Boeing Creek. Partial support for the successful conclusion of this project was provided by the U. S. Environmental Protection Agency's Waters and Watersheds Program Grant no. R825284-01-0, and the Center for Urban Water Resources Management at the University of Washington. The manuscript was greatly improved from careful reviews by Ellen Wohl, Jim Pizzutto, and Kelin Whipple.

REFERENCES

Armstrong, J. E., Crandell, D. R., Easterbrook, D. J., and Noble, J. B., Late Pleistocene stratigraphy and chronology in southwestern British Columbia and northwestern Washington, *Geological Society of America Bulletin*, 76, 321-330, 1965.

Arnold, C. L., Boison, P. J., and Patton, P. C., Sawmill Brook, an example of rapid geomorphic change related to urbanization, *Journal of Geology*, 90, 155-166, 1982.

Bird, J. F., Channel incision at Eaglehawk Creek, Gippsland, Victoria, Australia, Proceedings of the Royal Society of Victoria, 94, 11-22, 1982.

Booth, D. B., Runoff and stream-channel changes following urbanization in King County, Washington, in *Engineering Geology in Washington, Vol. II*, edited by R. Gallster, Washington Division of Geology and Earth Resources Bulletin 78, 639-650, 1989.

Booth, D. B., Stream-channel incision following drainage-basin urbanization, *Water Resources Bulletin*, 26, 407-417, 1990.

Booth, D. B., Urbanization and the Natural Drainage System--Impacts, Solutions, and Prognoses, *Northwest Environmental Journal*, 7, 93-118, 1991.

Booth, D. B., and Jackson, C. R., Urbanization of aquatic systems—degradation thresholds, stormwater detention, and limits of mitigation, *Journal of the American Water Resources Association*, 33, 1077-1090, 1997.

Burges, S. J., Stoker, B. A., Wigmosta, M. S., and Moeller, R. A., Hydrological information and analyses required for mitigating hydrologic effects of urbanization, *Water Resources Series Technical Report 117*, University of Washington, Department of Civil Engineering, Seattle, 1989.

Dinicola, R. S., Characterization and simulation of rainfall-runoff relations for headwater basins in western King and Snohomish Counties, Washington state, U. S. Geological Survey Water-Resources Investigation Report 89-4052, 1990.

Douglas, I., Urban sedimentology, *Progress in Physical Geography*, 9, 255-280, 1985.

Dunne, T., and Leopold, L. B., *Water in Environmental Planning*, W. H. Freeman and Company, New York, 1978.

Ebisemiju, F. S., The response of headwater stream channels to urbanization in the humid tropics, *Hydrological Processes, 3,* 237-253, 1989a.

Ebisemiju, F. S., Patterns of stream channel response to urbanization in the humid tropics and their implications for urban land use planning: a case study from southwestern Nigeria, *Applied Geography*, 9, 273-286, 1989b.

Graf, W. L., The impact of suburbanization on fluvial geomorphology, *Water Resources Research*, 11, 690-692, 1975.

Gregory, K. J., Davis, R. J., and Downs, P. W., Identification of river channel change due to urbanization, *Applied Geography*, 12, 299-318, 1992.

Hammer, T. R., Stream and channel enlargement due to urbanization, *Water Resources Research*, 8, 1530-1540, 1972.

Henshaw, P. C., Restabilization of stream channels in urban watersheds, American Water Resources Association, Annual Water Resources Conference on "Watershed Management to Protect Declining Species," Seattle, WA, 1999.

Hollis, G. E., and Luckett, J. K., The response of natural river channels to urbanization: two case studies from southeast England, *Journal of Hydrology*, 30, 351-363, 1976.

King County, 1990, Hylebos Creek and Lower Puget Sound Basins Current and Future Conditions Report, Department of Public Works, Surface Water Management Division, Seattle, WA, 4 sections.

Kondolf, G. M., Geomorphological stream channel classification in aquatic habitat restoration: uses and limitations, *Aquatic Conservation: Marine and Freshwater Ecosystems,* 5, 127-141, 1995.

Leopold, L. B., River channel change with time—an example, *Geological Society of America Bulletin*, 84, 1845-1860, 1973.

Leopold, L. B., Wolman, M. G., and Miller, J. P., *Fluvial Processes in Geomorphology,* W. H. Freeman and Co., San Francisco, 1964.

Miller, J. F., Frederick, R. H., and Tracey, R. J., Precipitation-frequency atlas of the western United States, U. S. Department of Commerce, Volume XI—Washington, Washington, DC, 1973.

Montgomery, D. R., and Buffington, J. M., Channel reach morphology in mountain drainage basins, *Geological Society of America Bulletin*, 109, 596-611, 1997.

Morisawa, M., and LaFlure, E., 1982, Hydraulic geometry, stream equilibrium and urbanization, *in Adjustments of the Fluvial System*, edited by D. D. Rhodes and G. P. Williams, Allen and Unwin, London, 333-350.

Moscrip, A. L., and Montgomery, D. R., Urbanization, flood frequency, and salmon abundance in Puget Lowland streams, *Journal of the American Water Resources Association*, 33, 1289-1297, 1997.

Mosley, M. P., The classification and characterization of rivers, in River Channels: *Environment and Process*, edited by K. Richards, Blackwell, Oxford, 294-320, 1987.

Nanson, G. C., and Young, R., W., Downstream reduction of rural channel size with contrasting urban effects in small coastal streams of southeastern Australia, *Journal of Hydrology*, 52, 239-255, 1981.

Neller, R. J., A comparison of channel erosion in small urban and rural catchments, Armidale, New South Wales, *Earth Surface Processes and Landforms*, 13, 1-7, 1988.

Odemerho, F. O., Limited downstream response of stream channel size to urbanization in a humid tropical basin, *Professional Geographer*, 44, 332-339, 1992.

Park, C. C., Channel cross-sectional change, *in Changing River Channels*, edited by A. Gurnell and G. Petts, John Wiley and Sons, Chichester, 117-145, 1997.

Rosgen, D. L., A classification of natural rivers, *Catena,* 22, 169-199, 1994.

Simon, A., A model of channel response in disturbed alluvial channels, *Earth Surface Processes and Landforms*, 14, 11-26, 1989.

Tinkler, K. J., and Parish, J., Recent adjustments to the long profile of Cooksville Creek, and urbanized bedrock channel in Mississauga, Ontario, *in Rivers Over Rock: Fluvial Processes in Bedrock Channels*, edited by K. J. Tinkler and E. E. Wohl, American Geophysical Union, Monograph 107, Washington, DC, 167-187, 1998.

Whitlow, J. R., and Gregory, K. J., Changes in urban stream channels in Zimbabwe, *Regulated Rivers: Research and Management*, 4, 27-42, 1989.

Wolman, M. G., A cycle of sedimentation and erosion in urban river channels, *Geografiska Annaler*, 49A, 385-395, 1967.

Development and Application of Simplified Continuous Hydrologic Modeling for Drainage Design and Analysis

C. Rhett Jackson[1], Stephen J. Burges[2], Xu Liang[3], K. Malcolm Leytham[4], Kelly R. Whiting[5], David M. Hartley[5], Curt W. Crawford[5], Bruce N. Johnson[6], and Richard R. Horner[7]

Protection of aquatic resources and property in urbanizing watersheds requires integrated mitigation efforts that include effective stormwater retention and detention systems (hereafter referred to simply as detention systems). In the Puget Sound region, detention systems designed with event models based on the SCS curve number method do not provide sufficient storage to prevent increases in peak flows and degradation of stream habitat. Continuous hydrologic analysis with Hydrologic Simulation Program-FORTRAN (HSPF) provides more accurate hydrologic analysis but is difficult. The authors have developed a runoff files system that provides design flows, conveyance analysis, and detention facility designs comparable to those obtained with HSPF but with much less effort. The system serves as the regulatory model for analyzing and designing stormwater projects in King County, Washington. The runoff files provide the user with a database of time-series of unit area land surface runoff (runoff files) pre-simulated with HSPF for a variety of land cover conditions and soil types and for two precipitation regions. Menu-driven software, the King County Runoff Time Series (KCRTS) program, accesses this flow database to create continuous hydrographs, estimate design flows, iteratively design detention facilities, and analyze conveyance systems. To speed analysis and design, the system uses an eight-year reduced hydrologic record that represents the statistical characteristics of runoff generated by the 40-year precipitation record available at the time of development. The tools and methodology developed here eliminate the difficulties of continuous flow analysis while delivering much of their power and versatility, allowing simplified continuous hydrologic analysis to replace and improve upon event-based methodologies. Furthermore, the facilities designed with this system should provide the flow reduction benefits ascribed to them, as opposed to facilities designed with event methodologies, which, at least in the Pacific Northwest, do not.

[1]Daniel B. Warnell School of Forest Resources, University of Georgia, Athens, GA, 30602-2152

[2]Dept. of Civil and Environmental Engineering, University of Washington, Seattle, WA, 98195

Land Use and Watersheds: Human Influence on Hydrology and Geomorphology in Urban and Forest Areas
Water Science and Application Volume 2, pages 39-58

[3]Dept. of Civil and Environmental Engineering, 631 Davis Hall, University of California at Berkeley, CA, 94720-1720

[4]Northwest Hydraulics Consultants, 16300 Christensen Road, Suite 350, Tukwila, WA, 98188-3418

[5]King County Department of Natural Resources, Water and Land Resources Division, 700 5th Avenue, Suite 2200, Seattle, WA, 98104

[6]Concept Engineering, 455 Rainier Blvd. N., Issaquah, WA, 98027

[7]Center for Urban Horticulture, University of Washington, Box 354115, Seattle, WA, 98195-4115

INTRODUCTION

A team of hydrologists, engineers, geologists, planners, and programmers worked for over 10 years to develop a hydrologic/hydraulic analysis and design system that integrated advanced and powerful rainfall-runoff modeling techniques with the practical considerations of a widely-used regulatory tool. The goal was to create an analysis system that was easy and efficient to use and that produced accurate assessments and design. The effort was motivated by the belief that current hydrologic design techniques created stormwater management systems that did not sufficiently protect aquatic resources. For obvious reasons, development of this tool had to balance scientific, practical, and political considerations. This paper describes the justification, development, and use of the resulting hydrologic analysis system.

Without mitigation measures, land development endangers property, water quality, and aquatic habitat conditions by a variety of mechanisms, the most important of which is alteration of runoff generation processes. Hollis [1975] found that urbanization of a watershed can increase peak flows by factors of two to five. Moderate amounts of urbanization in Puget Sound basins increased peak flow rates by factors of 1.5 to 2.75 [Moscrip and Montgomery, 1997]. Such flow increases can lead to dramatic channel incision [Booth, 1990]. Channel widths are greater in urbanized basins [Booth and Jackson, 1997; Booth and Reinelt, 1993; and Hammer, 1972], and dramatic channel changes can occur within two years of development of between 10 and 50 percent of a basin [Graf, 1975]. A threshold of urbanization of 10 percent effective impervious coverage of a watershed discriminated between stable and unstable channels in western King County, Washington [Booth and Jackson, 1997]. Klein [1979] found fish and invertebrate diversity and baseflows began to drop when impervious area in a basin exceeded 10 percent and were greatly reduced when impervious area exceeded 30 percent. Similarly, Steedman [1988] found that indices of biological integrity were lower in streams with larger amounts of urban development in the contributing area. Causality of aquatic system degradation is not clear, and it is known that the sensitivity of a watershed to development depends on soil characteristics, soil management, riparian management, stormwater management, and flow paths through the landscape.

On-site and regional detention and infiltration facilities can be effective tools for mitigating hydrologic impacts of development if properly sited and designed. On-site detention facilities can consume 10 to 15 percent of developable land, so the opportunity costs of these facilities are high. Therefore, it is important to develop design methodologies that accurately predict hydrologic behavior and generate near-optimal designs for stormwater management facilities.

Toward this goal, a runoff files implementation of the U.S. Environmental Protection Agency's Hydrologic Simulation Program FORTRAN model (HSPF) was developed as a tool to provide the accuracy and versatility of continuous flow modeling but eliminate many of the difficulties. The system provides a framework for efficient design of effective on-site stormwater detention facilities. This hydrologic modeling tool produces results (design flows, detention pond sizing, level-pool routing, etc.) comparable to those obtained with the HSPF model but with significantly less effort. This is achieved by providing the user with a set of time series files of unit area land surface runoff (called "runoff files") pre-simulated with HSPF for a range of land cover conditions and soil types. This system was built as a regulatory model for development projects in King County, Washington, and it is based on climate and soils data for this region. To expedite analysis and design, the runoff files depend on a reduced hydrologic record that is statistically representative of the available long record. Estimation of design flows and design of detention facilities is accomplished by accessing and manipulating the runoff file data by means of supporting software known as the King County Runoff Time Series (KCRTS). Such a runoff files system was first developed and implemented by Lumb and James [1976]. This effort extends their work to a different hydrologic regime and deals differently with the statistical problems of a condensed record.

The runoff files are a database of hourly and 15-minute time series data of simulated unit area land surface runoff and shallow groundwater flow (interflow) for two precipitation records and eight land use classifications. Each land use classification is defined by a soil type and a cover type. To take advantage of the runoff files, the analyst uses the KCRTS software to scale and combine the unit area runoff from a mix of land use types to create runoff time series for a basin and to size stormwater management facilities.

To determine reliable design flows, streamflow statistics, and facility designs, continuous hydrologic models must simulate long time series of flows on the order of 40 years or more. Use of long time-series of data imposes a computational burden on the user, so an important feature of the runoff files method is selection of a shorter sample of hydrologic data that is statistically representative of the full 40-year record of data generally available within King County. These runoff files were subjected to detailed examination and testing to identify seven individual years whose flow duration statistics nearly matched those from the full 40-year data record because the duration of high flows is a principal determinant of sediment routing and channel stability. Further testing was then carried out to allow flood frequency relationships to be developed from the seven years of data selected to represent the 40-year record. The pre-simulated runoff files are stored in an easily retrievable format on floppy disks and on a web server (http://splash.metrokc.gov/wlr/lds/manual.htm). Selection of a comparatively short record on which to base analysis greatly reduces the computational effort of using continuous simulation without losing the benefit of the information available in the longer 40-year record.

The runoff files include an eighth year in which a hypothetical 100-year event was embedded. This 100-year event was created by scaling up simulated runoff from a severe storm of January 9-10, 1990, to produce peak runoff rates and volumes comparable to those for a 100-year event. A hypothetical event was necessary because of its regulatory importance (especially for flood plain management) and because the available 40-year records did not include such an event.

Stormwater detention design using the KCRTS runoff files software involves three basic steps: 1) estimation of a pre-development continuous hydrograph from the parcel of land in question, 2) estimation of a post-development continuous hydrograph, and 3) stormwater facility design. The capabilities of the KCRTS software at present include the following:

1. Estimation of a pre-development continuous time series of flows for a specified land use and location within the County.
2. Estimation of a post-development continuous time series of flows for a specified land use and location within the County.
3. Level pool routing of time series through reservoirs including infiltration and up to two surface discharges.
4. Flow frequency and flow duration analyses for determining target releases, evaluating detention facility performance, and evaluating impacts to downstream problems.
5. Stage frequency and stage duration analyses for assessing water level fluctuations in lakes and wetlands.
6. Design of a variety of detention and infiltration ponds, vaults, and tanks.
7. Plotting of analysis results.

Software Design Goals

As a regulatory model, the KCRTS software had to balance technical capabilities with ease of use. While the primary goal was to produce accurate hydrologic analyses, the tool had to be acceptable to the development engineering community accustomed to working with event models. Designers of the software developed the following specifications believed necessary to make the tool worthwhile and acceptable:

1. Produce designs for effective detention and infiltration systems, i.e. designs that meet stated performance standards.
2. Yield accurate flow recurrence and duration statistics.
3. Perform routing calculations for eight-year time series in less than 30 seconds on a typical personal computer.
4. Operate on a personal computer with an Intel 80386 processor.
5. Fit the executable program and data files on two floppy disks.
6. Allow a user to design a detention facility in less than 30 minutes.

Some of these performance requirements betray the long development time of the runoff files and KCRTS software. Initial work on this system began in 1987, but the first working system was developed in 1993, and the first accepted regulatory version was not completed until 1998. Advances in computational hardware and data transfers have rendered some of these design goals quaint. Nevertheless, in its current form, KCRTS has achieved all of these objectives. The total file size of the runoff files and KCRTS software is 5.0 megabytes, and in compressed format, the files fit on two floppy disks. Rather than transfer by floppy disk, however, most users now download the runoff files from the King County design manual web site at http://splash.metrokc.gov/wlr/lds/manual.htm. The current version of KCRTS also supports long record runoff files (up to 99 years), and these long records are being used to assess large reservoirs with residence times in excess of a year.

Versatility of Analysis and Applicability to Volume-Driven Hydrologic Problems

Hydrologic problems in many climates, including that of lowland western Washington State, are driven by high volumes of flow from sequential winter storms rather than from peak flows from short duration and high intensity rainfall. Continuous hydrologic analysis tools are needed in these environments to provide useful analysis. Continuous models include algorithms that maintain a continuous water balance for a catchment to account for soil moisture and hydraulic conditions antecedent to each storm event [Linsley et al., 1982] whereas event models assume initial conditions and only address single storm events. As a result, continuous hydrologic models are more appropriate for evaluating runoff during periods of extended rainy weather.

The benefits of continuous hydrologic modeling with HSPF are summarized as follows:

1. A continuous model accounts for long-duration and high volume wet periods characterized by sequential, low-intensity rainfall events. Continuous simulation uses continuous long-term records of observed rainfall rather than short periods of data representing hypothetical "design" storm events. As a result, continuous simulation explicitly accounts for the long duration rainfall events typically experienced in the Pacific Northwest lowlands as well as the effects of rainfall antecedent to major storm events. Snow and rain-on-snow events are not important in the Puget Sound lowlands.
2. Continuous simulation allows direct examination of flow duration data for assessing the impacts of land use change (usually urban/suburban or industrial development) on stream erosion and morphology. An event model, whether using a 1-day or a 7-day storm, cannot provide such information.
3. A continuous model allows water level analysis for wetlands, lakes, and closed depressions whose water level regime is often dependent on seasonal runoff rather than 1-day or 7-day event runoff.
4. Continuous modeling with a long record or representative record produces retention/detention facility designs that achieve desired performance goals.
5. The continuous model HSPF has been shown to simulate runoff accurately from basins with a wide range of sizes and land-uses using the Puget Lowland generalized (regional) parameters developed by the United States Geological Survey (USGS) [Dinicola, 1989].

The importance of continuous modeling in the Puget Sound is illustrated in Figure 1, which shows a small basin's simulated runoff response to a series of winter storms and the outflow from a detention pond designed to control the peak annual flows from this basin. The largest outflow from the detention pond corresponds not to the peak inflow but rather to the high volume of flow from sequential storms.

Part of the need for versatility of hydrologic analysis arises from the multiple levels of stormwater flow control required by King County. The County has three tiers of flow control appropriate for different receiving water conditions: conveyance standard, duration control standard, and water level protection standard.

The conveyance standard is designed to control peak flood flow rates at their current levels and to maintain flows within the capacity of the conveyance system for most storm events. Specifically, the conveyance standard requires maintaining the pre-development peak flow rates for all recurrence intervals from the 2-year through the 10-year flow rates. This standard may be modified to ensure that downstream problems are neither created nor aggravated, and this may mean that control of a 100-year flood hydrograph is necessary as well. The conveyance standard is applied to basins with little habitat value or basins with resilient stream channels. This standard is also applied at the lower end of sensitive basins where additional flow attenuation provides little benefit to the receiving waters.

The duration control standard is designed to control the durations of geomorphically significant flows and thereby maintain existing channel and streambank erosion rates. A geomorphically significant flow is one that moves sediments on the channel bed. The flow that initiates transport of sediments varies from channel to channel, but one-half of the 2-year flow can be considered a general estimate of the erosion-initiating flow. A review of the literature indicates that erosion-initiating flows may range from 0.3 to 0.7 of the 2-year flow, but for gravel bed streams, one-half of the 2-year flow has some scientific backing [e.g., Pickup and Warner, 1976; Andrews, 1984; Carling, 1988; Sidle, 1988]. Specifically, the duration control standard requires maintaining the durations of high flows at their pre-development level for all flows greater than one-half of the 2-year flow up to the 50-year flow. This standard is applied to stream systems with erosion prone channels or important aquatic habitat. Application of this standard is most critical in headwater basins.

The most stringent detention standard applied in King County is the water level protection standard, which is motivated by the difficulty in mitigating water level changes in lakes, wetlands, and closed depressions that provide natural flood storage. Because such water bodies act as natural flow dampeners, it is difficult to detain collected stormwater beyond the natural residence time of these systems. Therefore, the increased volume of runoff from new development inevitably increases the peak water levels of these water bodies. The water level protection standard provides additional storage and increases the detention time to minimize these downstream impacts. It requires maintaining the durations of high flows at their pre-development level for all flows greater than one-half of the 2-year flow rate up to the 50-year flow rate and also attenuating the 100-year peak flow rate to its pre-development level.

Impetus for Developing a Continuous Flow Model for Hydrologic Design

Much of the impetus for developing KCRTS came from inferred problems with detention facilities designed with the Santa Barbara Urban Hydrograph Model (SBUH). SBUH is a version of the SCS curve number model [U.S. Soil Conservation Service, 1975], and it was the hydrologic and hydrau-

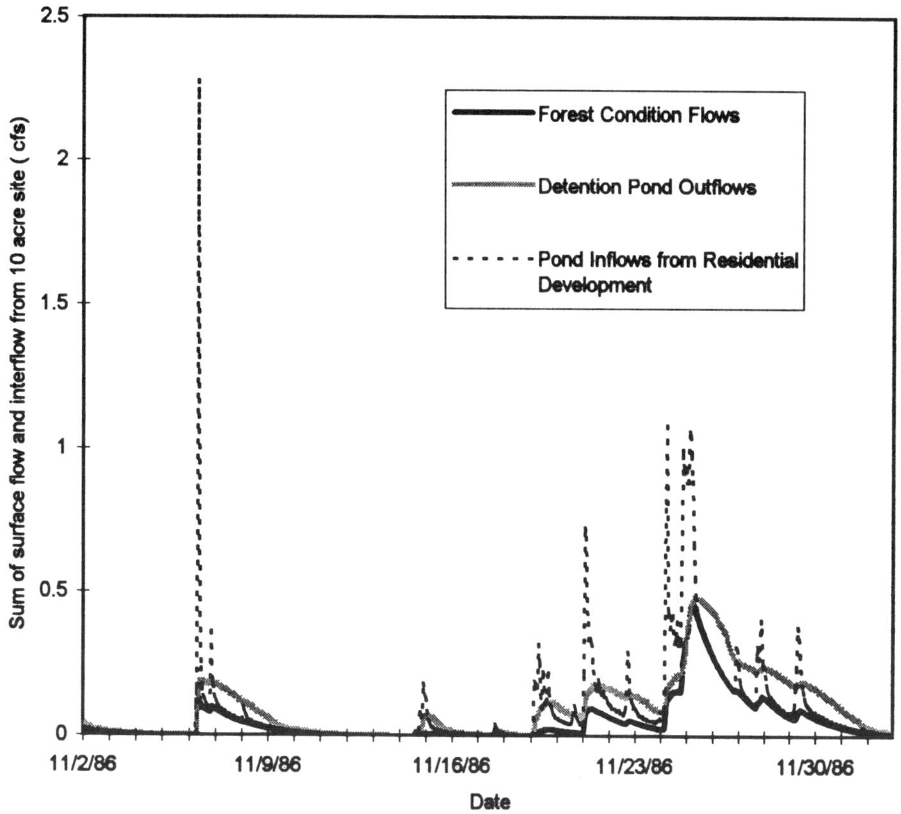

Figure 1. Runoff hydrographs from a 4.05-hectare (10-acre) site on till soils for three different conditions: 1) forest cover, 2) residential development (25 percent impervious and 75 percent lawn), and 3) the same residential development with a detention facility. The hydrographs show that the largest flows out of a detention facility are caused by high volume storms and do not necessarily correspond to the highest inflow peaks. The detention facility outflow hydrograph also shows the detention facility does not empty between sequential storms. Note: one cubic foot per second (cfs) equals 0.0283 cubic meters per second.

lic analysis model included in the 1990 version of the King County Surface Water Design Manual. Versions of the SCS curve number method, such as SBUH, are the standard tools around the country for stormwater analysis and design. King County Water and Land Resources Division personnel conducted extensive HSPF modeling of streamflows under future development scenarios for basins on the scale of hundreds of hectares to tens of square kilometers. In all cases, these simulations have shown that there was no way to regulate or apportion the total storage volume designed for new development with the SBUH model and achieve the stated goals for flow management. For example, HSPF simulations predicted that 2-year and 10-year flows would increase significantly in simulated basins in which all new developments included detention facilities designed to match the SBUH 2-year and 10-year predeveloped flows [Barker et al, 1991; King County, 1989; 1990a; 1990b; 1990c; 1991; 1992; 1993]. The scale of the increase depended on the intensity of the development and the pre-developed conditions. The detention volume was usually insufficient, even though a 30 percent safety factor was applied to the detention facility volumes after design. Figure 2 shows flow increases predicted for a King County watershed after high density residential development with SBUH designed detention facilities.

SCS-based models are inappropriate for detention pond sizing in the Puget Sound climate for several reasons. Gage data and observations demonstrate that SBUH greatly overpredicts peak flow rates for pasture conditions and slightly overpredicts flows for forest conditions. As a result, release rates for detention facilities designed with these models are too high. SBUH-designed detention facilities were often observed nearly empty or at the dead storage level during large rainstorms because the ponds released water too quickly. King County first attempted to address these problems by calibrating local curve numbers, by setting the release rate for a pre-development flow that was more frequent than the design inflow, and by using a seven-day

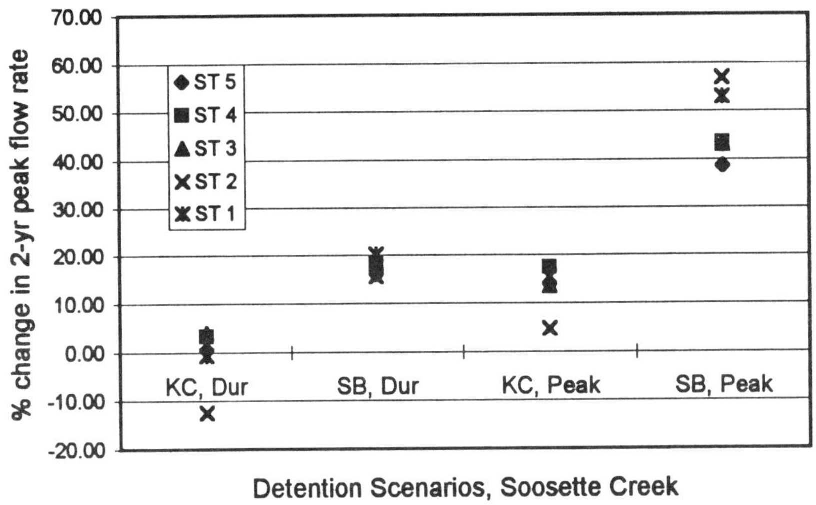

Figure 2. Percent change in the two-year peak flow rate for the five sub-basins (ST1 through ST5) in the Soosette Creek basin under four detention scenarios: 1) KCRTS duration control standard (KC, Dur), 2) SBUH duration control standard (SB, Dur), 3) KCRTS conveyance standard (KC, Peak), and 4) SBUH conveyance standard (SB, Peak). All simulations assume high density residential development and are compared to flows simulated from 1985 land use conditions.

design storm. These repairs were not considered as robust or useful as the solutions provided by the runoff files. These modifications were also controversial since it was perceived that the County was tinkering with an established engineering design method to achieve larger detention ponds.

The original purpose for the SCS method was for the design of small irrigation or agricultural drainage projects in ungaged catchments. The propensity to overestimate peak flows was not of paramount concern since it led to conservative design of spillways, conveyances, and other hydraulic structures. The apparent conservatism in estimation of peak flows leads to over-estimation of pre-development peak flow rates. As a result, target detention facility release rates (pre-development flow rates) are over-estimated and storage requirements are underestimated.

The single 24-hour event used in SBUH cannot represent the sequential storm characteristics of Puget Sound winters where most flooding is driven by storms lasting six to seven days with wet antecedent conditions. The SBUH method assumes that detention facilities are empty at the start of a design event, whereas an effective detention facility is often partially full as a result of preceding storms. Aware of these problems, King County personnel developed several methods to fix the problems of SBUH [e.g. Barker et al, 1991; King County, 1990d], but these manipulations led to multiple and confusing design standards and did not directly address flow duration problems. Another problem with the SBUH methodology was the large amount of judgement allowed and required in determining times of concentration and curve numbers. This latitude meant developers could bias calculations to minimize the sizes of their detention ponds and reduce the loss of developable land and agency personnel had no consistent methodology for evaluating submitted designs. The many problems associated with the use of "design-storm" methods have been long known and documented [Guo and Adams, 1999].

DEVELOPMENT OF RUNOFF FILES

Eight steps were involved in the development of the runoff files and KCRTS: 1) selection of HSPF model parameters for a range of land cover conditions and soil types, 2) quality assurance and correction of rainfall data, 3) selection of a short climate record that accurately substitutes for the long record, 4) generation of runoff files using HSPF, 5) determining plot positions for peak annual flows so that the short record could be used for flow recurrence estimation, 6) creation of 100-year flood hydrographs, 7) model verification (against HSPF long-term simulations) and modification, 8) development of supporting software (KCRTS), 9) evaluating and justifying differences between KCRTS-designed facilities and those designed with an event methodology, and 10) training the engineering community to use the new system.

Ungaged Catchment Problem - Regional Parameterization

The first step in development of the runoff files method was selection of land cover and soil types for which runoff files were to be generated together with selection of appropriate HSPF model parameters for these land cover and soil types. The USGS had previously calibrated HSPF against streamflow data from 21 stream gage sites in western King and Snohomish Counties to develop generalized model parameters for common combinations of surficial geology, land cover (soil/cover types), and slope [Dinicola, 1989]. While hillslope runoff characteristics depend on many soil and landform attributes, lumping of soil/cover types is necessary to achieve a manageable number of flow records for inclusion in the runoff files. The USGS generalized parameters also separated land surfaces into three slopes, flat, moderate, and steep, but these slope categories were not used in the runoff files because subsequent evaluations against gage data questioned their reliability. Therefore, all slopes in the runoff files were categorized as moderate.

The gaged basins used to develop the generalized HSPF parameters ranged from 3.36 to 173 square kilometers. The generalized parameters have subsequently been tested against gaged runoff from small watersheds for several development project applications and in a research project. Accounting for soil moisture, groundwater, and streamflow interactions becomes more difficult in the small watersheds, but the generalized parameters still produce adequate simulations.

The generalized parameters are used in HSPF models in the Puget Sound basin when local gage data for calibration are not available. The generalized parameters are also used as the starting parameters when calibrating HSPF flows to gage data. While the generalized parameters are poorly suited to simulate runoff from certain geologic conditions found in the area, experience has shown that the regional parameters provide robust simulations for the large majority of the land in King County. The flows in the runoff files were created with the generalized parameters. The runoff files cannot be calibrated to streamflow data, and if calibration is needed for a project, HSPF must be used.

Representation of Local Hydrologic Characteristics

Under the runoff files method, the land surface hydrologic response (represented by a time series of unit area land surface runoff) was generated by HSPF for eight soil/cover types and for two long-term (40+ years) hourly rainfall stations, one representing the western lowlands of King County and the other representing the foothills and valleys to the east. The far eastern part of the County is covered by the Cascade Mountains, which are largely uninhabited, so no system was developed for this area. Runoff files were generated with precipitation data from these stations for the following eight soil/cover types: impervious, till-forest, till-pasture, till-grass, outwash-forest, outwash-pasture, outwash-grass, and wetland. West of the Cascades, King County was overridden by a continental glacier 12,000 years ago, and almost all surficial soils in this area are derived from either glacial till or glacial outwash. The till soils have relatively small amounts of soil storage, and their runoff rates are much greater than those of outwash soils, which typically feature deep, high-porosity soils. Dinicola [1989] had previously defined soil/cover types for the region, but his system was simplified here by eliminating the slope classifications for till soils.

While HSPF simulates surface runoff, interflow, and groundwater flow, only the surface and interflow components of runoff are included in the runoff files because these files have been created as an analysis tool for engineers designing stormwater systems for new developments and roads. The large majority of developments are relatively small, and it is usually not appropriate to include groundwater flows in estimates of the surface or near-surface runoff from a site. For example, a detention facility serving a small development on till soils is likely to receive surface runoff and interflow (shallow groundwater flow) from the site. Groundwater generated on the site would seep through the underlying till and would likely reappear in springs or seeps some considerable distance from the site.

An interflow component of runoff is not computed for outwash soils because there is assumed to be no low-permeability subsurface layer. Runoff files for on-site detention facility design were thus generated with only the surface flow component for outwash soils and impervious surfaces and with both surface flow and interflow for till soils and wetlands.

Initial Selection of Precipitation Gages

The orographic effects of the Olympic Mountains, the Cascade Range, and the Cascade foothills strongly affect precipitation patterns in King County, and annual precipitation varies by 50 percent across the urban and suburban areas of the County. The developers of KCRTS originally intended to use as many as nine hourly precipitation gages to simulate runoff for use in the runoff files to capture the geographic variability of the precipitation and runoff regimes. Findings from preliminary data analysis reduced the number of records to be analyzed in detail to six. The names and locations of precipitation records considered are shown in Figure 3. The precipitation records were evaluated in 1991, and data through 1990 were used. Four of the precipitation records covered time periods in excess of 30 years: Seattle Tacoma International Airport (Seatac), Landsburg , Carnation, and Snoqualmie Falls. The remaining records, Auburn and Seattle EMSU (North Seattle), were less than 20 years in length. Each of these precipitation records was evaluated for data reliability, period of record, and reasonableness of simulated runoff.

The precipitation records were checked for data gaps and anomalies. Data gaps and anomalies were numerous, which is not surprising to persons accustomed to working with

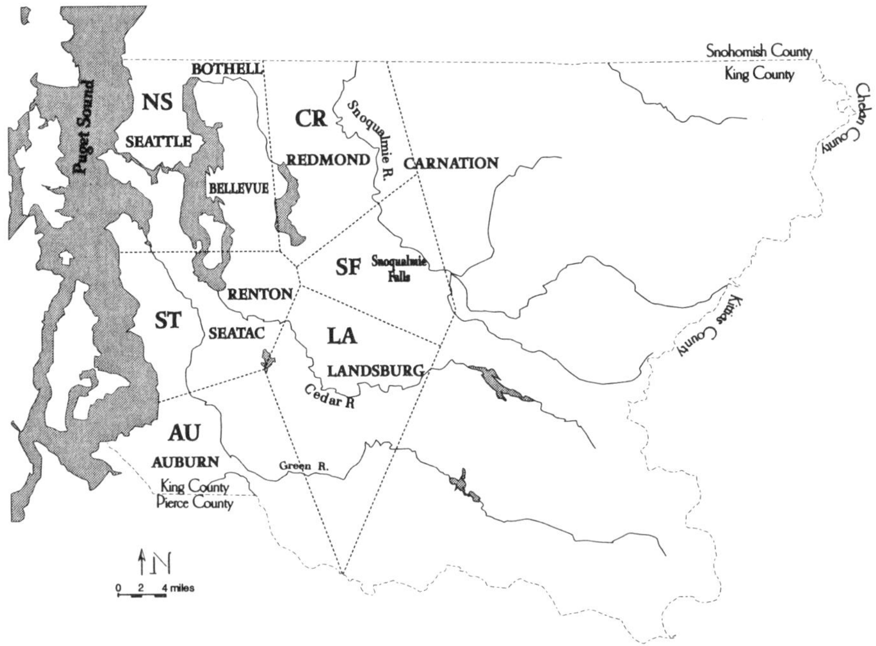

Figure 3. The six precipitation stations and their corresponding runoff files regions that were intensively analyzed for quality of data record and runoff characteristics. NS is North Seattle, CR is Carnation, SF is Snoqualmie Falls, LA is Landsburg, AU is Auburn, and ST is Seatac. The eastern part of the County is mountainous, largely uninhabited, and its precipitation is dominated by snowfall, so no attempt was made to analyze precipitation records or develop runoff files for that area.

precipitation records. Data gaps and errors were identified by manually checking electronic records against printed records, and problem data were replaced by transposition of data from nearby gages. Runoff was then simulated with the edited data, and the simulated runoff was evaluated for anomalies. These anomalies pointed to other errors in the precipitation records, which were edited again. The precipitation data were not corrected for wind effects, so it is likely that rainfall is underestimated. Editing the precipitation records was a time-consuming and labor-intensive process. In a similar effort to ensure the quality of precipitation records, the Miami Conservancy District conducted four separate rounds of editing before achieving acceptable precipitaton data quality for the design of dams and channels that have been in place since 1922 [Morgan, 1951].

Creation of Short Record and Selection of Water Years

To minimize the computational burden of conducting hydrologic analysis with a long record, the development team tried to identify a shorter sample of data that was statistically representative of the existing 40 year record and that could be used as a surrogate for the full record. The full Seatac precipitation record was searched for any combination of seven water years that together would produce flow duration statistics for common soil/cover types that matched the flow duration statistics from a full record simulation. In other words, running this combination of seven water years sequentially in HSPF produces flow duration statistics close to those from the long-term Seatac record. Selection of the short record began in the fall of 1987 and used 40 years of data (water years 1948-1987). Because of the labor-intensive effort required to select seven representative water years, this process was not repeated for the other precipitation records, but rather it was assumed that the same seven water years would be appropriate. This assumption was tested subsequently and determined to be invalid, as will be discussed in a later section.

Selection of the seven water years was a laborious trial and error process that was not guided by a defined algorithm. The flow duration curve was not robust over a seven-water-year sample, meaning that the flow duration curve varied considerably based on the selection of the water years, even when the average annual precipitation for the selected seven-year sample was near the long-term average. The following seven water years were identified as producing runoff with flow duration curves that match those produced from a 40-year simulation: 1951, 1966, 1972, 1977, 1983, 1986, and 1987. These water years include relatively wet and relatively dry years, and there is no characteristic common to these water years. It is not known whether these seven years constitute a unique optimal solution to the short record problem since there was no systematic effort to try every combination of seven years.

Evaluation and Elimination of Gage Records

The runoff files were tested and modified based on their performance relative to long-term HSPF simulations. For each of the three long-term (30+ years) local precipitation records other than the Seatac record, simulations were conducted to test whether the seven water years selected from the Seatac record served as reliable surrogates for the full record at each station. Flow duration statistics generated from the short seven-year record were compared with those from the full period of record.

Analysis of simulation results quickly revealed great differences in runoff bchavior predicted from the Snoqualmie Falls precipitation record and the other local precipitation records. Although these problems were revealed by the flow duration test simulations, the problems were not related specifically to flow durations. Previous HSPF modeling in the region, including the regional calibration effort by Dinicola [1989] and basin planning modeling done by King County, had all used precipitation records that recorded rainfall in hundredths of an inch (Note: most rain gages used in the United States still record rainfall in inches and fractions thereof. A hundredth of an inch is equal to 0.254 millimeters). The Snoqualmie Falls record, however, recorded rainfall in tenths of an inch (or units of 2.54 millimeters). It was determined that the HSPF parameters calibrated for precipitation records recording hundredths of an inch would not work well with rainfall data recorded in tenths of an inch. The coarser precipitation record artificially punctuated and temporarily intensified the low intensity drizzle common in the Puget Sound. Rather than ten hours each receiving one hundredth of an inch, for example, the rainfall would be recorded as nine hours with no rainfall and one hour with one tenth of an inch. The result was that HSPF would simulate surface runoff when none had occurred. The coarse precipitation records were therefore incompatible with the previous continuous flow modeling work that had been done in the region. For this reason, the Snoqualmie Falls record was eliminated from the runoff files database.

The assumption that the short record selected from Seatac data would suffice for the other gages held only for the Landsburg gage. Comparisons of the short- and long-term flow statistics for the Carnation record showed that the chosen water years did a poor job of representing flow durations in the Carnation area. The Carnation record was dropped from the runoff files because there was insufficient manpower to select a representative short record for this gage. Furthermore, the Carnation record demonstrated that the chosen subset of seven years did not replicate long-term flow statistics for every local record. Since the validity of the selected set of years could not be sufficiently tested for the relatively short local precipitation records at Auburn and Seattle EMSU, these records were also dropped from the runoff files. After starting with six potential precipitation records for the County, only the Seatac and Landsburg records were retained in the runoff files system because their short record statistics compared favorably with their long record statistics. As will be discussed in a later section, additional tests were conducted to evaluate the validity of using only the Seatac and Landsburg short records to simulate the entire County.

The statistical agreement between long-term HSPF flows and KCRTS flows are illustrated in Figures 4a-4c, which show that the runoff files overestimated flow durations in almost all cases. Because the exaggeration was not extreme and was quite uniform, so that till forest flows, till pasture flows, and till grass flows were all overestimated by the same degree, duration-control facilities designed with KCRTS were equivalent to those designed with HSPF as will be discussed later. The overestimation of durations for high flows resulted because the original water year selection efforts matched flow durations in linear space, so that flow durations matched for the lowest 99 percent of flows. In stormwater design, however, it is also necessary to match durations for the highest 1 percent of flows in log exceedance space. Because the selected subset of water years contain several high flow winters (1951, 1986, and 1987), the flow durations for the highest 1 percent of flows are exaggerated. Subsequent tests showed that, because the exaggeration of high flows was consistent among soil/cover types, the effect on facility design was not significant.

Creation of Synthetic 100-year Runoff Hydrograph

Many regulations concerning hydraulic design and hydrologic control require analysis and design considering the 100-year flow, and therefore the runoff files needed to include a 100-year flow. Creating a 100-year flow hydrograph, however, is difficult. Because King County's precipitation and streamflow records only extend about 45 years, estimates of 100-year flow rates and volumes are not reliable. In addition, creating 100-year flows poses a multiple probability problem; in reality, no single storm will create 100-year flows for each of the land types. A storm that generates a 100-year flow from forested outwash soils is very different from a storm that generates a 100-year flow from grassed till soils. If flow from a single event is scaled up so that every land type produces a 100-year peak flow rate, then the flow rate for a collection of land types will be significantly larger than the true 100-year flow rate for that combination.

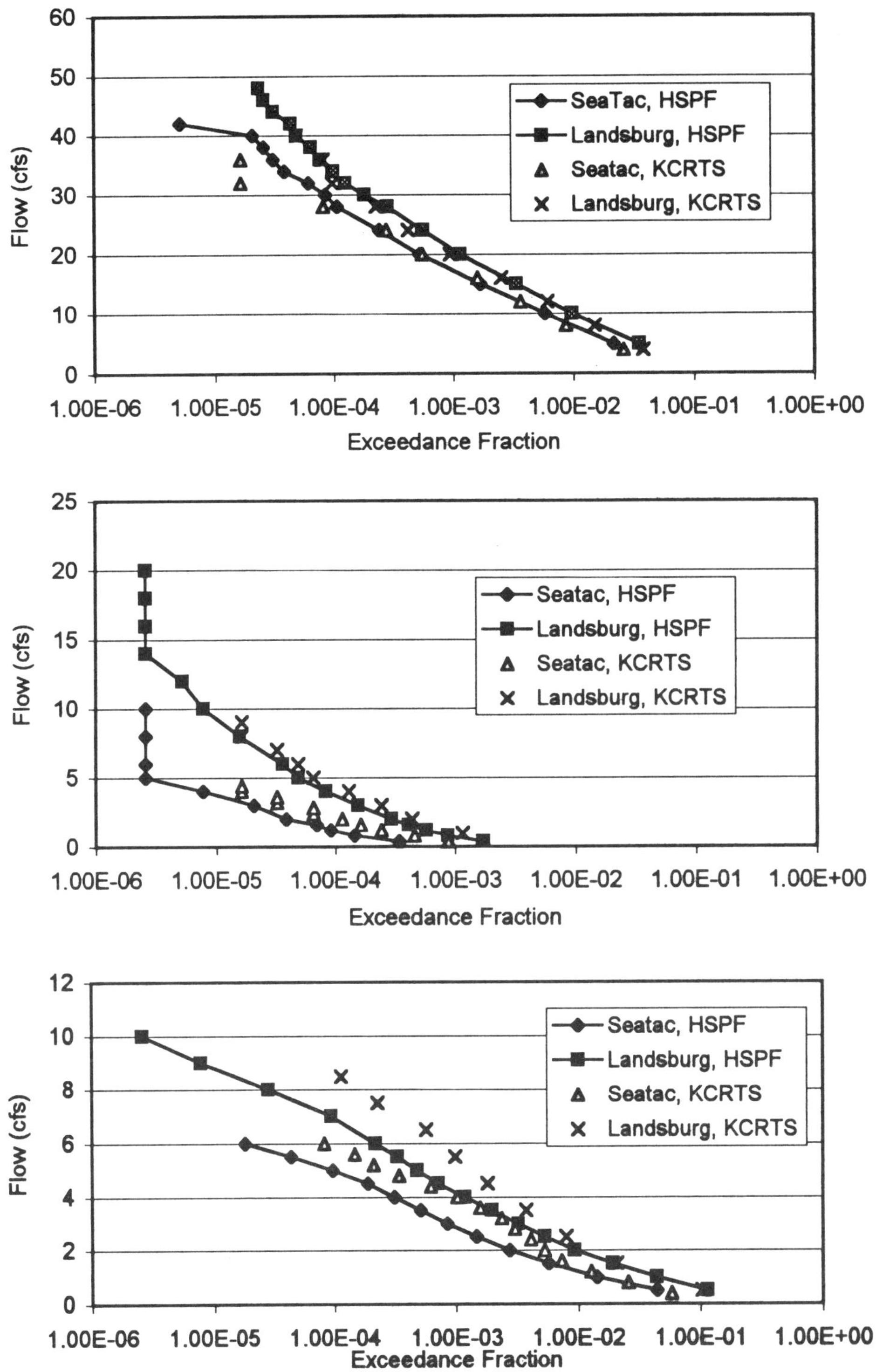

Figures 4a-4c. Comparison of HSPF-generated and KCRTS-generated flow durations for the highest 1 percent of flows. Durations are shown for three cover types, all located in the Seatac precipitation region: a) impervious, b) till soils with forest cover, and c) outwash soils with grass cover. KCRTS simulated flows consistently over-estimate the durations of flows for the top 1 percent of flows. The original selection of the abbreviated meteorologic record was based on matching the lowest 99 percent of flows. Note: one cubic foot per second (cfs) equals 0.0283 cubic meters per second.

Recognizing that not all of these problems could be solved, the creation of the 100-year runoff hydrograph series was simplified to two tasks: 1) selecting an appropriate storm period for scaling, and 2) determining scale factors.

In King County, the January 1990 storm was unusual in that it produced high peak flows for all land types because of wet antecedent conditions. The storm volume and duration was such that it produced high flows from outwash soils and the peak intensity was such that it produced high flow rates from till soils and impervious surfaces. For this reason, the January 1990 storm event could be scaled so that it produces "100-year" flows for aggregations of land types while still producing high flows (greater than or equal to any in the seven-year runoff files record) for each individual land type.

For each land type, 100-year peak flow rates were estimated by fitting a Log Pearson Type III distribution to peak annual flows generated with 42-year (1948-1990) HSPF simulations. The Log Pearson Type III fitting scheme, however, overestimates long return period flows from the wetland and outwash soils because of the attenuating effects of surface storage in wetlands and the high amount of soil storage in outwash soils. Dynamic water storage within a hydrologic system tends to flatten the hydrologic response at higher return periods [Wolff and Burges, 1994]. Therefore, 100-year peak flow rates for outwash and wetland soils were determined graphically by eyeball fitting of a flow frequency curve to peak flows plotted using the Gringorten plotting scheme [Stedinger et al., 1993]. Peak annual flows from outwash and wetland soils plot linearly on a semi-log graph.

Scale factors were selected to produce a weighted-average scaling factor that increases the January 1990 peak flow from a mixture of land types by about 20 to 30 percent. This is approximately what is necessary to raise observed January 1990 peak flow rates to statistically-determined 100-year rates. The first step in choosing appropriate scaling factors was determining the ratio between the 100-year flow rate and the January 1990 flow rate for each land type. For most land types, the final scaling was less than this ratio. Because a uniform scaling factor over-scales flows from some land types and under-scales others, individual scaling factors were determined for each land type. Obviously, creation of a 100-year design event was a trial-and-error process without standard protocols. The resulting time series produced an approximate 100-year flow hydrograph for mixes of land types and at least produced the largest flow in the eight-year record for each land type.

The 100-year flow was added to the runoff files by embedding flows from the scaled January 1990 event into flows generated from the last water year (1987) and creating a fictional eighth year, which was appended to the runoff files. Data from the eighth year are not used in the flow duration statistics. The importance of the eighth year of flow data is to provide a synthetic 100-year hydrograph, which in some cases is the key design consideration. No analysis was done to assess whether the choice of water year appended to the runoff files affected the facility performance with respect to the 100-year flow.

Flow Recurrence Estimation

While information on flow durations is indispensable for assessing the impacts of development on stream morphology, flow frequency data are also needed to assess the impacts of development and to design detention and infiltration facilities. Fitting a heavy-tailed extreme value distribution, such as a Log Pearson Type III distribution, to the uncommonly large floods contained in the eight annual peaks from the runoff files provided terribly inaccurate flow recurrence estimates. Eight years is far too short a record for flow frequency estimation, and these years were not selected because of their annual peak distribution. To conduct flow frequency analyses with the eight-year runoff file record, it was necessary to assign fixed return periods to the ranked annual peak flows from the eight-year record. Plotting positions for the eight annual peaks were chosen so that the peaks fell on the flow recurrence curves derived from 42-year simulations. This approach produced frequency curves that matched curves from the full record reasonably closely with the best agreement for till soils and the poorest agreement for outwash soils (Table 1 and Figures 5a and 5b).

TESTING AND MODIFYING THE RUNOFF FILES

Project-Scale Tests

Extensive tests of KCRTS-designed detention facilities were conducted by routing long-term HSPF generated flow series through the facilities to determine whether discrepancies in flow statistics from the short runoff files records caused faulty facility design. Tests were run at two scales: the scale of the development project (1 to 50 hectares), and the basin scale (hundreds to thousands of hectares). At the project scale, County staff initially selected eight residential and commercial building projects that had been permitted in the preceding 10 years and that represented a range of pre-development and post-development conditions. Detention facilities were designed with KCRTS, and then the design's performance was checked using HSPF by routing a full 42-year time series of simulated hourly flows through the facility and comparing pre-development and post-development flood frequency and flow duration curves derived from the 42-year simulation with those produced using KCRTS. The HSPF simulations used the same meteorological stations, model parameters, and areas used in the KCRTS designs.

TABLE 1. Return periods (years) assigned to ranked KCRTS peak annual flows generated from the Seatac and Landsburg precipitation records

Rank of peak flow	1	2	3	4	5	6	7	8
Seatac Precipitation Record	100	25	10	5	3	2	1.3	1.1
Landsburg Precipitation Record	100	25	10	8	5	3	2	1.1

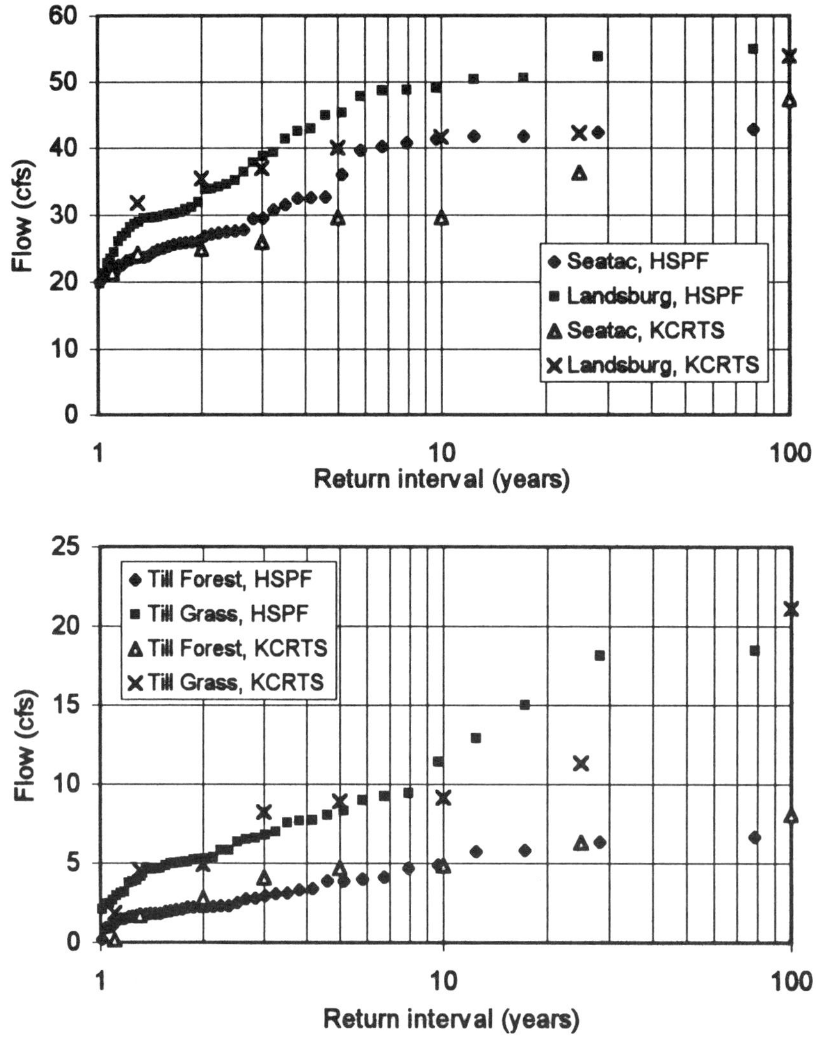

Figures 5a and 5b. Comparison of HSPF-generated and KCRTS-generated flow recurrence curves for several cover types: a) impervious surfaces in the Seatac and Landsburg precipitation regions, and b) till soils with forest cover, till soils with pasture cover, and till soils with grass cover, all within the Seatac precipitation region. Note: one cubic foot per second (cfs) equals 0.0283 cubic meters per second.

The principal differences between the KCRTS and HSPF analyses resulted from the length of the meteorological record used. These tests hence provide a check on: a) the degree to which the eight-year sample of data used in the runoff files is representative of the available 42-year record, and b) the validity of the pre-development flow frequencies (and hence detention pond target releases) assigned to peak flows extracted from the runoff files.

These analyses demonstrated that almost all designs using KCRTS met or came close to meeting their performance standards when tested with HSPF. Without the use of volume safety factors, conveyance protection ponds maintained post-development peak flow recurrence estimates at pre-developed levels for all flows from the 2-year pre-development flow through the 10-year pre-development flow. Facilities designed to match flow durations from one-half of the 2-year through the 50-year flow generally achieved this goal and matched the level of the 100-year flow. The latter accomplishment was serendipitous and not a design intention. However, flow durations were not always matched as well as had been hoped at the low end of the geomorphically significant range. One case study facility only provided flow duration control down to about 85 percent of the 2-year flow (rather than 50 percent as targeted) because the pre-development 2-year flow from KCRTS was about 35 percent greater than the estimate from HSPF. At the other end of the spectrum, another case study facility controlled flow durations down to about 40 percent of the 2-year flow, suggesting that this facility might be larger than actually required. The case study test results also suggested that the runoff files overestimated peak 100-year flow rates for all conditions (pre- and post-developed), as determined from HSPF, by up to 25 percent. Given the inherent uncertainties in estimating 100-year peak flows, this discrepancy was of minimal concern. Generally, the case study test results suggested KCRTS-designed facilities could be expected to meet stated performance standards reasonably well and much better than facilities designed with single storm methodologies and resulting "design hydrograph."

Basin-Scale Tests

KCRTS-designed facilities were also tested with respect to their control of basin-scale flows in simulations of complete build-out (development of a basin to its zoning potential) of two suburban watersheds. The watersheds analyzed were Soosette Creek, which covers 1404 hectares and includes five sub-basins, and Upper Bear Creek, which covers 4706 hectares and includes 15 sub-basins. These two basins feature very different current conditions, zoning, and detention requirements. The Soosette Creek basin was chosen as a case study to represent an already impacted stream system expected to accommodate a high degree of development. At the time of testing, Soosette Creek maintained a marginal but still viable salmonid population including coho salmon, but the coho population could not withstand much more hydrologic disturbance. Upper Bear Creek was chosen to represent a relatively undisturbed and resource-rich system expected to accommodate future low density residential development throughout the basin. The creek still supports several wild salmonid runs, and the basin included a mix of 1-acre (0.405 hectare) and 5-acre (2.02 hectare) zoning and development projects ranging from single family residences, small plats, and large subdivisions.

Soosette Creek simulations. The "pre-developed" or base case simulations were based on land coverage determined from 1985 aerial photographs, and the undeveloped areas at this time included a large amount of pasture land. The post-developed simulations assumed that all "developable" land was built as high density residential (10 to 15 dwelling units per hectare, or four to six dwelling units per acre). The high density residential simulations assumed 25 percent effective impervious surface coverage and 75 percent lawn. In future conditions, 20 percent of the forest cover that existed in 1985 was retained as open space, and all of the wetland areas were assumed to be undisturbed. None of the runoff from forest lands or wetlands was routed to detention facilities. In all future scenarios, 91 percent of the new development was assumed to build detention facilities, and the remaining 9 percent of the development was assumed to consist of projects that are not large enough to merit drainage review by the County. Future scenarios were run with KCRTS and SBUH designed detention facilities for both the conveyance standard (control of peak flows between the 2-year and 10-year flow) and the duration control standard (control of flow durations from one-half of the 2-year flow up to the 50-year flow). All detention facilities included volume safety factors (volume added at all stages after facility design), and these factors were 20 percent for KCRTS-designed facilities and 30 percent for SBUH designed facilities. A volume safety factor is incorporated by increasing the area of the pond so that the water volume is a certain percentage larger at every water level. In other words, the storage part of the stage-storage-discharge relationship is increased without altering stage or discharge. Due to political and policy considerations, the final implemented version of KCRTS did not include safety factors. The basin-scale tests were not re-run after the decision to drop the safety factor. Small-scale tests suggest the average performance of detention facilities should not change significantly.

Facility performance was evaluated partly by use of a normalized channel stability index defined as follows:

Normalized Channel Stability Index=

$$\frac{\text{2-year}_{dev} - \text{2-year}_{pre}}{\text{10-year}_{pre} - \text{2-year}_{pre}}$$

This ratio is equal to zero when there is no increase in the 2-year flow, it is equal to one when the developed 2-year flow equals the pre-developed 10-year flow, and it is greater than

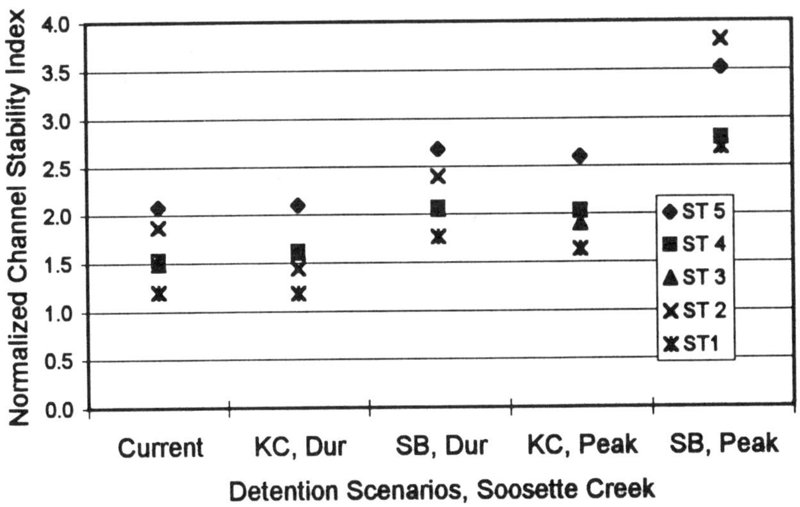

Figure 6. Normalized channel stability indices for the five sub-basins in the Soosette Creek basin under current conditions and four future detention scenarios: 1) KCRTS duration control standard (KC, Dur), 2) SBUH duration control standard (SB, Dur), 3) KCRTS conveyance standard (KC, Peak), and 4) SBUH conveyance standard (SB, Peak). All simulations assume high density residential development and are compared to flows simulated from 1985 land use conditions.

one when the developed 2-year flow exceeds the pre-developed 10-year flow. Observations of channel morphology, habitat characteristics, and fish usage throughout King County have indicated that channels with values greater than one are unstable and unable to support salmonid populations [Booth and Reinelt, 1993; Booth and Jackson, 1997]. Systems for which values of this hydrologic index are near zero usually have excellent habitat and healthy fish populations unless some other deleterious factor, such as riparian zone destruction, water pollution, sediment input, or blockage to fish passage, is present. This stability index was developed from observations of streams lacking appreciable engineering mitigations of hydrologic change. In such systems, increases in flood frequency are inevitably associated with increases in flow durations. If flow durations increase without any increase in flow frequency, as can happen when detention systems are designed to control peaks only, this stability index is not useful and will not identify a system in possible jeopardy from increased flow durations.

As evidenced by changes in the 2-year flow, by the predicted stability indices, and by increases in the durations of the 2-year flow (Figures 2, 6, and 7), the KCRTS duration control standard was the only one of the four detention standards that could protect the habitat quality and remaining salmonid fishery of Soosette Creek. The KCRTS duration control standard achieved its goal of maintaining current modeled durations of high flows in all five modeled reaches of Soosette Creek. The SBUH duration control standard allowed 60 percent increases in the duration of the 2-year flow. It also allowed the channel stability index to increase to values for which salmonids have rarely been observed in other King County streams [Booth and Jackson, 1997; Lucchetti and Fuerstenburg, 1993].

The failure of the SBUH duration control standard was mostly due to the large amount of conversion of pasture land in this basin. SBUH substantially overpredicts pre-developed flows for pasture land and thus sizes detention facilities with overly large release rates and insufficient volume. As will be shown in the Bear Creek basin study, the SBUH duration control standard worked well for conversions of forest land. In the Soosette Creek basin simulations, the KCRTS conveyance standard allowed small increases in the 2-year flow, but it maintained the 10-, 25-, and 100-year flows at their current levels. The SBUH conveyance standard allowed substantial increases in peak flows: 50 percent increases for the 2-year flow and 30 percent increases for the 10-, 25-, and 100-year flows. Overall, the KCRTS-designed detention facilites achieved their stated design goals in this basin while the SBUH-designed facilities did not.

Upper Bear Creek simulations. The "pre-developed" or base case simulations were based on land coverage determined from 1985 aerial photographs when most of the undeveloped area was covered by forests. The post-developed conditions assumed that eight sub-basins were completely developed to 5-acre zoning and six sub-basins were developed to 1-acre zoning. For 5-acre zoning, the assumed effective impervious area was 4 percent, and for 1-acre zoning it was 10 percent. In all sub-basins, 35 percent of the forest land in 1985 was retained as open space in future conditions and all wetland

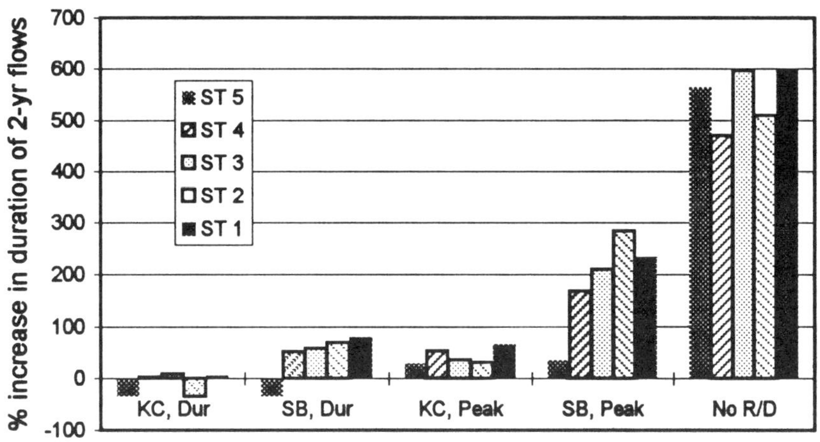

Figure 7. Percent change in the duration of the two-year flow for the five sub-basins in the Soosette Creek basin under current conditions and four future detention scenarios: 1) KCRTS duration control standard (KC, Dur), 2) SBUH duration control standard (SB, Dur), 3) KCRTS conveyance standard (KC, Peak), and 4) SBUH conveyance standard (SB, Peak). All simulations assume high density residential development and are compared to flows simulated from 1985 land use conditions.

areas were assumed undisturbed. Runoff from forest lands and wetlands was not routed to detention facilities. In all future scenarios, 40 percent of the new development was assumed to be single family development, which would not construct detention facilities, and 30 percent was considered to be small plats, whose runoff was detained in some scenarios and not detained in others. This is because King County historically exempted small projects from detention requirements, and one of the regulatory decisions being considered was maintenance or elimination of the so-called peak flow exemption, which exempts from detention requirements all development projects that do not increase the 100-year flow from the site by more than 0.5 cfs. The remaining 30 percent of new development was assumed to consist of large projects, which included detention facilities in all cases. Five future scenarios were modeled: 1) KCRTS duration control with no peak flow exemption, 2) KCRTS duration control with the exemption, 3) SBUH duration control with the exemption, 4) SBUH conveyance standard with the exemption, and 5) no detention.

Most of the development in this basin would occur on previously forested till soils, and the SBUH method does not exaggerate flows from this land type, so the SBUH duration control standard worked well here. There was not a major difference in either the facility volumes or performance between the KCRTS and SBUH duration control standards in this basin, although KCRTS-designed detention facilities provide slightly better performance with slightly smaller facilities. For this basin, impacts did not depend much on design methodology but critically depended on the peak rate exemption. Eliminating the peak rate exemption and designing facilities with KCRTS maintained flow durations at essentially their pre-developed levels. In all other scenarios, durations of 2-year flows increased significantly. With no mitigation, durations of 2-year flows increase 75 to 240 percent. With either KCRTS or SBUH duration control, about half of the unmitigated flow duration increases oc curred if the peak flow exemption was maintained so that small plats did not build detention facilities (Figures 8, 9, and 10).

Geographical Scaling of Runoff Time Series

Since the condensed meteorologic record could not be used for the Carnation gage, it was important to determine how much regional differences in precipitation affected stormwater facility design. Using the long-term precipitation records from Seatac, Landsburg, and Carnation, detention facility performance was tested across regions to determine whether there was any value to having multiple precipitation regions in the runoff files database. For instance, it was possible that facilities designed with Seatac precipitation would work fine for runoff generated from either Landsburg or Carnation precipitation. If this were the case, it would only be necessary to use flows generated from one meterologic station. These tests were also used to determine whether scale factors for flows were needed to eliminate abrupt discontinuities in flow statistics and detention facility volumes across the boundaries of precipitation regions.

These tests indicated that detention facilities designed with Seatac precipitation worked satisfactorily when subjected to

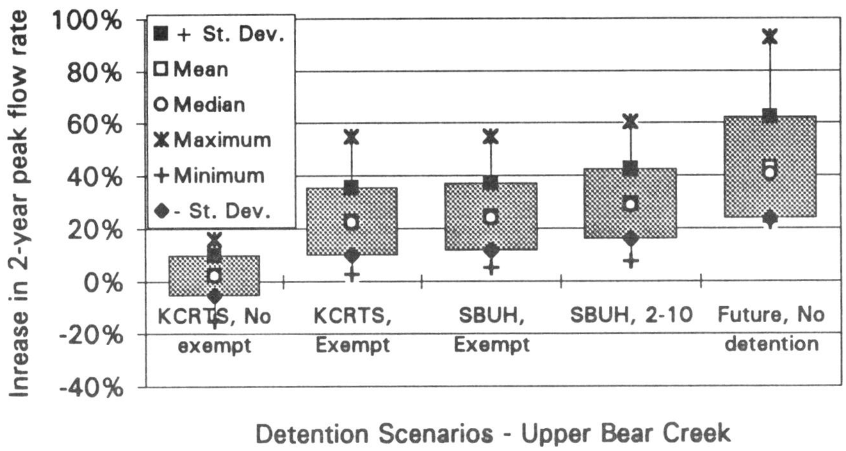

Figure 8. Percent change in the two-year peak flow rate for the 15 modeled stream reaches in the Upper Bear Creek basin under five future scenarios: 1) KCRTS duration control standard and no peak flow exemption (KCRTS, No exempt), 2) SBUH duration control standard with the peak flow exemption (SBUH, Exempt), 3) KCRTS duration control with the peak flow exemption (KCRTS, exempt), 4) SBUH conveyance standard with the peak flow exemption (SBUH, 2-10), and 5) no detention. The peak flow exemption eliminates detention requirements for the projects that do not increase the 100-year flow from the site by more than 0.5 cfs. All future simulations assume low-density residential development and are compared to flows simulated from 1985 land use conditions.

flows generated from Carnation precipitation. Detention facility volumes and flow statistics were similar between the two regions. This finding eliminated the impetus to correct the problems with the abbreviated Carnation record since the Seatac record could be used in lieu of the Carnation record. Detention facilities designed with Seatac precipitation did not perform well when subjected to flows generated from Landsburg precipitation. Because of the strong orographic effects on precipitation caused by the topography of King County, precipitation depths vary considerably across the densely inhabited areas of the County. Annual precipitation at the Landsburg gage, located in a stream valley in the foothills of the Cascade range, is approximately 30 percent greater than at the Seatac gage located adjacent to the Puget Sound. Flow peaks and durations were much larger in the Landsburg region than they were at Seatac. Furthermore, detention volumes determined with the Seatac and Landsburg records differed by 15 to 20 percent.

As a result of these tests, the Landsburg record was chosen to generate runoff time series for the eastern half of the residential area of the County, and the Seatac record was used in the western half of the residential area of the County. Geographically variable scaling factors were developed to provide a smooth transition in runoff and detention volumes across the County. Application of KCRTS without geographic scaling would result in a geographically abrupt regulatory discrepancy. At some locations in the County, a development on one side of a street would have to build 20 percent larger detention facilities than an identical project on the other side of the street. To eliminate this volume discrepancy, scale factors were developed so that detention volumes are equivalent at the border of the Seatac and Landsburg runoff files regions. In addition, the Landsburg flows were scaled upward to the east to account for the progressively wetter climate climbing into the Cascade foothills. A map of the geographically variable scaling factors for the flow time series is shown in Figure 11.

Engineering Community Educational Outreach

Because analysis and design with a continuous model is very different from analysis and design with an event model, training of members of the engineering community (both designers and development review staff) was necessary before the runoff files and KCRTS were accepted. Most engineers are trained in school to use SCS-based event models, but few are exposed to continuous hydrologic models. A KCRTS manual was developed, and multiple training workshops were held to introduce and elucidate the key differences between event-based and continuous hydrologic analysis.

Similar to design using an event model, preliminary design of detention facilities with KCRTS is based on matching flows from several month-long test hydrographs. Final design is not achieved, however, until the statistics of the outflow time series for the full eight years match the predeveloped flow statistics for the full eight years. Somewhat surprising to people unaccustomed to continuous simulation

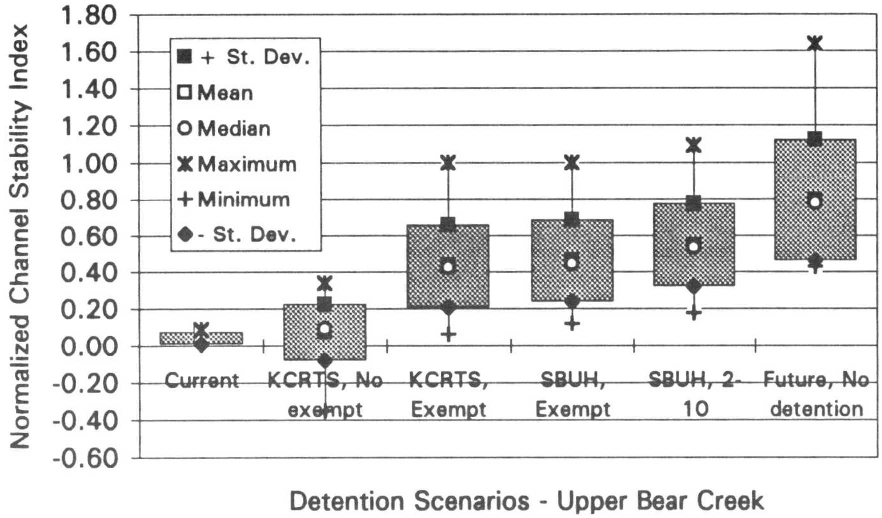

Figure 9. Normalized channel stability indices for the 15 modeled stream reaches in the Upper Bear Creek basin under five future scenarios: 1) KCRTS duration control standard and no peak flow exemption (KCRTS, No exempt), 2) SBUH duration control standard with the peak flow exemption (SBUH, Exempt), 3) KCRTS duration control with the peak flow exemption (KCRTS, exempt), 4) SBUH conveyance standard with the peak flow exemption (SBUH, 2-10), and 5) no detention. The peak flow exemption eliminates detention requirements for the projects that do not increase the 100-year flow from the site by more than 0.5 cfs. All future simulations assume low-density residential development throughout the basin.

models, the return period of the peak flow leaving a detention facility may not have the same return period as the peak flow entering the facility for the same event. The ability of detention ponds to control peak flows depends on both the volume and the peak of the inflow hydrograph and the amount of water stored in the pond at the start of the storm. Generally, it is high volume storms, rather than high intensity storms, that cause detention ponds to fill and overtop. This lack of correspondence in the return periods of peak inflows and outflows complicates the design process and often requires changing the choice of test hydrographs. Similarly, the pre-development peak annual flows may not occur during the same storms as the post-development peak annual flows. The type of storm that causes high peak flows differs between soil/cover combinations. Peak runoff rates from impervious surfaces depend on rainfall intensity and are

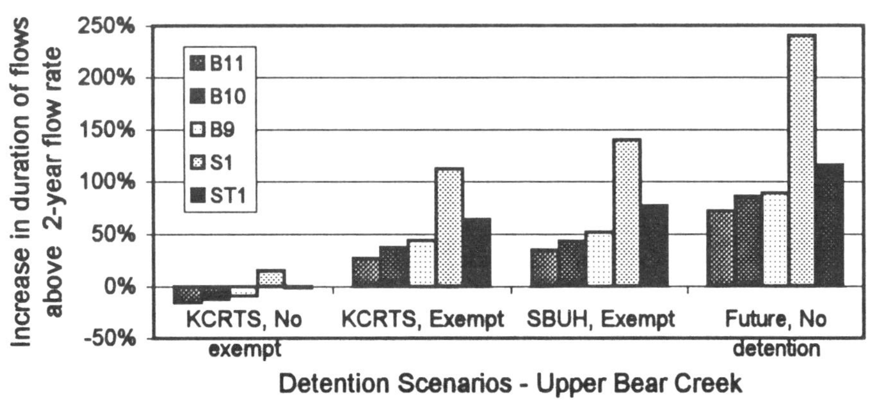

Figure 10. Percent change in the duration of the two-year flow for the 15 modeled stream reaches in the Upper Bear Creek basin under five future scenarios: 1) KCRTS duration control standard and no peak flow exemption (KCRTS, No exempt), 2) SBUH duration control standard with the peak flow exemption (SBUH, Exempt), 3) KCRTS duration control with the peak flow exemption (KCRTS, exempt), 4) SBUH conveyance standard with the peak flow exemption (SBUH, 2-10), and 5) no detention. The peak flow exemption eliminates detention requirements for the projects that do not increase the 100-year flow from the site by more than 0.5 cfs. All future simulations assume low-density residential development and are compared to flows simulated from 1985 land use conditions.

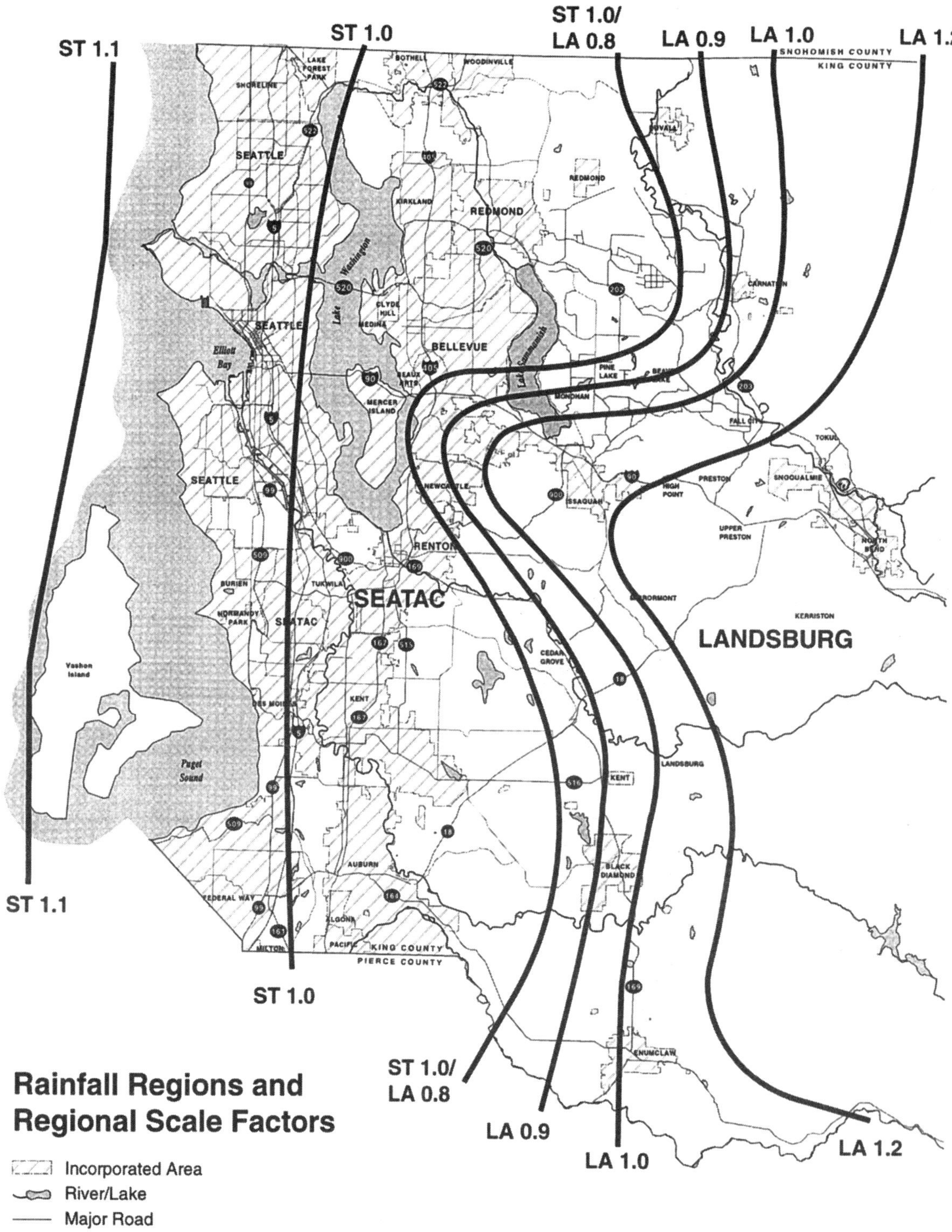

Figure 11. Scaling factors applied to flows from the Landsburg (LA) and Seatac(ST) runoff files to account for orographic effects on precipitation and to eliminate discrepancies in flow statistics and detention facility volumes across precipitation region boundaries.

insensitive to antecedent conditions, whereas peak runoff rates from a forested outwash soil depend on the storm volume and the antecedent conditions and are relatively insensitive to rainfall intensity. For all these reasons, facility design with a continuous model is an iterative, trial-and-error process that requires thinking differently about hydrologic statistics than is encouraged by event models.

Educating the engineering community on the justification for and benefits of this system was necessary to garner political support to change the County's regulatory model to the runoff files and KCRTS. The County Council had to vote to adopt the KCRTS system as the regulatory model for development projects before County staff could require its use, and it was unlikely that the Council would have voted in favor of the runoff files system if the engineering community opposed it. The Council voted to adopt the runoff files system as the County's regulatory model in the spring of 1998.

CONCLUSIONS

Creating an easy-to-use, fast-operating, continuous hydrologic and hydraulic model for stormwater analysis and design was a laborious process that had to be guided at several junctures by professional judgement since standard methodologies did not exist for all the necessary tasks. While the effort required to develop this system was large, so are the potential benefits. In many climate regimes, of which the Puget lowlands and the foothills of the Cascade mountains are two, continuous hydrologic analysis is necessary to design effective stormwater detention systems that meet their desired and stated performance goals. Truth in advertising is one of the benefits provided by the runoff files system. Detention facilities designed with event methods to maintain certain frequency flows at their pre-development levels usually fail to do so, but tests indicate that KCRTS-designed facilities reliably achieve their goals as tested by long-term HSPF simulations. The runoff files and KCRTS software system also provide versatility of analysis, allowing for assessment of flow durations and water level fluctuations in lakes and wetlands. The runoff files system also facilitates permit review because there are no factors such as times of concentrations or curve numbers to estimate or manipulate. KCRTS-designed facilities are easy for permit reviewers to check.

The ability to analyze and control flow durations is critical when managing streams with important aquatic resources. The listing of several Pacific salmonid species under the Endangered Species Act has given more impetus to land managers in the Pacific Northwest to protect habitat conditions in urbanizing stream basins. The versatility of the KCRTS package assists this effort. Without effective control of flow durations, habitat quality is doomed in many streams even if peak flow rates are maintained at their pre-development levels.

Acknowledgments. In addition to the authors, many people contributed significantly to the design, development, and testing of the runoff files system and KCRTS. Jim Kramer, who directed the King County Surface Water Management Division until 1994, provided essential political support, and the King County Water and Land Resources Division funded the development of the runoff files system. Current and former King County staff members important to the development of KCRTS included Linda Holden, Jeff Stern, Jeff Burkey, Derek Booth, Gino Lucchetti, Bob Fuerstenburg, Joe Miles, and Jeff O'Neill. Previous King County hydrologists Bruce Barker, Mark Wigmosta, and Ralph Nelson laid much of the groundwork for this effort. Important consultants included Doug Beyerlein and Rick Schaefer. Evan Twombly of Northwest Hydraulics Consultants programmed the KCRTS supporting software. Many people in the engineering community provided feedback and suggestions after working with prototypes of the software. This project was greatly aided by the HSPF modeling work done by Rick Dinicola of the USGS. We dedicate this work to the memory of the late Dr. Robert C. Johanson (1941-1998) who contributed so much to the development of HSPF and its use worldwide.

REFERENCES

Andrews, E.D. 1984. Bed-material entrainment and hydraulic geometry of gravel-bed rivers in Colorado. *Geol. Soc. Am. Bull.* 95:371-378.

Barker, B.L, R.D. Nelson, and M.S. Wigmosta. 1991. Performance of detention ponds designed according to current standards. In: *Puget Sound Research '91. Puget Sound Water Quality Authority, Olympia, WA.* pp. 64-70.

Booth, D.B. 1990. Stream-channel incision following drainage basin urbanization. *Water Resour. Bull.* 26:407-417.

Booth, D.B., and C.R. Jackson. 1997. Urbanization of aquatic systems, degradation thresholds, stormwater detention, and the limits of mitigation. *J. Am. Water Resour. Assoc.* 33(5):1077-1090.

Booth, D.B., and L.E. Reinelt. 1993. Consequences of urbanization on aquatic systems – measured effects, degradation thresholds, and corrective strategies. Watersheds '93, Conference sponsored by the U.S. Environmental Protection Agency, Alexandria, VA, March 21-24. pp. 545-550.

Carling, P. 1988. The concept of dominant discharge applied to two gravel-bed streams in relation to channel stability thresholds. *Earth Surf. Process. Landforms.* 13:355-367.

Dinicola, R.S. 1989. Characterization and simulation of rainfall-runoff relations for headwater streams in western King and Snohomish Counties, Washington State. U.S. Geological Survey Water-Resources Investigation Report 89-4052. 52 pp.

Graf, W.L. 1975. The impact of suburbanization on fluvial geomorphology. *Water Resour. Res.* 11(5):690-692.

Guo, Y., and B.J. Adams. 1999. An analytical probabilistic approach to sizing flood-control detention facilities. *Water Resour. Res.* 35(8):2457-2468.

Hammer, T.R. 1972. Stream and channel enlargement due to urbanization. *Water Resour. Res.* 8:1530-1540.

Hollis, G.E. 1975. The effects of urbanization on floods of different recurrence intervals. *Water Resour. Res.* 25:135-140.

King County. 1989. Bear Creek current and future conditions analysis. Department of Public Works, Surface Water Management Division, Seattle, Washington (six sections).

King County. 1990a. Soos Creek basin plan and final environmental impact statement. Dept. of Public Works, Surface Water Management Division, Seattle, Washington. 330 pp.

King County. 1990b. Hylebos Creek and Lower Puget Sound basins current and future conditions report. Dept. of Public Works, Surface Water Management Division, Seattle, Washington (four sections).

King County. 1990c. East Lake Sammamish basin conditions report – preliminary analysis. Dept. of Public Works, Surface Water Management Division, Seattle, Washington. 148 pp.

King County. 1990d. King County surface water design manual, updated November 1995. King County Department of Natural Resources, Water and Land Resources Division, Seattle, Washington.

King County. 1991. Issaquah Creek basin current/future conditions and source identification report. Dept. of Public Works, Surface Water Management Division, Seattle, Washington (twelve sections).

King County. 1992. East Lake Sammamish basin and nonpoint action plan. Dept. of Public Works, Surface Water Management Division, Seattle, Washington. 362 pp.

King County. 1993. Issaquah Creek basin and nonpoint action plan. Dept. of Public Works, Surface Water Management Division, Seattle, Washington. 279 pp.

Klein, R.D. 1979. Urbanization and stream quality impairment. *Water Resour. Bull.* 15(4): 948-963.

Linsley, R.K., M.A. Kohler, and J.L.H. Paulhus. 1982. *Hydrology for engineers*. McGraw Hill, New York.

Lucchetti, G., and R. Fuerstenburg. 1993. Relative fish use in urban and non-urban streams. Proceedings of the Conference on Wild Salmon. Vancouver, BC.

Lumb, A.M., and D. James. 1976. Runoff files for flood hydrograph simulation. ASCE, *J. Hydraul. Div.* 102:1515-1531.

Morgan, A.E. 1951. *The Miami Conservancy District.* McGraw Hill, New York.

Moscrip, A.L., and D.R. Montgomery. 1997. Urbanization, flood frequency, and salmon abundance in Puget Lowland streams. *J. Am. Water Resour. Assoc.* 33(6):1289-1297.

Pickup, G., and R.F. Warner. 1976. Effects of hydrologic regime on magnitude and frequency of dominant discharge. *J. Hydrol.* 29:51-75.

Sidle, R.C. 1988. Bed load transport regime of a small forest stream. *Water Resour. Res.* 24:207-218.

Stedinger, J.R., R.M. Vogel, and E. Foufoula-Georgiou. 1993. Chapter 18: Frequency analysis of extreme events. In: *Handbook of hydrology,* David Maidment, Ed. McGraw-Hill Inc., New York.

Steedman, R. J. 1988. Modification and assessment of an index of biotic integrity to quantify stream quality in Southern Ontario. Can. *J. Fish. Aquat. Sci.* 45:492-501.

U.S. Soil Conservation Service. 1975. Urban hydrology for small watersheds. Technical Release Number 55. Washington, D.C.

Wolff, C.G., and S.J. Burges. 1994. An analysis of the influence of river channel properties on flood frequency. *J. Hydrol.* 153: 317-337.

Sliding in Seattle: Test of a Model of Shallow Landsliding Potential in an Urban Environment

David R. Montgomery and Harvey M. Greenberg

Dept. of Geological Sciences, University of Washington Seattle, Washington

William T. Laprade and William D. Nashem

Shannon & Wilson, Seattle, Washington

Compilation of a 100-year record of 1,358 landslide locations allows testing of a process-based model for shallow landslide initiation in the City of Seattle, Washington. The relative slope stability model is based on coupling a topographically-driven model for shallow throughflow with the infinite-slope stability model. Three digital elevation models (DEMs) were used to generate predicted patterns of potentially unstable ground: the standard US Geological Survey (USGS) 30 m DEM; a 10 m DEM created from USGS 7.5' topographic contours; and a 1.5 m DEM created from Seattle Engineering Department contours. Model performance varied with DEM grid size, but areas identified as high risk occupy less than 1% of the area of the City. The map pattern of historic landslide locations corresponds well to areas predicted to be at risk for shallow landslide initiation in spite of the extensive hydrologic modifications typical of urban environments and the strong influence of glacial stratigraphy and groundwater flow on near-surface hydrologic processes in Seattle. In addition, the unusually long-term record of landslide locations suggests that areas predicted to be potentially unstable but that have not yet failed can be interpreted as at risk of failure, as landslides have occurred in proximity to approximately half of the area of potentially unstable ground over the period of record. Comparable performance of a slope based hazard assessment indicates that in Seattle gradient is more important than drainage area as a control on potentially unstable ground. Our analysis indicates that landslide hazards in Seattle are strongly associated with a small but dispersed area of the City that can be objectively identified in spite of the hydrologic complexity of the urban environment.

INTRODUCTION

Landslides pose a significant hazard in steep urban areas around the world (Jones, 1973; Ellen and Wieczorek, 1988; Brabb and Harrod, 1989). While the influence of land use on the initiation of shallow landsliding is widely recognized (Sharpe, 1938; Sidle et al., 1985) many re-

Land Use and Watersheds: Human Influence on Hydrology and Geomorphology in Urban and Forest Areas
Water Science and Application Volume 2, pages 59-73

searchers in the past several decades have focused primarily on landslide processes in rural or urbanizing areas, and in particular on the role of forestry practices (Swanson and Dyrness, 1975; O'Loughlin and Pearce, 1976; Gray and Megahan, 1981). Nonetheless, the problem of predicting areas prone to slope failure is important for setting public policy in urban areas (Nilsen et al., 1979). Hazards include the potential for damage to private property, the cost of repairing and maintaining public infrastructure in slide-prone areas and loss of life.

There are many approaches to the problem of predicting areas prone to shallow landsliding. The simplest approach, which is widely used in land use planning, is based on a critical slope angle to designate areas of high hazard. Such an approach, however, does not use the effects of land form and local geology on landslide potential. A number of more complex approaches to predicting landslide hazards are based on correlations with slope, lithology, land form and/or geologic structure (Campbell, 1975; Hollingsworth and Kovacs, 1981; Seely and West, 1990; Montgomery et al., 1991; Ellen et al., 1993; Derbyshire et al., 1995). Another approach for predicting areas prone to shallow landsliding relies on combining topographically-driven hydrologic models with slope stability models (Okimura and Ichikawa, 1985; Dietrich et al., 1993; 1995; van Asch et al., 1993; Montgomery and Dietrich, 1994; Wu and Sidle, 1995; Montgomery et al., 1998). Tests of such coupled models of near-surface runoff and slope stability against mapped landslide locations reveal that the topographic control of drainage area and local slope on shallow landsliding can provide a reasonable measure of relative landslide potential in rural areas (Montgomery and Dietrich, 1994; Dietrich et al., 1995; Montgomery et al., 1998). However, the profound effects of urbanization on near-surface hydrologic processes suggest the potential for significant problems in applying such models in developed areas, as runoff may not follow topographically defined pathways and patterns of soil saturation may not correspond to topographically-driven predictions.

Issues surrounding management of land use on steep slopes are often contentious in urban areas, as the desires of developers and private land owners can conflict with governmental interest in public safety and minimizing costs to repair and maintain public infrastructure (such as bridges, roads and utilities). A recurring question relevant to such conflict is how to evaluate sites that are classified as high risk, but that have not yet failed. Are such sites places where site-specific conditions provide stability greater than that implied by the hazard rating, or are they areas poised to fail in the future and therefore hazardous to develop on or below? Here we test the application of a physically-based model of the topographic control on shallow landsliding to an urban environment and use an extraordinary long-term record of landslide locations to examine the relation of landsliding to areas predicted to be potentially unstable.

MODEL OF SHALLOW LANDSLIDING

The model that we used is discussed in detail elsewhere (Dietrich et al., 1993; 1995; Montgomery and Dietrich, 1994; Montgomery et al., 1998), so here we provide only an overview of the model and its assumptions. Our approach is based on coupling a hydrologic model to a limit-equilibrium slope stability model to calculate the critical steady-state rainfall necessary to trigger slope instability at any point in a landscape. The hydrologic model assumes that flow infiltrates to a lower conductivity layer and follows topographically-determined flow paths to map the spatial pattern of equilibrium soil saturation based on analysis of upslope contributing areas, soil transmissivity, and local slope (O'Loughlin, 1986). Specifically, local wetness (W) is calculated as the ratio of the local flux at a given steady-state rainfall (Q) to that upon complete saturation of the soil profile, which may be expressed as

$$W = \frac{Q\,a}{bT\sin\theta} \tag{1}$$

where a is the upslope contributing area (m^2), b is the contour length across which flow is accounted for (m), T is the soil transmissivity (m^2/day), and θ is the local ground slope (degrees). Adopting the simplifying assumption that the saturated conductivity does not vary with depth results in W=h/z for W 1 (Dietrich et al., 1995), where h is the thickness of the saturated soil above the impermeable layer and z is the total thickness of the soil.

Combining this hydrologic model with the infinite-slope stability model (see Selby, 1993) provides a simple model for failure of shallow soils where the critical steady-state rainfall required to cause slope instability (Q_c) is given by

$$Q_c = \frac{T\sin\theta}{(a/b)} + (\rho_s/\rho_w)[1-(\tan\theta/\tan\phi)] \tag{2a}$$

for cohesionless soils where ρ_s is the saturated bulk density of the soil, g is gravitational acceleration, ρ_w is the density of water, and ϕ is the friction angle of the soil (Montgomery and Dietrich, 1994). For soils with an apparent cohesion (C'), Q_c is given by (Montgomery et al., 1998)

$$Q_c = \frac{T\sin\theta}{(a/b)}\left[\frac{C'}{\rho_w g z\cos^2\theta\tan\phi} + (\rho_s/\rho_w)[1-(\tan\theta/\tan\phi)]\right] \tag{2b}$$

Values of W greater than 1.0 imply that excess water runs off as overland flow, as there is no mechanism in this model for generating pore pressures greater than hydrostatic. Slopes that are stable even when W=1.0 are interpreted to be unconditionally stable and to require excess pore pressures to generate slope instability. Similarly, slopes predicted to be unstable even when dry (i.e., when W=0) are considered to be unconditionally unstable areas where soil accumulation would be difficult. Critical rainfall values can be calculated for locations with slopes between these criteria, but the combined influence of the steady-state hydrologic assumption, lateral reinforcement by roots that extend across the side of potential failures (Burroughs, 1984; Reneau and Dietrich, 1987) and systematic variations in soil thickness (Dietrich et al., 1995) mean that without calibration or further modifications the approach to landslide hazard assessment embodied in equation (2) can simply identify areas with equal topographic control on shallow landslide initiation.

To date, tests of the model embodied in equation (2) have revealed that, as predicted, shallow landslides preferentially occur in areas with low modelled critical rainfall. Montgomery and Dietrich (1994) used high resolution digital elevation models (DEMs) to compare the locations of field-mapped landslides with Q_c values predicted by equation (2a) for 3 small catchments in the western United States. In the Tennessee Valley, California, and Split Creek, Washington, catchments they found that 96% and 100% of the mapped landslides overlapped Q_c categories less than 100 mm/day (using site-specific values of T, $\tan\phi$ and ρ_s). At the third study site (Mettman Ridge, Oregon) 73% of the mapped slides overlapped areas with Q_c 100, and the remaining 27% of the slides were associated with road drainage concentration or were in subtle topographic hollows not resolved in the digital topography. In applications to larger watersheds, Pack and Tarboton (1997) found the model to be useful for landslide hazard prediction in British Columbia. At the regional scale, Montgomery et al. (1998) compared 3,224 landslide locations mapped from sequential aerial photographs with Q_c values predicted from 30 m DEMs for 14 watersheds that cover 2,993 km^2 in Oregon and Washington. In each of the watersheds the frequency of mapped landslides (i.e., slides/km^2) was inversely related to Q_c. However, model performance varied widely between watersheds, with the best performance generally in steep watersheds underlain by shallow bedrock and the worst performance in low-gradient watersheds underlain by thick glacial deposits. Montgomery et al. (2000) tested the discriminatory power of the model by comparing the landslide frequencies in high hazard classifications (i.e., low Q_c categories) with a comparable number of randomly placed "landslides" for a series of watersheds in Oregon and Washington. They found that the model provides a substantial improvement over the random sampling of the landscape for low Q_c categories, but that for high Q_c values (i.e., lower hazard) the model provided no clear improvement over a random model. To date, such tests of the model's performance indicate that it provides a reasonable prediction of areas at high risk for shallow landsliding in a wide range of lithologies and environments. The actual rates of landsliding associated with high hazard categories, however, vary widely among drainage basins, and therefore use of the model in specific risk assessments requires local calibration.

LANDSLIDES IN SEATTLE

We chose to investigate urban applications of the model in the city of Seattle because: i) landslides cause extensive damage during large storms in the Seattle metropolitan area; ii) Seattle is a geologically heterogeneous environment that would provide a harsh test of the model; iii) a field-checked, long-term record of landslide locations was recently compiled; and iv) we have local knowledge of the area.

Seattle lies on and around a series of north trending linear ridges carved by the Puget lobe ice sheet during its last advance in the late Pleistocene (Plate 1). The glacial stratigraphy of the Seattle area influences landslide processes (Waldron, 1962; Galster and Laprade, 1991). Many of the hill tops are capped by Vashon till, a basal lodgment till that ranges from a gravelly, sandy silt to silty sand with clay and scattered cobbles and boulders. The Esperance Sand, which underlies the Vashon till, represents advance outwash and is composed of fine to medium sand with local silt beds and channel deposits of gravel. The lower contact with the Pleistocene lake bed deposits of the Lawton Clay is gradational over several meters. Other pre-Vashon glacial and nonglacial deposits are exposed locally throughout the city. Infiltration of rainfall through the Vashon till and highly-conductive Esperance Sand to the very low conductivity Lawton Clay leads to the widespread occurrence of springs and seeps along the trace of the Esperance/Lawton contact (Figure 1). Many landslides in Seattle are associated with this contact and Tubbs' (1975) map of high hazard landslide zones drew heavily upon its outcrop pattern.

The steep bluffs that border Puget Sound are unstable in many places, but they are also popular sites for expensive view homes. Many of the bluffs are steeper than 40° and

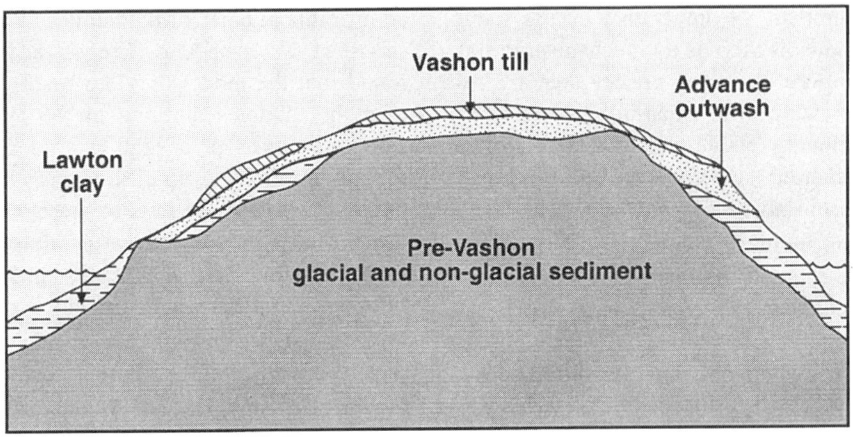

Figure 1. Typical cross-section through a hillslope in the city of Seattle showing the influence of late-glacial stratigraphy on slope hydrology and landslide processes [modified by Laprade from a figure in Tubbs (1974)]. Geologic units are Vashon till overlying advance outwash (Esperance Sand), Lawton clay, and Pre-Vashon glacial and non-glacial sediments.

have a sharp break in slope at their head and relatively linear profiles down to the toe of the slope. Such features suggest that most of the steep waterfront bluffs are actively maintained by shallow landsliding and bluff retreat over geomorphic time scales (i.e., 100's to 1000's of years). Deep-seated sliding also occurs in some locations along the coastal slopes of Puget Sound.

Past Landsliding

Landslides have repeatedly caused damage in Seattle. Tubbs (1974; 1975) studied landslides that resulted from severe storms during the winter of 1971-1972, during which at least 100 landslides occurred in the City of Seattle. He documented that human influences, such as concentrating water, cutting or filling on a slope, were involved in more than 80% of the landslides examined. Tubbs (1975) also compiled a record of landslide occurrences reported in *The Seattle Times* during the period from 1933 to 1972. During this period, landslides were reported in 30 out of 39 years (77%), and 5 years had more than 10 reported landslides. The date on which 160 of the reported landslides initiated could be determined to within a period of 2 days and based on this compilation Tubbs developed a relationship between the number of landslides reported on a given day (N) and the daily precipitation (P) that can be approximated as $N=10^{(P-64)}$, where P is in millimeters. Although Tubbs (1975) also developed relations for other rainfall durations, the relation with daily rainfall suggests that the incidence of landsliding in Seattle increases systematically once a daily threshold of 64 mm (2.5 inches) of rainfall is reached. Implicit in this relation, however, is the influence of antecedent rainfall.

Landslides during the winter storms of 1996-1997 caused more than $34 million in property damage to City facilities alone, and extensive damage to private property, spawning efforts to reevaluate Seattle's landslide policies (Schell, 1998). As part of Seattle's landslide response effort, the geotechnical firm Shannon & Wilson compiled a map of 1,358 landslide locations from city records. To generate this extraordinary record of landsliding, information in city records was compiled to create a record of the street address of landslides reported since 1890. Primary data sources for this compilation were records from the city engineering department, consultant reports, damage claims, and infrastructure crews. The address of each property that experienced a landslide was georeferenced and compiled in ARC/INFO. Field verification revised each mapped landslide location to within approximately 10 to 30 m of its headscarp. A total of 171 reported "landslides" were eliminated from the data base; slides were eliminated because they could not be reasonably field verified or field inspection revealed that reports of their occurrence actually referred to toppled retaining walls, collapsed trench walls and the like. Because the landslide locations were stored as points, rather than as polygons that defined landslide boundaries, we could not conduct the same tests of model performance that we have conducted in previous studies in which we analyzed the predicted potential for instability within sites of landslide initiation (Montgomery and Dietrich, 1994; Dietrich et al., 1995; Montgomery et al., 1998). Instead, we examined model performance based on

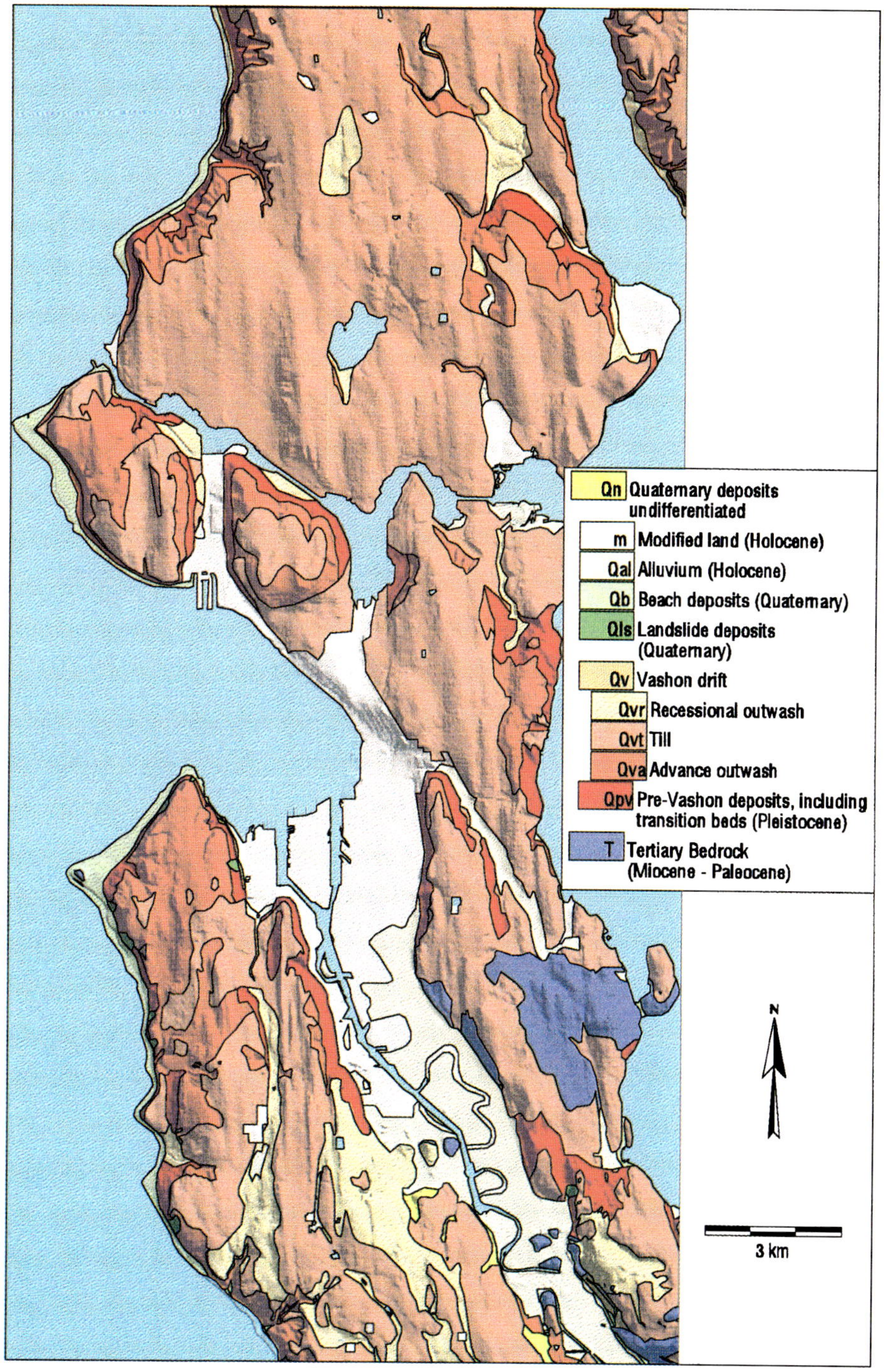

Plate 1. Shaded digital relief map of Seattle with overlaid geologic map [digital data from Booth and Sacket (1997)].

TABLE 1. Percent of Seattle predicted to be in each Q_c category.

DEM Grid Size	Critical Rainfall Values (mm/day) Q_c<50	Q_c<100	Q_c< 200	Q_c< 400	Total of Potentially Unstable Ground
30 m	0.1	0.3	0.5	0.7	0.7
10 m	0.3	0.6	0.9	1.3	1.6
1.5 m	0.6	0.8	1.2	1.8	4.0

the distance from each landslide record to slopes predicted to be potentially unstable, and vice-versa, in a broad comparison of city-wide landslide distributions with predicted patterns of landslide susceptibility.

MODEL TEST

We ran SHALSTAB using equation (2b) on three DEMs of the City of Seattle, the coarsest of which was the standard US Geological Survey (USGS) 30 m coverage. We also used the USGS 10 m DEM generated by gridding a vectorized contour coverage from the USGS 7.5' topographic quadrangles, and created a very fine-resolution, 1.5 m grid size DEM from 2 foot contour interval maps provided by the City. For all three model runs, we used a single set of parameters that reasonably approximate general properties of glacial deposits in Seattle (Koloski et al., 1989) to isolate the predicted topographic control on shallow landsliding: (ρ_s/ρ_w)=2.0, C=2 kPa, ϕ=33°, and T=65 m^2/day. We then compared landslide locations in the Shannon and Wilson data base to the distribution of potentially unstable slopes by evaluating the distance from each landslide location to potentially unstable slopes grouped by Q_c categories. We also calculated the cumulative distribution of landslides within a variable buffer diameter of potentially unstable ground.

DEM grid size strongly affects representation of drainage area and slope, with coarser grids depicting gentler slopes (Zhang and Montgomery, 1994). Consequently, the grid size of the DEM used to drive SHALSTAB influences the predicted extent of potentially unstable ground. At a 30 m grid size <1% of the city is predicted to be potentially unstable (i.e., all Q_c categories), with only 0.3% of the city in the highest hazard categories (i.e., Q_c<100 mm/day) (Table 1). At a 10 m grid size (Plate 2) 1.6% of the City is predicted to be potentially unstable, but there is more spatial coherency to the pattern of predicted instability, and the areas of potential instability extend farther along the coastal bluffs and include areas at the heads and margins of incised canyons, and the steep margins of linear ridges. At the very fine grid size of 1.5 m, the entire coastline and the outline of all canyons and linear ridges are identified as potentially unstable. However, most of the potentially unstable ground is in the lowest hazard category (Q_c>400 mm/day), which previous studies (Montgomery and Dietrich, 1994; Montgomery et al., 1998) have found to represent a very low risk for landslide initiation. Without this lowest hazard category the total area of potentially unstable slopes is less than 2% of the City (Table 1). All of the models predict that most of the potentially unstable ground is located along the steep bluffs that border Puget Sound. Although DEM grid size affects the hazard rating for those portions of the city that lie within each Q_c category, areas at high risk for shallow landslide initiation occupy a very small portion of the City.

The composite map of historic landslide locations in Seattle (Figure 2) bears a strong resemblance to the maps of potentially unstable ground. Landslides are concentrated along the coastal bluffs, in and around canyons, and on the steep margins of linear ridges. The conspicuous gap in landsliding along the coast at the north end of Elliot Bay reflects that records of landslides in Discovery Park were not included in the City's records because nothing was damaged by the slides—and hence no records were kept. We do not know how many more landslides occurred that did not cause sufficient damage to merit recording by City personnel. Although only a fraction of the potentially unstable ground fails during any given storm, the long-term record of slide locations implies that, over time, failures may be expected to occur throughout the areas predicted to be potentially unstable.

The fine-grid DEMs have a better spatial correpondence between landslides and potentially unstable ground (Figure 3). The uncertainty inherent in registering mapped landslides to a DEM can be addressed by using a variable radius within which to consider a slide associated with grid cells predicted to be potentially unstable. For both the 30 m and 10 m DEMs, almost 70% of the landslides are within 30 m of potentially unstable ground. Using the 10 m DEM, over 50% of the landslide locations are within 10 m of potentially unstable ground, whereas 45% of the landslides are within 30 m of potentially unstable ground using

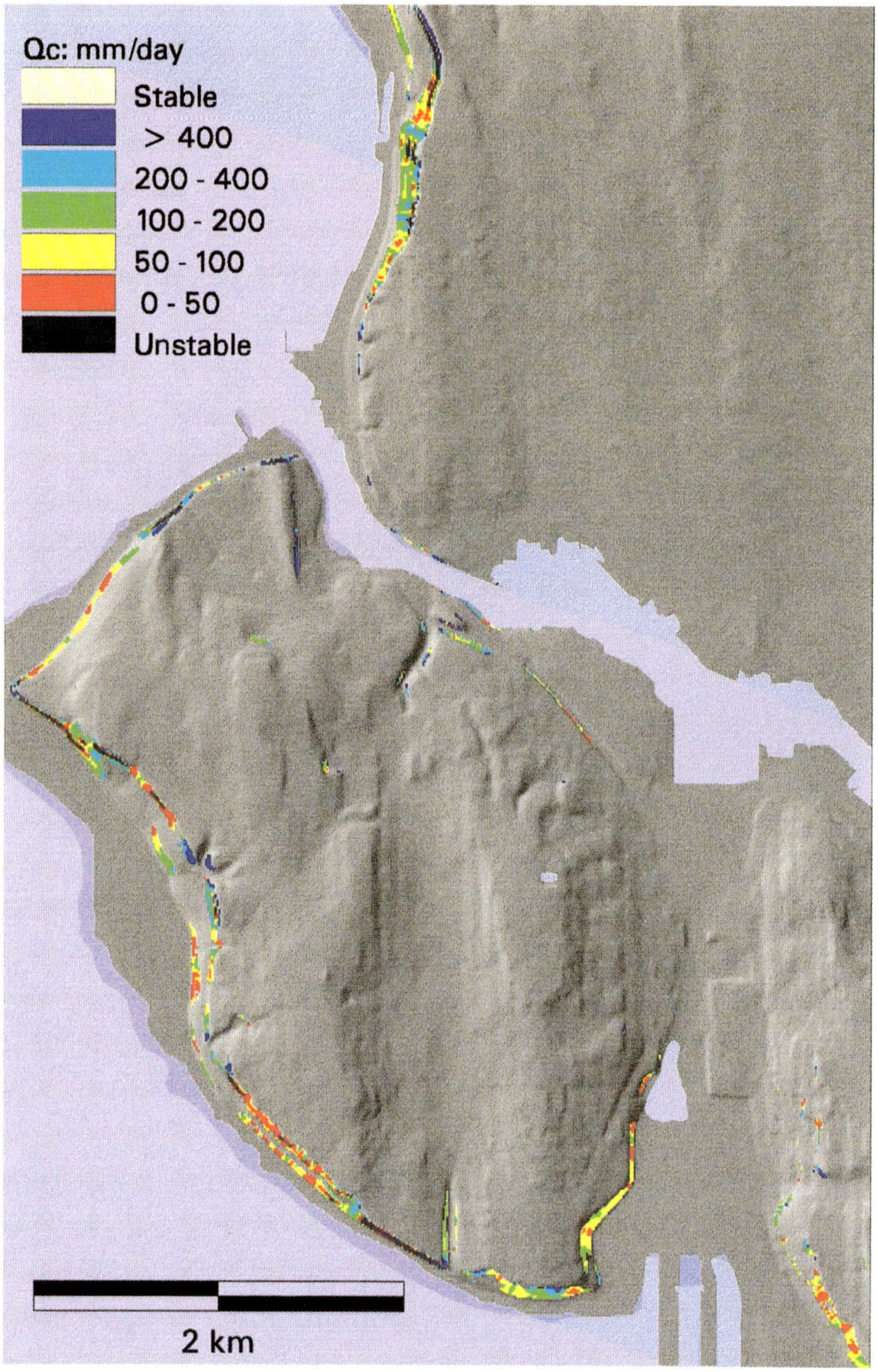

Plate 2. Predicted critical rainfall values for portions of the City of Seattle predicted using a 10-m grid DEM.

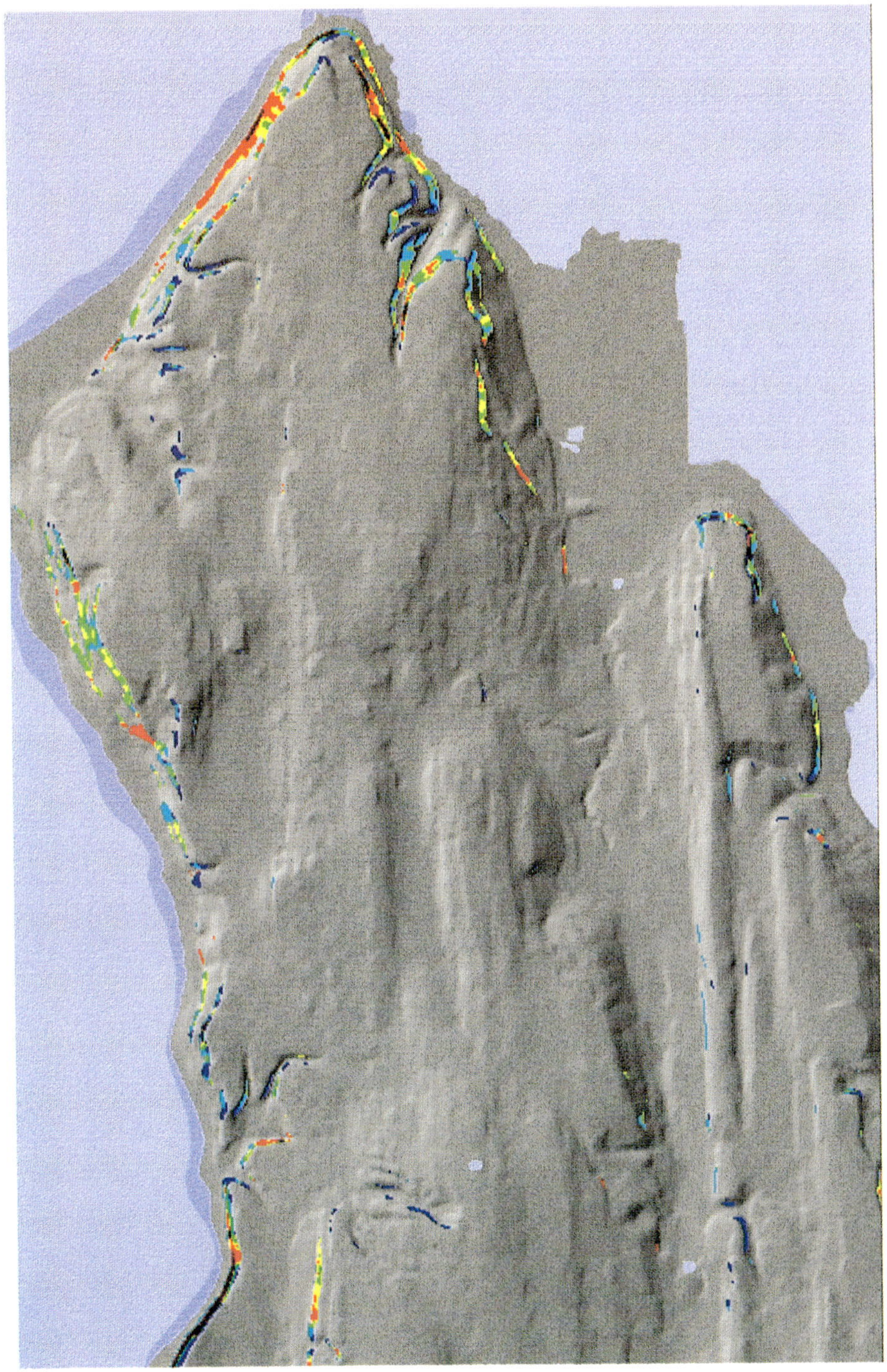

Plate 2 (continued)

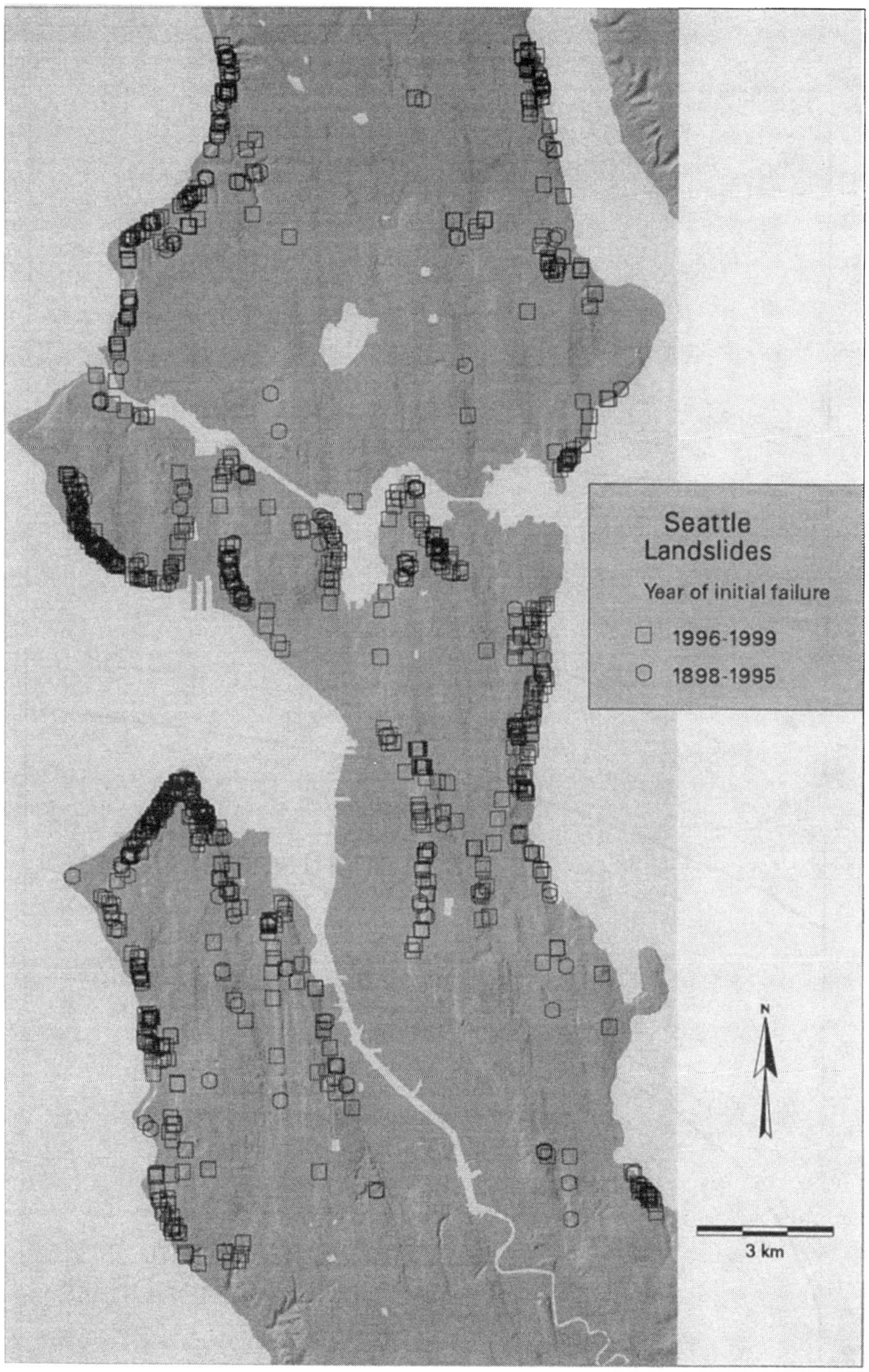

Figure 2. Map of in-city landslide locations reconstructed from City records extending from 1897 through 1997.

the 30 m DEM. Over 95% of the mapped landslide locations are within 10 m of potentially unstable ground predicted using the 1.5 m grid DEM. Recall that the mapped landslides have no dimension themselves, as they were mapped as point locations.

The proportion of the high hazard areas in proximity to locations that experienced landsliding over the period of record varies as a function of both Q_c values and DEM grid size (Figure 3). We generated a point at the center of each grid cell predicted to be potentially unstable and measured

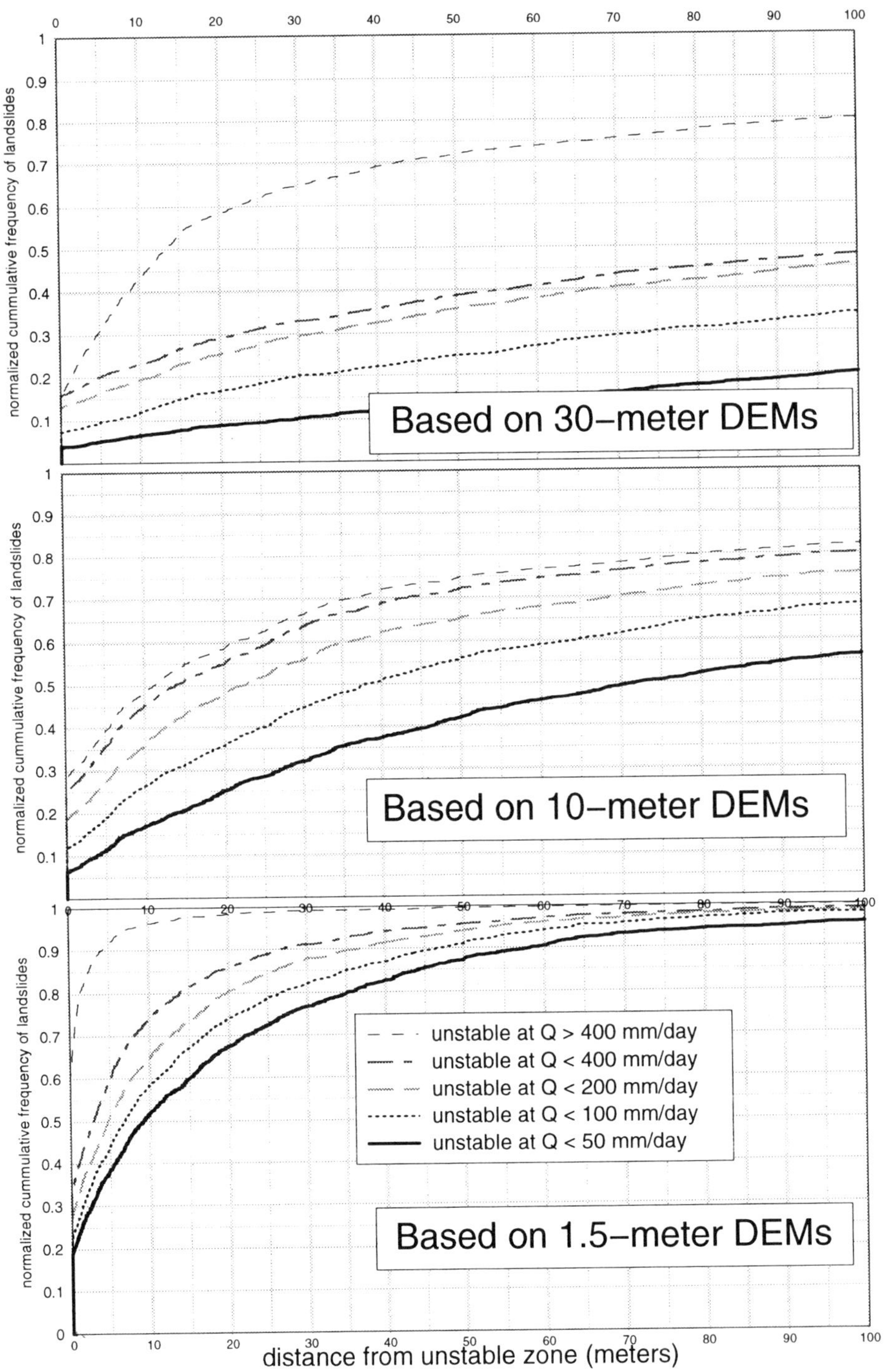

Figure 3. Cumulative percentage of recorded landslides that fall within a specified distance of potentially unstable ground for 30 m, 10 m, and 1.5 m DEMs.

the distance to the nearest point representing a mapped landslide. For the 30 m grid size DEM, 20% of the potentially unstable grid cells had a landslide that plotted within 30 m of the centerpoint of the cell, and over 50% of all of the potentially unstable grid cells were within 100 m of a mapped landslide (Figure 3). For the 10 m DEM, just 15% of the potentially unstable grid cells were within 30 m of a landslide, and 45% of all of the potentially unstable grid cells were within 100 m of a mapped landslide (Figure 4). For the 1.5 m DEM, less than 15% of the potentially unstable grid cells were within 10 m of a landslide, and less than 40% of all of the potentially unstable grid cells were within 100 m of a mapped landslide (Figure 4). A higher proportion of the potentially unstable grid cells were associated with landsliding in the coarser-grid DEMs. Hence, a finer resolution grid size provides more accurate depiction of where landslides occurred, but finer resolution DEMs also generate additional zones of predicted instability that are farther from past landslides.

DISCUSSION

We find the reasonable model performance surprising given that the fundamental hydrologic assumption of shallow topographically driven throughflow that drives the model is not necessarily met in urban environments with substantial impervious area and extensive drainage alterations. The reasonable model performance is particularly intriguing given the strong degree of stratigraphic control on landsliding in Seattle (Tubbs, 1974; Galster and Laprade, 1991). Apparently, the topographic attributes of steep slopes with large drainage areas characterize locations where stratigraphic conditions strongly influence the frequency or rates of sliding and bluff retreat, indicating a strong correlation between geological materials and slope form.

The strong model performance in an urban environment where glacial stratigraphy strongly influences landslide processes suggests that either: shallow throughflow is more important than generally recognized in Seattle (as suggested by the updated stratigraphic cross sections of Figure 1); redirection of water onto high hazard slopes dicates those places that do fail; or the specific location of slope failures in Seattle is primarily controlled by the distribution of steep slopes rather than by generalizable patterns in the near-surface hydrology. The latter interpretation is supported by a comparative analysis of the proportion of the total area of the city needed to cover a given proportion of mapped landslides in hazard zones defined by 30 m buffers determined by SHALSTAB and by using a slope-driven hazard model defined by simply taking the steepest slope in the city and progressively assessing the proportion of landslides accounted for within the buffers as the critical slope decreases (Figure 5). Although the hazard zones defined by SHALSTAB account for a greater proportion of the mapped landslides than the slope-based model, the difference is small. Hence, it is the small area of very steep slopes in Seattle that dominates the historic record of landsliding.

The long-term record of landsliding provides guidance for how to interpret areas predicted to be potentially unstable, but in which landslides have not been observed historically or over the period covered by aerial photography. This issue is important for addressing how to interpret "high hazard" areas that have not slid in the decades prior to an evaluation. The record of Seattle landsliding shows that over time landslides may be expected to eventually occur throughout the areas predicted to be potentially unstable. Of course, some areas identified as at high risk may not be as hazardous as predicted due to local hydrogeological factors not accounted for in the model or to misrepresentation of actual topography in the DEM. Moreover, the particular locations that fail in a given storm may reflect both local conditions that change over time (e.g., root strength, soil thickness, and anthropogenic drainage modifications) and variability in storm-specific patterns of high-intensity rainfall cells that can drive landsliding. Nonetheless, our analysis supports interpreting "type II error", in which the model identifies slopes that have not yet failed to be potentially unstable, as identifying potential sites for future instability.

The highest densities of landsliding occur in the highest hazard areas as defined by simple topographic characteristics. These high hazard areas include a few clearly-defined geomorphic environments: coastal cliffs and bluffs that border Puget Sound, the steep sides of linear ridges, and the head and margins of Holocene canyons incised into late glacial deposits. Hence, the pattern of high risk areas across the city reflects topography, even though site-specific storm and geological conditions may control the timing and locations of specific failures. The approximately 15 to 30% of the landslides that did not plot within 30 to 100 m of potentially unstable ground is comparable to results from different analyses of forested areas using 30 m and 10 m DEMs (Montgomery and Dietrich, 1994; Montgomery et al., 1998).

Landslide hazard evaluation in urban environments also involves other processes not incorporated in the model, in particular deep-seated slides and the delineation of runout and deposition zones downslope of initiation sites. Addition of a simple runout algorithm (e.g., Montgomery and Dietrich, 1994) could address the issue of routing and

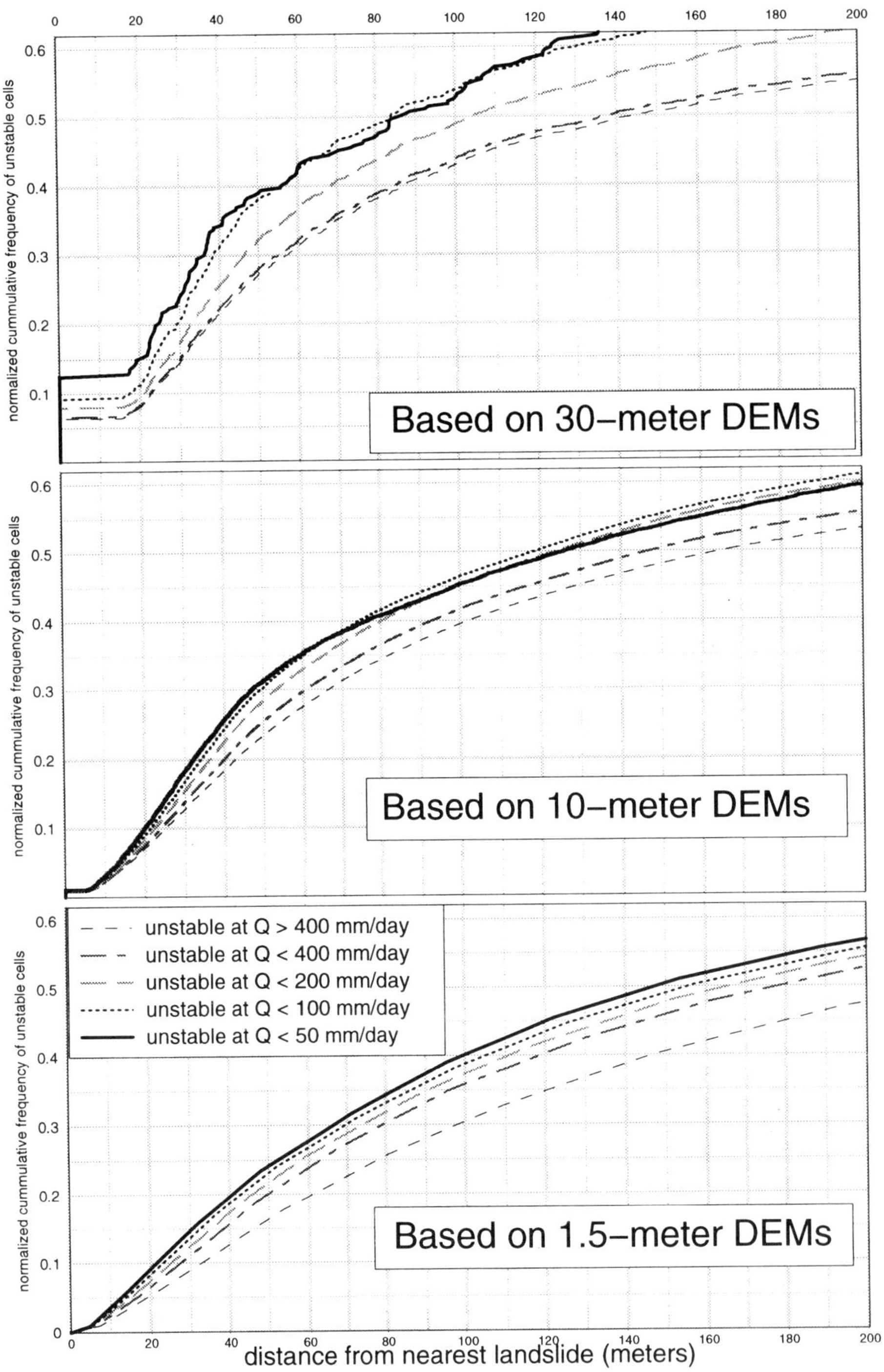

Figure 4. Cumulative percentage of areas mapped as potentially unstable that fall within a specified distance of recorded landslides for 30 m, 10 m, and 1.5 m DEMs.

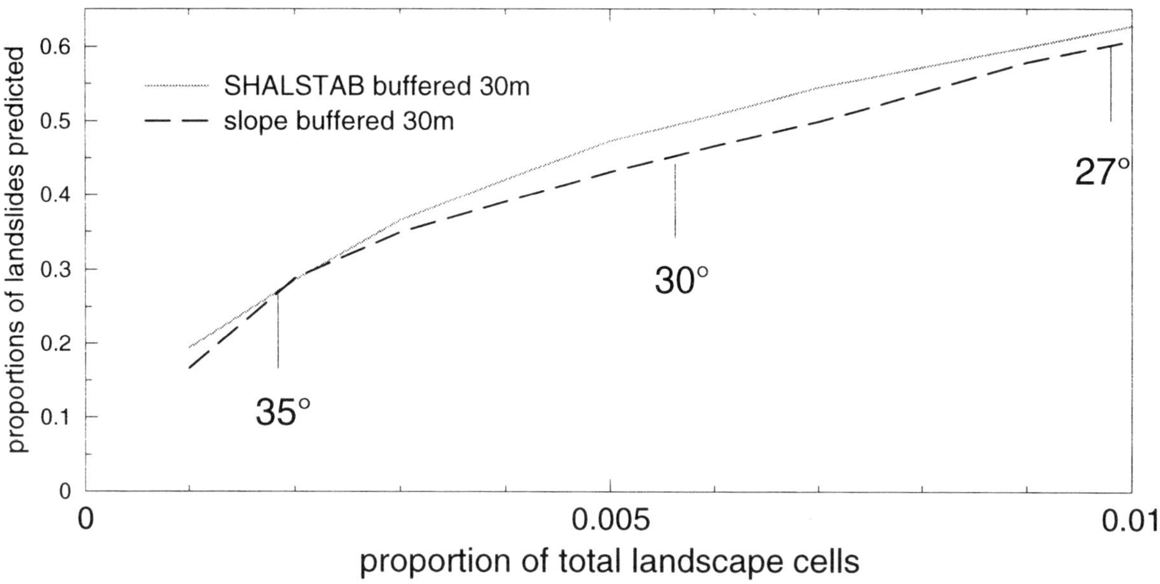

Figure 5. Proportion of mapped landslides (y-axis) with hazard zones that cover a given portion of Seattle (x-axis) for both SHALSTAB and a simple threshold slope model.

downslope hazards, which are important public safety issues. Avoidance of development in high hazard zones is the only sure way to minimize public exposure to impacts from landslides. The problem of whether to restrict development in "high hazard" zones involves both the time scale over which we may be able to rely on such models to predict failures and the definition of acceptable risk; private individuals or developers seeking to construct a view home in a hazardous location may be more willing to accept risk than the public agency charged with providing ongoing access and utilities to potentially unstable locations. Although models can help to provide an objective framework within which to develop policies, it is important to test model performance against available data on observed landslide locations—especially in environments where fundamental assumptions of the model may be poor representations of field conditions and processes. Furthermore, in some applications simple models such as a critical-slope may suffice for hazard delineation, and prove more attractive for regulatory purposes due to their inherent simplicity.

The small area of potentially unstable ground and the strong spatial correspondence between areas predicted to be potentially unstable and the landslides that have occurred over almost a century indicate that landslide hazard areas in Seattle are relatively predictable. No one can predict which of the potentially unstable areas will fail in a particular storm, but our results indicate that over the course of a century a substantial portion of ground predicted to be potentially unstable did experience instability. In addition, the strong fidelity of landslides to predicted high hazard areas implies that landslide hazards could be managed in much the same manner as flood hazards, only over longer planning horizons due to the less frequent impact of landsliding on specific areas within hazard zones (flood plains are innundated frequently, as flow in typical rivers rises overbank every year or two). Both flood plains and potentially unstable ground can be defined on a topographic basis, and the degree of acceptable risk or subsidy for those dwelling on or building in such areas is an appropriate topic for consideration by local, state and federal agencies. Should development be allowed in such areas? Should houses and businesses be rebuilt in slide-prone areas? Should disaster aid be channeled to assist in maintaining development in such areas? The small proportion of the City of Seattle that is at risk for shallow landsliding certainly motivates asking whether the public at large should bear the cost of repairing private property or maintaining public infrastructure to serve private interests on potentially unstable ground. While there are many ways to address these questions in the political arena, the objective delineation of areas potentially at risk to landslide initiation provides a means for developing solutions likely to achieve whatever policy objectives are defined by the political process.

Acknowledgments. This research was supported by the National Science Foundation through grant CMS96-10269 and the University of Washington's Puget Sound Regional Synthesis Model (PRISM) project. We thank Derek Booth and two anonymous reviewers for comments on a draft manuscript, and the City of Seattle for providing their GIS data.

REFERENCES

D. B. Booth and J. Sacket, compilers, 1997, Geologic map of King County, Washington: King County, Department of Natural Resources, digital map and database.

Brabb, E. E., and Harrod, B. L. (Eds.), 1989, *Landslides: Extent and Economic Significance*, A. A. Balkema, Rotterdam, 385p.

Burroughs, E. R., Jr., 1984, Landslide hazard rating for portions of the Oregon Coast Range: in *Proceedings of the Symposium on Effects of Forest Land Use on Erosion and Slope Stability*, Honolulu, Hawaii, p. 265-274.

Campbell, R. H., 1975, *Soil Slips, Debris Flows and Rainstorms in the Santa Monica Mountains and Vicinity, Southern California*, US Geol. Surv. Prof. Paper 851, 51p.

Derbyshire, E., van Asch, T., Billard, A., Meng, X., 1995, Modelling the erosional susceptibility of landslide catchments in thick loess: Chinese variations on a theme by Jan de Ploey: Catena, v. 25: 315-331.

Dietrich, W. E., Wilson, C. J., Montgomery, D. R., McKean, J., and Bauer, R., 1992, Channelization thresholds and land surface morphology: *Geology*, v. 20, p. 675-679.

Dietrich, W. E., Wilson, C. J., Montgomery, D. R., and McKean, J., 1993, Analysis of erosion thresholds, channel networks and landscape morphology using a digital terrain model: Journal of *Geology*, v. 101, p. 259-278.

Dietrich, W. E., Reiss, R., Hsu, M.-L., and Montgomery, D. R., 1995, A process-based model for colluvial soil depth and shallow landsliding using digital elevation data: *Hydrological Processes*, v. 9, 383-400.

Ellen, S. D., Mark, R. K., Cannon, S. H., and Knifong, D. L., 1993, *Map of Debris-Flow Hazard in the Honolulu District of Oahu, Hawaii*, US Geol. Surv. Open-File Report 93-213, 25p.

Ellen, S. D., and Wieczorek, G. F. (Eds.), 1988, *Landslides, Floods and Marine Effects of the Storm of January 3-5, 1982, in the San Francisco Bay Region, California*, US Geol. Surv. Prof. Paper 1434, 310p.

Galster, R. W., and Laprade, W. T., 1991, Geology of Seattle, Washington, United States of America: *Bulletin of the Association of Engineering Geologists*, v. 28, p. 235-302.

Gray, D. H., and Megahan, W. F., 1981, Forest vegetation removal and slope stability in the Idaho Batholith, Res. Pap. INT-271, 23pp., US Dept. of Agriculture, Forest Service, Ogden, Utah.

Hollingsworth, R., and Kovacs, G. S., 1981, Soil slumps and debris flows: Prediction and Protection: *Bulletin of the Association of Engineering Geologists*, v. 18, 17-28.

Jones, F. O., 1973, *Landslides of Rio de Janeiro and the Serra das Araras Escarpment, Brazil*, U.S. Geol. Surv. Prof. Paper 697, 42p.

Koloski, J. W., Schwarz, S. D., and Tubbs, D. W., 1989, Geotechnical properties of geologic materials, in *Engineering Geology in Washington*, Volume I, Washington Division of Geology and Earth Resources Bulletin 78, p. 19-26.

Montgomery, D. R., Wright, R. H., and Booth, T., 1991, Debris flow hazard mitigation for colluvium-filled swales: *Bulletin of the Association of Engineering Geologists*, v. 28, p. 299-319.

Montgomery, D. R., and Dietrich, W. E., 1994, A Physically-Based Model for the Topographic Control on Shallow Landsliding: *Water Resources Research*, v. 30, p. 1153-1171.

Montgomery, D. R., Sullivan, K., and Greenberg, H., 1998, Regional test of a model for shallow landsliding: *Hydrological Processes*, v. 12, p. 943-955.

Montgomery, D. R., Schmidt, K. M., Greenberg, H. M., and Dietrich, W. E., 2000, Forest clearing and regional landsliding, *Geology*, v. 28, p.311-314.

Nilsen, T. H., Wright, R. H., Vlasic, T. C., and Spangle, W. E., *Relative slope stability and land-use planning in the San Francisco Bay Region, California*, US Geol. Survey Prof. Paper 944, 96p.

Okimura, T., and Ichikawa, R., A., 1985, Prediction method for surface failures by movements of infiltrated water in a surface soil layer: *Natural Disaster Science*, v. 7, p. 41-51.

O'Loughlin, C. L., and Pearce, A. J., 1976, Influence of Cenozoic geology on mass movement and sediment yield response to forest removal, North Westland, *New Zealand: Bulletin of the International Association of Engineering Geologists*, v. 14, p. 41-46.

Pack, R. T., and Tarboton, D. G., 1997, New developments in terrain stability mapping in B. C., Paper presented at the 11th Vancouver Geotechnical Society Symposium - Forestry Geotechnique and Resource Engineering, 12p.

Reneau, S. L., and Dietrich, W. E., 1987, Size and location of colluvial landslides in a steep forested landscape: in *Proceedings of the International Symposium on Erosion and Sedimentation in the Pacific Rim*, R. L. Beschta, T. Blinn, G. E. Grant, G. G. Ice, and F. J. Swanson (eds.), Int. Assoc. Hydrol. Sci. Pub. 165, p. 39-48.

Schell, P., 1998, *Landslide Policies for Seattle: Preliminary Recommendations to the Landslide ad hoc Committee*, Report submitted to the Seattle City Council by Mayor Paul Schell, 23p.

Seeley, M. W., and West, D. O., 1990, Approach to geologic hazard zoning for regional planning, Inyo National Forest, California and Nevada: *Bulletin of the Association of Engineering Geologists*, v. 27, p. 23-35.

Selby, M. J., 1993, *Hillslope Materials and Processes*, Oxford Univeristy Press, Oxford, 451p.

Sharpe, C. F. S., 1938, *Landslides and Related Phenomena: A Study of Mass-Movements of Soil and Rock*, Columbia University Press, New York, 137p.

Swanson, F. J., and Dyrness, C. T., 1975, Impact of clearcutting and road construction on soil erosion by landslides in the western Cascade Range, Oregon, *Geology*, v. 3, p. 393-396.

Tubbs, D. W., 1974, *Landslides in Seattle*, Washington Division of Geology and Earth Resources, Information Circular 52, Olympia, 15 p.

Tubbs, D. W., 1975, *Causes, Mechanisms and Prediction of Landsliding in Seattle*, Ph.D. dissertation, Dept. of Geological Sciences, Univ. of Washington, Seattle, 94p.

van Asch, T., Kuipers, B., van der Zanden, D. J., 1993, An information system for large scale quantitative hazard analyses of landslides: *Zeitschrift für Geomorphologie, Supplementband* 87, p. 133-140.

Waldron, H. H., Liesch, B. A. Mullineaux, D. R., and Crandell, D. R., 1962, *Preliminary Geologic Map of Seattle and Vicinity, Washington*: US Geol. Surv. Misc. Geologic Investigations Map I-354.

Wu, W., and R. C., Sidle, 1995, A distributed slope stability model for steep forested basins: *Water Resources Research*, v. 31, p. 2097-2110.

Zhang, W., and Montgomery, D. R., 1994, Digital elevation model grid size, landscape representation, and hydrologic simulations: *Water Resources Research*, v. 30, p. 1019-1028.

Section 2

Influence of Forest Management Activities on Hydrology and Geomorphology

Introduction to Section 2—Problems in Measuring and Modeling the Influence of Forest Management on Hydrologic and Geomorphic Processes

Thomas Dunne

Donald Bren School of Environmental Science and Management, University of California, Santa Barbara, California

Hydrologic and geomorphic changes resulting from alterations to the land surface and canopy cover in forest lands are subtler than those occurring in urban areas. Changes in infiltration capacity that alter the partitioning of runoff between overland and subsurface routes are smaller in managed forests than in urban areas, except locally on roads and following rare, intense burns in some biomes. Sedimentation rates and water quality degradation are not as intense. However, the changes in forests are more extensive and they diminish wildland resources that require prodigious amounts of space and high-quality water and the occasional disturbance of habitats (through fire and flood) in ways that are inconvenient for human inhabitants or resource managers. Thus, the influence of forest activities on runoff, erosion, and the form and biophysical functioning of landscapes have long been a source of contention between people who place a high value on wild living resources and the landowners and technical specialists most closely associated with timber harvest (from which, of course, most of us also profit).

For most people and their opinion leaders, forested mountain lands have been out of sight and out of mind for decades—the realm of timber managers—until recent debates about endangered species, water quality, and flood hazards to communities that are spreading into forested mountains have ignited a broader interest. Alteration of forest cover through inadvertent land-use experiments has highlighted the various functions of a forest cover (canopy interception, evapotranspiration, root strength), the various influences of timber harvest (road construction), and how they interact with topography, rainfall, and snowmelt to control the magnitude and tempo of water and sediment delivery to channels. Fewer studies of the influences of soil type on watershed response to timber management have been made so far. However, relevant studies of the disturbance at landscape scale have suffered from a variety of constraints because the per-unit-area value of these wildlands is low compared to the high-value, intensively engineered, urban lands, where concerns about safety and sanitation established, early in the 20th century, the social expectation of intensive investment in site characterization and long-term monitoring. We have not yet established the social value of these operations in forested lands convincingly enough to garner support for thorough, long-term empirical studies.

Monitoring programs in most forest lands (with a few exceptions in carefully managed, uncharacteristic, experimental watersheds) are typically too brief to sample the variability of natural and disturbed hydrologic regimes, and thus have a high probability of missing critical events such as large floods, landslides and debris flows. Sparse monitoring rarely captures the intersection of large rainstorms with the transient effects of timber harvest, such as the removal of canopy or root structure, or the installation or grading of road fills and drainage systems. Records are gradually lengthening at a few sites, as represented by the paper in this volume by Lewis et al, and the regional-scale statistical survey of floods by Jones and Grant [1996], but providing purely empirical resolution to questions such as whether timber harvest increases the magnitude of large floods or the risk of landsliding will always suffer from the difficulties of small sample size in the critical range of large events.

Constraints on the number and size of monitored sites, and the need to make collateral measurements in order to understand the occurrence of destructive processes, also result in most measurement programs being too local to obtain accurate spatial averaging of, say, storm rainfall, snow accumulation or melt rates, landsliding, or other erosion processes. Thus, it is difficult to use monitored watersheds to estimate the probability density functions of various processes that trigger floods and that load materials (sediment, nutrients,

Land Use and Watersheds: Human Influence on Hydrology and Geomorphology in Urban and Forest Areas
Water Science and Application 2

woody debris) into channels and re-shape their ecosystems. It would be difficult, for example, for an engineer with a design problem or a water-quality target to use monitored watersheds in the Pacific Northwest to estimate probable sediment yields from wild and managed forest watersheds.

The papers in this section begin to redress some of these difficulties. Most of them utilize explicit models of processes to formulate hypotheses that can be tested against the sparse empirical record, and they take advantage of the largest source of new hydrologic data in the history of the science: digital spatial databases of topography and, to a lesser extent, soil properties. Despite the progress reported herein, however, the message that resonates throughout the papers is the constraint of data quality and availability, and thus the need for investment in field data collection programs to define the fluctuating condition of forested wildlands in various regions of the United States. Indeed, the condition in which we now find ourselves (as agencies facing lawsuits and other instruments of policy debate, or as individual scientists being asked questions about issues of rising public concern with inadequate data in our hands) indicates that society would probably have been better served by investing more funds and effort in measuring the properties of wildlands and monitoring their changes in the intervening years. We would now be in a better position to make decisions about acceptable ranges and even required changes of habitat condition or about water quality variability, ecosystem restoration, and sustainable development. It would be wise to avoid continuance of the same mistake, and hydrologists can help by articulating the need and designing the solution more effectively than has been our record to date.

The papers collected here, which are among the best examples published in the field, testify to the continuing need for well-supported programs of field data collection and field process studies. Even the carefully designed and relatively long statistical study reported by Lewis et al. highlights the brevity of rainfall and runoff records and the complexity of interactions that affect peak storm flows and suspended sediment concentrations in very small logged watersheds. Although Lewis et al. demonstrate the extraordinary care that must be taken to control for various effects such as rainstorm size, time since disturbance, and antecedent wetness, the presence of many transient effects such as sediment storage and varying sediment sources still make the data set difficult to interpret and to extrapolate regionally. And this is for a relatively straightforward experiment with a simple road network and harvest practices, no devastating flood in the short record, and single ownership and timing of harvest! Dietrich et al illustrate how innovations in modeling the spatial distribution of landsliding are limited by the coarse nature of most digital topographic products in forested mountains. Modeling the timing of landsliding through timber cycles (as in the paper by Sidle and Wu), even for conceptual and planning purposes, requires information that is rarely available. The data-quality issue results from the lack of interest in data gathering by some resource management agencies, and the lack of inter-agency cooperation in field studies and data gathering (on, for example, streams that cross agency land boundaries or watersheds that don't coincide with a single agency's land). The lack of high-quality data is sustained by a tradition among hydrologists of utilizing whatever fragmentary records we can get from hydrometric networks that were not originally designed for scientific purposes.

Hydrologists have been so grateful for years for the scraps of data that we obtain from measuring stations, mostly located not for the purpose of understanding river flow or sediment sources, but for an agency's need to monitor contaminant loads or water quality standards, or a power or irrigation company's need to define its resource, or a local government's desire to keep an eye on their resources to deflect criticism or law suits. If we are lucky, there is a rain gauge in the vicinity of the monitored watershed, and we can locate those data in another agency. We have not invested time, as a professional community, in the design of (and garnering of support for) hydrometric networks that would be useful for testing theories, examining whether transferable watershed-scale parameterization is possible, etc. We have not aggressively investigated the possibilities of technology for efficient data collection - e.g. rapid installation and calibration of flow recorders at many locations in a region; sediment samplers calibrated to total suspended load; and remote sensing of basin-wide channel morphology, wood loading, or water temperature. Various remote sensing initiatives promise to assist forest hydrology through the enhancement of hydrometeorological databases. For example, the need to calibrate spaceborne radiometers has motivated support for well-endowed surface radiometer sites [Augustine et al., 2000]. A vital challenge for hydrologists desiring to improve rainfall-runoff predictions is to combine ground networks of rain gauges (after appropriate densification) with radar to define spatial and temporal fields of rainfall, which is, after all, the largest component of the hydrologic balance equation outside of the snow zone. Hydrologists need to participate in exploring and improving calibrations under a range of raindrop size and topographic ruggedness, exploiting data handling systems for rapid data processing, and examining the hydrologic implications of improved knowledge of precipitation.

Limitations of data add significance to cooperative efforts in sharing of old records within and between agencies and the academic community. Fortunately, US Geological Survey streamflow data (though not raw data such as sediment concentrations, gauging station morphology, channel hydraulics, or discharge rating curves) have long been routinely available, and the USGS has greatly improved access to flow records and digital topography using the worldwide web. The academic community, through the NSF-sponsored Long Term Ecological Research network, is also promoting the sharing of hydrologic data along with experience and insights gained about model applications. The USGS has also promoted the development and dissemination of models to assimilate data and transfer it to unmonitored sites [Leavesley et al., 2000].

It is generally acknowledged that mathematical models must play an ever-greater role in answering increasingly detailed questions about hydrologic mechanisms and conditions. It is impossible to collect and present sufficient measurements of processes over large areas and multiple years, or to sample and understand the many interactions between processes and the factors that control them, without resorting to complex sequences of mathematical expressions, and computing the implications of the limited range of conditions or the fragments of process understanding that we are able to capture empirically. However, models immediately raise the need for high-quality input and other data for testing. These data have to be obtained quickly and at low cost.

During the past two decades, of course, the most important increase in data availability has arisen through digital representations of watershed condition, especially in topography and cover, and the capacity for updating and refining such representations. In particular, the availability of digital topography, first through digitization of aerial photographs, and more recently from side-looking airborne radar and laser altimetry, has expanded the application of spatially explicit, topographically driven, environmental models in mountain watersheds. Important innovations have occurred in: conceptual and mechanistic runoff models [e.g. Beven and Kirkby, 1979; O'Loughlin, 1986; Moore et al., 1988]; hydroecological models [e.g. Band et al., 1991]; erosion and landform predictions [e.g. Dietrich et al., 1993]; and snow pack energy balance models [Marks and Dozier, 1992].

The papers in this section continue the developments referred to above, and focus on management and policy issues that have become particularly contentious in the Pacific Northwest of the United States, and are emerging in other timber regions. Despite increasing regulation of timber harvest in the past three decades (involving rates of cutting, method of logging, constraints on distributing various activities, and requirements for best management practices), society's sensitivities to the effects of timber harvest have simultaneously escalated. There is greater concern for aquatic ecology, the condition of streams, and for the preservation of biodiversity, particularly for certain charismatic species such as salmon and other endangered species. Details of questions have increased to levels such as: "What is the role of small to medium-sized floods in disturbing channel beds?" or "What is the effect of timber harvest on stream turbidity?" The issues are also intensified because of population moving into areas formerly reserved for logging, thereby intensifying concern about whether rainfall or rain-on-snowmelt floods are increased by timber harvest. Although qualitative ideas exist about the sign of the resulting changes, hydrology has not had the capacity for sufficiently quantitative predictions that can be used for defensible analyses of risk and of cumulative effects of multiple changes at the watershed scale.

Several of the papers establish that spatially explicit models, incorporating new spatial data, modern computing power, and insights from process studies from the 1960s-1980s, can capture relevant hydrologic processes at a resolution and a geographic scale sufficient for policy making and forest engineering plans, as long as data of high quality are available. The papers by Wigmosta and Perkins and by Bowling and Lettenmaier both illustrate how the Distributed Soil-Hydrology-Vegetation Model (DSHVM), a DEM-grid-cell-based model of hydrologic processes characteristic of maritime, mountainous environments, can be augmented to examine the role of road networks in intercepting subsurface flow and routing it quickly to a watershed outlet. The model was originally formulated by Wigmosta et al. [1994]. Bowling and Lettenmaier also examine the synergies between enhancing the capacity of the watershed to evacuate runoff in this way and increasing the volume of runoff due to the removal of canopy interception and evaporation. However, despite the availability of several spatial databases for these intensively studied basins of high commercial value, it was still necessary to augment the characterization of the road network through GPS surveys, including the localization of about 20 culverts per km2 of watershed in order to capture some runoff processes at the scale at which they express themselves.

Both modeling efforts pay attention to using observable processes and measurable quantities to constrain parameter values as far as possible. They verify predictions wherever possible (as in the case of Bowling and Lettenmaier's observations of the altitudinal limits of snow coverage.) However, it was still necessary to calibrate the volumes and timing of runoff by adjusting the transmissivity of the soil, and the geometry of the subsurface flow and water-holding fields. The need for calibration on each watershed raises the question of whether transmissivity (and perhaps other parameter values) can be regionalized and correlated with some identifiable controlling factor. It would be reassuring to know whether a set of catchments in the same geologic-physiographic region calibrate with the same transmissivity, or whether the representation of material properties by that parameter is so occluded with geometric effects, with typical rainstorm characteristics, or with recent moisture conditions that it is essentially not transferable from one basin or period to another. Effects of the calibration period may be particularly critical in the case of non-equilibrium forests.

The need for better definition of soil and root-zone dimensions and of precipitation fields in complex terrain arises again in the modeling of landslides later in the volume. However, testing of the landslide predictions is facilitated by spatially explicit observations of the absence, presence and timing of the critical events. Rainfall-runoff predictions are verified against point measurements of streamflow, and cannot be tested to any significant degree against spatial data. In fact, monitoring of runoff at a resolution needed to test hypotheses about mechanisms [Moore and Thompson, 1996; Anderson et al., 1997; Hutchinson and Moore, 2000] and attempts to measure hydraulic conductivity independently of the "watershed permeameter" [Davis et al., 1999] indicate just how much spatial averaging and conceptual revision are required in our application of familiar constitutive equations at the watershed scale.

Although it is difficult to imagine verifying or falsifying a complex model such as DHSVM, which includes such innovations as the use of a predicted two-dimensional wind field over complex terrain, snow accumulation and metamorphism, canopy processes affecting snow and liquid water, a multi-layer root zone, as well as overland, subsurface, and channelized runoff, the model provides a vessel for our communal understanding of the hydrologic processes that are believed to be important in maritime mountains. It also constitutes a way of systematically analyzing the various interactions that can only be sampled in a very sparse manner by even enhanced field monitoring programs. Such a model provides a template for field studies to investigate various components of the hydrologic cycle and to compute the significance of their interactions. The limiting resource appears to be the lack of a tradition in hydrology for organizing large-scale, long-running tests of modeling capability in well-instrumented field sites or regions. Such tests would require funds for watersheds instrumented to test hydrologic hypotheses and decisions about whether to concentrate process studies in a relatively few, heavily studied catchments to capture the expected synergies, or to distribute the field sites widely in order to expand the range of processes, environmental conditions, and parameter values. Another use to which models of this type could be put involves forensic studies of extraordinarily large and damaging floods for which adequate data, especially on rainfall or snowmelt, are available. This would avoid the problem of extrapolating parameter values derived from small floods to large ones. It might also make use of the rare, adequate data sets (e.g. on canopy interception processes) from the few intensively instrumented watersheds that have sampled very large events. Increasing transfer of information from experimental watersheds by means of spatially explicit modeling remains to be done. Given society's recent focus on pollution and water quality rather than floods, there would be greater support for such forensic studies if the flood triggers major pollution events, such as tailings dam failures or nitrate pollution of rivers. Another opportunity for well-instrumented studies of rainfall and runoff is presented by each large conflagration in western forests, but the academic hydrology community and the land management agencies have not yet coordinated their activities and resources to take advantage of such opportunities.

Hydrologic interest in forest roads also extends to their effects on sediment delivery to streams. Luce and Black have developed an extraordinary data set, monitoring sediment loss from 74 road segments that sample a range of gradient, plot length, cutslope height, and soil texture. The problem of modeling to assimilate and interpret the results initially seems simpler than watershed-scale erosion prediction, but the strong sensitivity of sediment production to small local differences in environmental characteristics and recent history of a road segment constrains process modeling. The results are satisfying in this sense: they can be rationalized in terms of qualitative conceptual models of field observations, and provide useful calibrations of the effects of time on sediment availability. On the other hand, transferable predictions are severely limited by sensitivity to transient effects such as sediment availability as conditioned by the recent history of grading and ditch clearing. Surprisingly, traffic intensity did not emerge as a factor controlling sediment availability.

Papers by Dietrich et al. and Sidle and Wu continue the theme of sediment production. Taken in combination, these papers provide a review of current methods for prediction of landslide sources, particularly of locations vulnerable to timber harvest effects. Sidle and Wu begin with a review of terrain mapping with multi-factor overlays, which is still the best way of integrating the effects of material properties such as strength, stratigraphy, anisotropy, heterogeneity, and of taking into account historical records and the experience of the delineator, even if the technique is not very useful for delineating specific landslide sites or their timing. Both papers explore the leverage to be gained from application of theories of subsurface flow and slope stability to predicting the temporal and spatial occurrence of landslides using digital topographic data.

Both spatially distributed models combine the infinite-slope stability model with a subsurface flow model for steady-state conditions (Dietrich et al.) or finite-duration storms (Sidle and Wu). Iverson [2000] has recently presented a subsurface flow theory that has the capacity for representing the different timescales of transient pore-pressure response involved in both short, intense rainstorms that induce rapid, vertical percolation in permeable soils and the slower, downslope percolation responsible for the drainage-area effect on shallow landslides and for triggering the instability of deep landslides in low-permeability soils long after rainfall. Coupled with digital elevation models, this theory will allow an extension of current topographically driven models of landsliding, --- at the expense, once again, of the need for transferable parameterization based on field studies since the model is also based on the Richards equation. All of these methods can take advantage of automated procedures for processing digital topography in hydrologic calculations, and of removing some operator variance in recognizing sites with topography that could promote failure. They can also efficiently transfer results to unsurveyed basins, using thresholds from a few basins in which detailed landslide mapping has occurred soon after a large rainstorm or in which there has been a detailed historical audit of landsliding throughout the air-photo history.

The Sidle and Wu model adds the transient effect of root decay and recovery resulting from timber harvest, fire and disease and the stochastic occurrence of rainstorms to calculate the temporal and spatial distribution of landsliding, and thereby to map failure probabilities across a disturbed landscape. However, our general ignorance of soil depths, especially in hollows, propagates uncertainties through these predictions of landslide occurrence, and can only be compensated by intensive site investigation [Dengler et al., 1987], the assumption of a probability distribution of depths [Ward et al., 1982], or stochastic simulations of very long series of events that includes the evacuation and filling of hollows with mobile colluvium [Benda and Dunne, 1998; Benda et al., 1998].

Not surprisingly, both the Sidle and Wu and Dietrich et al. modeling efforts indicate that leaving trees in hollows reduces the risk of landsliding, at least over the multi-decadal time scale. Root cohesion is particularly difficult to represent in such stability models, since the mechanics are not well understood and the effect is time-dependent immediately after harvest or fire through at least a few decades of succession, during the period of species replacement and changing stem density. However, root reinforcement will probably become the focus of intense investigation now that there is a consensus about the effects of decreasing root strength on slope stability after years of the debate being diverted towards the (now discounted) destabilizing effects of removing the normal stress due to tree weight. The timber industry will probably now encourage investigations of the role of the understory, and the possibility of designing root systems through manipulation of forest-age structure during harvest cycles [Sidle, 1992]. However, as it is presently formulated in slope stability models, a single threshold value of "soil strength" may not be measurable at the relevant scale. It exhibits strong local variations, from the distance between tree perimeters to the half-widths of hollows, and between deep soils and shallow soils. Yet, if one tries to back-calculate "strength" from the discrimination between failure sites and stable sites, a large range of combinations of relevant controls (assumed values of pore pressure, friction angles, cohesion, and failure geometry) can interact to affect the derived values, even in an apparently homogeneous landscape. The Dietrich et al. paper, for example, demonstrates the difficulty of calibrating a spatial model of landslide occurrence, in which small uncertainties in parameter values (probably in the range of unknowability for most extensive applications) cause several- fold differences in the predicted spatial density of failures. The Sidle and Wu model (dSLAM) concentrates our thinking and formal analysis on the multiplicity of interacting factors affecting landslide occurrence, but the parameter estimation problem is daunting and measurement programs are in their infancy.

Rather than attempting a full prediction of landslide occurrence under transient conditions, Dietrich et al. concentrate on illustrating and (most importantly) evaluating the capability of their model, SHALSTAB, for recognizing the topographic attributes that localize shallow landsliding and debris-flow generation. The authors discuss the significance of ignoring other attributes (e.g. soil depth, root strength, and subsurface conductivity) that are essentially unknowable at present, at least without intensive site investigation. Once all of these quantities are fixed, the model can be run to test the hypothesis that the topographic attributes represented in the model (area drained per unit of contour width and local hillslope gradient) are good predictors of where in the landscape landslides will be concentrated. Considerable attention is paid to the role of topographic resolution in DEMs, and the limits of accuracy in mapping landslide sites.

Dietrich et al. suggest that their model predictions be interpreted as qualitative rankings of risk, and they illustrate why in the application of such a model for decision-making there is no substitute for intelligent interaction between model and interpreter. It seems unlikely that there will ever be (the dream of managers and regulators) a prediction tool that can be used routinely, objectively, and reproducibly by people with little or no hands-on experience of the phenomenon being modeled. This has long been the assumption underpinning the use of handbooks of engineering practice, and their modern incarnations in planning models, which are downloadable from the worldwide web and can be used as a cookbook, for example, to estimate Total Maximum Daily Loads of pollutants such as sediment. Instead, the accuracy of predictions is likely to depend on judgment and skill. Simply having applied the model many times does not constitute "experience". Moreover, effective use of the predictions will depend on value-based decision-making (the precautionary principle, maximum utility, or alternatives). Dietrich et al. provide valuable examples of applications of their model for: site-specific prescriptions by landowners; regional planning by government agencies; and hazard mapping. The model's capability lies in indicating that portion of the landscape in which most landslides are likely to occur. Site-specific conditions that dictate when landsliding might occur depend on knowledge of management and storm history.

What makes models difficult for managers and policy makers to comprehend and use is the concept that models are approximations, limited to certain types of representation by the resolution at which they are designed and supplied with input data. Once a model prediction is made, it is viewed by a scientist as a hypothesis to be checked by measurements (monitoring), and probably refuted or refined. Managers usually need an answer, with some urgency, even if it is obtained through methods that do not represent processes or conditions found on the watershed. Instead, they need an objective calculation device, relatively free from operator variance, from which all users (including contentious stakeholders or technical analysts hired by litigating opponents) will obtain more or less the same answer. The paper by Jackson et al. in Section 1 of this volume reports a modeling and decision-making approach for the urban environment in which all model users obtain essentially the same answer. Once the calculation (or a consequent decision) is made, it tends to be embraced and defended to the death as the basis for a decision that had to be made in uncertain times. Often, the agency embracing the prediction does no monitoring, in case the prediction proves to be inaccurate or even incorrect. These difficulties in using modeling and prediction are antithetical to scientific hydrology, but they limit support for improvements in model construction and in measurement systems for testing models under field conditions where there is a chance of reducing uncertainty or discovering new hydrologic principles.

The papers collected in this section provide grounds for optimism and challenge. Progress is being made in the application of hydrologic and geomorphic theory and in the use of an extensive new data source: digital topography. However, the progress demands improvements in other parts of our knowledge, including the geometry and material properties of various environmental features, and the best ways to think about and use these modeling procedures in our contributions to environmental problem solving.

REFERENCES

Anderson, S. P., W. E. Dietrich, D. R. Montgomery, R. Torres, M. E. Conrad, and K. Loague, Subsurface flow paths in a steep, unchanneled catchment, *Water Resour. Res.*, 33, 2637-2654, 1997.

Augustine, J. A., J. J. DeLuisi, and C. N. Long, SURFRAD—A national surface radiation budget network for atmospheric research, *Bull. American Meteorological Society*, 81, 2341-2357, 2000.

Band, L. E., D.L. Peterson, S.W. Running, J.C. Coughlan, R. Lammers, J. Dungan and R. R. Nemani, Ecosystem processes at the watershed level: Basis for distributed simulation, *Ecological Modeling*, 56, 171- 196, 1991.

Benda, L. and T. Dunne, Stochastic forcing of sediment supply to channel networks from landsliding and debris flow, *Water Resour. Res.*, 33, 2836-2849, 1998.

Benda, L., D. J. Miller, T. Dunne, G. H. Reeves, and J. K. Agee, Dynamic Landscape Systems, in *Ecology and Management of Streams and Rivers in the Pacific Northwest Coastal Ecoregion,* edited by R. Naiman and R. Bilby, pp. 261-288, Springer Verlag, New York, NY, 1998.

Beven, K. and M. J. Kirkby, A physically based, variable contributing area model of basin hydrology, *Hydrol. Sci. Bull.*, 24, 43-69, 1979.

Davis, S. H., R. A. Vertessy, and R. P. Silverstein, The sensitivity of a catchment model to soil hydraulic properties obtained by using different measurement techniques, *Hydrol. Processes,* 13, 677-688, 1999.

Dengler, L, A.K. Lehre, and C. J. Wilson, Bedrock geometry of unchannelized valleys, *Proc. Symp. Erosion and Sedimentation in the Pacific Rim,* Internat. Assoc. Hydrol. Sciences Pub. 165, 81-90, 1987.

Dietrich, W. E., C.J. Wilson, D.R. Montgomery, J. McKean, and R. Bauer, Analysis of erosion thresholds: channel networks and landscape morphology using a digital terrain model, *Journal of Geology*, 101, 259-278, 1993.

Hutchinson D. G, and R. D. Moore, Throughflow variability on a forested hillslope underlain by compacted glacial till, *Hydrol. Processes,* 14, 1751-1766, 2000.

Iverson, R. M., Landslide triggering by rain infiltration, *Water Resour. Res.*, 36 (7), 1897-1910, 2000.

Jones, J. A., and G. E. Grant, Peak flow responses to clearcutting and roads in small and large basins, Western Cascades, Oregon, *Water Resour. Res.*, 32, 959-974, 1996.

Leavesley, G.H., Restrepo, P.J., Stannard, L.G., Frankoski, L.A., and Sautins, A.M, The modular modeling system (MMS)—A modeling framework for multidisciplinary research and operational applications, in, *GIS and Environmental Modeling: Progress and Research Issues*, edited by M. Goodchild, L. Steyaert, B. Parks, M. Crane, M. Johnston, D. Maidment, and S. Glendinning, S., pp. 155-158, GIS World Books, Ft. Collins, CO., 2000.

Marks, D., and J. Dozier, Climate and energy exchange at the snow surface in the alpine region of the Sierra Nevada: 2. Snow cover energy balance, *Water Resources Res.*, 28, 3043-3054, 1992.

Moore, I. D., E. M. O'Loughlin, and G. J. Burch, a contour-based topographic model for hydrological and ecological applications, *Earth Surface Processes,* 13, 305-320, 1988.

Moore, R. D. and J. C. Thompson, Are water table variations in a shallow forest soil consistent with the TOPMODEL concept?, *Water Resour. Res.*, 32, 663-669, 1996.

O'Loughlin, E. M., Prediction of surface saturation zones in natural catchments by topographic analysis, *Water Resources Research,* 22, 794-804, 1986.

Sidle, R. C., A theoretical model of the effects of timber harvesting on slope stability, *Water Resour. Res.*, 28, 1897-1910, 1992.

Ward, T.J., R. Li, and D. B. Simons, Mapping landslide hazards in forest watersheds, *J. Geotechnical Division, American Society of Civil Engineers*, 108(GT2), 319-324, 1982.

Wigmosta, M. S., L. W. Vail, and D. P. Lettenmaier, A distributed hydrology-vegetation model for complex terrain, *Water Resour. Res.*, 30, 1665-1679, 1994.

Thomas Dunne, Donald Bren School of Environmental Science and Management, University of California Santa Barbara, Santa Barbara, CA 93106; tdunne@bren.ucsb.edu

Impacts of Logging on Storm Peak Flows, Flow Volumes and Suspended Sediment Loads in Caspar Creek, California

Jack Lewis[1], Sylvia R. Mori[2], Elizabeth T. Keppeler[1], Robert R. Ziemer[2]

Models are fit to 11 years of storm peak flows, flow volumes, and suspended sediment loads on a network of 14 stream gaging stations in the North Fork Caspar Creek, a 473-ha coastal watershed bearing a second-growth forest of redwood and Douglas-fir. For the first 4 years of monitoring, the watershed was in a relatively undisturbed state, having last been logged prior to 1904, with only a county road traversing the ridgetops. Nearly half the watershed was clear-cut over a period of 3 years, and yarded primarily using uphill skyline cable systems to spur roads constructed high on the slopes. Three tributaries were maintained as controls and left undisturbed. Four years of data were collected after logging was completed. Exploratory analysis and model fitting permit characterization and quantification of the effects of watershed disturbances, watershed area, antecedent wetness, and time since disturbance on storm runoff and suspended sediment. Model interpretations provide insight into the nature of certain types of cumulative watershed effects.

INTRODUCTION

This paired-watershed study in the North Fork of Caspar Creek was motivated by a desire to understand how a particular logging system affects storm peak flows, flow volumes, and suspended sediment loads in a second-growth coastal redwood forest. The logging system consisted of clear-cutting with streamside buffers, and yarding primarily by skyline to spur roads located on upper slopes and ridges. Primary objectives were to quantify how impacts vary with different levels of disturbance and how the effects of a given disturbance vary downstream. Pursuant to these objectives, a statistical model was developed for a treatment-and-control experimental design involving multiple watersheds. The study was also an opportunity for testing new technologies, and demonstrates two new automated schemes for suspended sediment sampling. Techniques for estimating sediment loads from these samples are tested and applied.

Storm Peaks

Throughout much of the Pacific Northwest, a large soil moisture deficit develops during the dry summer. With the onset of the rainy season in the fall, the dry soil profile begins to be recharged with moisture. In the H.J. Andrews Experimental Forest in the Oregon Cascades, the first storms of the fall produced streamflow peaks from a 96-ha clear-cut watershed that ranged from 40% to 200% larger than those predicted from the pre-logging relationship [*Rothacher*, 1971; 1973]. In the Alsea watershed near the Oregon coast, Harris [1977] found no significant change in the mean peak flow after clear-cutting a 71-ha watershed or patch cutting 25% of an adjacent 303-ha watershed. However, when Harr [1976] added an additional 30 smaller early winter runoff events to the data, average fall

[1]Pacific Southwest Research Station, U. S. Forest Service, Arcata, California

[2]Pacific Southwest Research Station, U. S. Forest Service, Albany, California

Land Use and Watersheds: Human Influence on Hydrology and Geomorphology in Urban and Forest Areas
Water Science and Application Volume 2, pages 85-125

Published in 2001 by the American Geophysical Union

peak flow was increased 122%. In Caspar Creek, Ziemer [1981] reported that selection cutting and tractor yarding of an 85-year-old second-growth redwood and Douglas-fir forest increased the first streamflow peaks in the fall about 300% after logging. The effect of logging on peak flow at Caspar Creek was best predicted by the percentage of area logged divided by the sequential storm number, beginning with the first storm in the fall. These first rains and consequent streamflow in the fall are usually small and geomorphically inconsequential in the Pacific Northwest. The large peak flows, which tend to modify stream channels and transport most of the sediment, usually occur during mid-winter after the soil moisture deficits have been satisfied in both the logged and unlogged watersheds.

Studies of large peak flows in the Pacific Northwest have not detected significant changes after logging. Rothacher [1971, 1973] found no appreciable increase in peak flows for the largest floods attributable to clear-cutting. Paired watershed studies in the Oregon Cascades [*Harr* et al., 1979], Oregon Coast Range [*Harr* et al., 1975; *Harr*, 1976; *Harris*, 1977], and at Caspar Creek [*Ziemer*, 1981; *Wright* et al., 1990] similarly suggested that logging did not significantly increase the size of the largest peak flows that occurred when the ground was saturated.

Using longer streamflow records of 34 to 55 years, Jones and Grant [1996] evaluated changes in peak flow from timber harvest and road building from a set of three small basins (0.6 to 1 km^2) and three pairs of large basins (60 to 600 km^2) in the Oregon Cascades. In the small basins, they reported that changes in small peak flows were greater than changes in large flows. In their category of "large" peaks (recurrence interval greater than 0.4 years), flows were significantly increased in one of the two treated small basins, but the 10 *largest* flows were apparently unaffected by treatment. Jones and Grant [1996] reported that forest harvesting increased peak discharges by as much as 100% in the large basins over the past 50 years, but they did not discuss whether the largest peak flows in the large basins were significantly affected by land management activities. Two subsequent analyses of the same data used by Jones and Grant concluded that a relationship could not be found between forest harvesting and peak discharge in the large basins [*Beschta* et al., 1997; *Thomas* and *Megahan*, 1998].

There are several explanations why relationships between land management activities and a change in storm peaks have been difficult to document. First, the land management activity may actually have no effect on the size of storm peaks. Second, because major storms are infrequent, the range of observations may not adequately cover the range of interest. Third, if the variability in response is large relative to the magnitude of change, it may be difficult to detect an effect without a large number of observations. Fourth, land-use changes in a large watershed are often gradual, occurring over several years or decades. The use of an untreated control watershed whose flows are well-correlated with the treated watershed can greatly increase statistical power, if both watersheds are monitored for an adequate number of years before and after the treatment is applied. The variability about the relation between the two watersheds can be critical. For example, when the South Fork (pre-treatment RMSE = 0.232) was used as the control, no change in peak streamflow was detected at the North Fork Caspar Creek weir after about 50% of the 473-ha watershed had been clear-cut logged. However, when the uncut tributaries within the North Fork (pre-treatment RMSE = 0.118) were used as the controls, an increase in peaks was detected [*Ziemer*, 1998]. In the analyses described in this paper, uncut tributaries in the North Fork will be used as controls for treated subwatersheds in the North Fork.

Sediment Loads

Paired watershed studies have been utilized to study the effects of logging activities on sediment loads as well as peak flows. Detecting changes in sediment loads is even more difficult than for peak flows, because sediment loads are more variable and more costly to measure. Studies are often dominated by a single extreme event [*Grant* and *Wolff*, 1991; *Rice* et al., 1979; *Olive* and *Rieger*, 1991], making the results more difficult to interpret. Most studies have utilized annual sediment loads [*Harris*, 1977; *Rice* et al., 1979; *O'Loughlin* et al., 1980; *Grant* and *Wolff*, 1991; *Megahan* et al., 1995], usually determined by surveys of settling basins behind impoundments. Sediment passing over a spillway is typically determined using sediment rating curves that relate suspended sediment concentration and water discharge.

Only one of these studies has been conducted in the redwood region. Rice et al. [1979] reported the suspended sediment load was 270% above that predicted for 1 year following roading of the South Fork of Caspar Creek, and the debris basin deposit 50% above that predicted. Lewis [1998] estimated an increase of 212% in suspended load in the 6 years following logging of the South Fork, despite a 3300 m^3 landslide contributing directly to the stream in the control watershed.

In the Alsea watershed in coastal Oregon, Brown and Krygier [1971] found a doubling of sediment loads in the year after roading in two different watersheds. In the watershed that was completely clear-cut and burned to the mineral soil the next year, sediment loads increased more than 10-fold the first year, then gradually declined in 7 years to near pretreatment levels [*Harris*, 1977]. In the watershed that was 25% clear-cut in three small units and remained mostly unburned, the road effect diminished in the second year, and measured increases in loads were not statistically different from the pretreatment relationship. Differences between sediment yields from the two treated watersheds were attributed primarily to the burning.

Sample sizes are necessarily rather limited in analyses using annual loads, an unfortunate situation, considering the variability in response. It is rare to find studies with more than 5 years of pretreatment measurements of sediment on both control and treated watersheds. Exceptions are the experiments in the Alsea [*Harris*, 1977] and the Silver Creek [*Megahan* et al., 1995] watersheds, which had 7 and 11 years' pretreatment data, respectively. Many studies have used no pretreatment measurements at all [*Plamondon*, 1981; *O'Loughlin* et al., 1980; *Leaf*, 1970]. These must rely on unproven assumptions about the relation between control and treated watersheds. Post-treatment sample sizes are limited by the rapidly changing conditions that usually follow a disturbance. In analyses based on annual loads, conditions might return to pretreatment levels before enough data are available to demonstrate a change occurred. Even if a change can be detected, it is difficult to establish reasonable bounds on the magnitude of change in the face of such high variability and small sample sizes.

Some paired watershed studies have attempted to look at changes in sediment concentrations. In the Alsea watershed study, an analysis of changes in sediment rating curves was less effective than an analysis of annual loads [*Brown* and *Krygier*, 1971]. Such analyses will usually be limited by the inadequacy of models relating sediment concentration to flow. Olive and Rieger [1991] were unable to establish a useful calibration using sediment concentrations, attributing the failure to the highly variable hydrologic environment. Fredricksen [1963] used paired specimens (collected within 1 hour of each other) to analyze changes in the H.J. Andrews concentrations, but found it necessary to discard 8 of 83 data points that represented "unpredictable events" and "sudden movements of soil". Considering the episodic nature of sediment transport, it is not surprising that simul-taneous specimens from adjacent watersheds are poorly related. Such episodic events should probably be focused upon rather than discarded.

Utilizing storm sediment loads circumvents the problems of properly pairing concentration data and permits much larger sample sizes than are possible in analyses of annual loads. Larger sample sizes permit more powerful statistical analyses and construction of confidence limits and prediction limits for responses. Because of the cost of reliably estimating storm loads, studies based upon them are rare. Miller [1984] estimated storm loads from three control and three treated watersheds using pumped specimens triggered at regular time intervals. Although no pretreatment data were collected, the replication of both treatment and control permitted an analysis of variance on storm ranks each year following the treatment. But sampling at regular time intervals will tend to miss peak concentrations in flashy watersheds unless the intervals are very short, in which case more field and lab work is required. In our study we used schemes that increased the probability of sampling during high flows and turbidities.

Cumulative Effects

A great deal of concern has been focused on the cumulative watershed effects of forest harvesting activities. This study design includes multiple gaging stations in the same watershed in order to evaluate cumulative effects. According to the U.S. Council on Environmental Quality's interpretation of the National Environmental Policy Act, a "cumulative impact" is the impact on the environment which results from the incremental impact of the action when added to other past, present, and reasonably foreseeable future actions, regardless of what agency...or person undertakes such other actions [CEQ guidelines, 40 CFR 1508.7, issued 23 April 1971]. An activity's importance may depend heavily upon the context of historic and future land use. A wide variety of interactions is imaginable. We attempt to answer three questions that arise with regard to cumulative watershed effects of logging activities :

1. How are impacts related to the total amount of disturbance? In particular, were the effects of multiple disturbances additive in a given watershed?
2. How do impacts propagate downstream? In particular, were downstream changes greater than would be expected from the proportion of area disturbed?
3. Can activities that produce acceptable local impacts result in impacts that are unacceptable by the same standard at downstream locations? In particular, were sediment loads in the lower watershed elevated to higher levels than in the tributaries?

The scope of these questions is limited here in order to permit scientific investigation. For example, question (2) does not consider that larger watersheds may experience different types of impacts than contributing watersheds upstream, and question (3) does not consider that different standards may be appropriate downstream because different resources may be at risk. Nevertheless, partial answers to these questions can be provided with regards to storm peak flows, flow volumes, and suspended sediment loads through watershed experiments and mathematical modelling.

Environment and History

The Caspar Creek Experimental Watersheds are a pair of rain-dominated forested catchments in the Jackson Demonstration State Forest on the coast of northern California. The 473-ha North Fork and the 424-ha South Fork are both located in the headwaters of the 2,167-ha Caspar Creek, which discharges into the Pacific Ocean near the town of Caspar. Uplifted marine terraces, to 320 m in elevation, are deeply incised by antecedent drainages resulting in a topography composed of steep slopes near the stream channel and broad rounded ridgetops. About one third of the basin's slopes are less than 17° and only 7% are greater than 35°. The watershed receives an average of 1200 mm of rainfall each year, 90% falling in the months of October through April. The forest is composed mainly of redwood (*Sequoia sempervirens* [D.Don.] Endl.), Douglas-fir (*Pseudotsuga menziesii* [Mirb.] Franco), grand fir (*Abies grandis* [Dougl. ex D.Don] Lindl.), and western hemlock (*Tsuga heterophylla* [Raf.] Sarg.). The well-drained clay loam soils developed in sandstone and shale units of the Franciscan assemblage [*Bailey* et al., 1964] and are highly erodible.

Streamside landslides, gully erosion, and debris flows are the major erosional processes delivering sediment to the channel system. Soil pipes, common in the unchannelized swales, and steep ephemeral tributaries discharge to the Caspar Creek main stems. Based on debris basin surveys and suspended sediment measurements, the perennial, gravel-bed North Fork channel typically transports about 70% of its sediment load in suspension, and sand rarely exceeds 50% of the suspension. Gravel bars associated with woody debris jams and debris-induced bank erosion furnish the bulk of bedload transported during peak flows. Finer sediments cap the highest gravel bars and are stored in pools for transport during modest storm flows [*Lisle* and *Napolitano*, 1998].

Between 1860 and 1904, the old-growth forest in the Caspar Creek watersheds was clear-cut and burned. Log drives were triggered by opening the spillway gates of log crib along the main-stem reaches of both the North Fork and the South Fork, profoundly affecting channel morphology during the earliest logging effort [*Napolitano*, 1998]. These gave way to semi-mechanized yarding of tributary catchments using railway inclines (tramways) and steam donkeys [*Henry*, 1998]. A historic stage coach route and a mid-1900's era forest road totaling 11.4 km in length follow the watershed divide along the north and east of the North Fork.

In 1962, Caspar Creek became the site of a paired watershed experiment. In 1968, the South Fork watershed was roaded, and from 1971-1973, it was selectively logged by tractor, while the North Fork watershed was maintained as an undisturbed control [*Rice* et al., 1979; *Ziemer*, 1981; *Wright* et al., 1990; *Keppeler* and *Ziemer*, 1990]. In 1985 and 1986, 59 ha of an ungaged tributary basin in the lower North Fork was clear-cut. The present study of cumulative impacts began in 1985 in the 384-ha Arfstein subwatershed (ARF), gaged on the North Fork's main stem just above the confluence with the ungaged tributary (Figure 1). When the stability of ARF's discharge rating equation recently came into question, we decided to use the larger North Fork watershed (NFC) in place of ARF for the analysis of storm peaks and flow volumes. ARF was retained, however, for the sediment analyses because roughly 40% of the suspended sediment settles in a debris basin immediately above the North Fork weir and thus is not measured at the NFC gaging station.

METHODS

Treatment

The treatment design was based on compliance with the California Forest Practice Rules in effect in the late 1980's, except that the proportion of the watershed cut in a 3-year period was atypically high for a watershed of that size. Streams bearing fish or aquatic habitat were protected with selectively logged buffer zones 15 to 46 m in width, depending on stream classification and slope steepness.

Logging began in the headwaters of the North Fork in May 1989 and ended in the lower watershed in January 1992 (Figure 1). Clear-cuts totalled 169 ha (43% of ARF) in blocks of 9 to 60 ha and occupied 30% to 98% of treated subwatersheds. Total logged areas, including timber selectively removed from stream buffer zones, are slightly larger (Table 1). The 60 ha cutblock was composed of two adjacent subwatersheds (CAR and GIB), and an exemption was required from the maximum clear-cut size permitted under California Forest Practice Rules in effect at the time. Of the clear-cut areas, 81% was skyline yarded to landings on spur roads built on the upper hillslopes away from the creeks. Logs only had to be suspended at one point, but in most cases full suspension was achieved by setting the chokers near the middle of the log. This prevented ground dragging except near landings and convex slope breaks. The remaining 19% of the clear-cut area was tractor yarded and was limited to ridgetop areas where slopes were generally less than 20°. In addition, about 34% of the

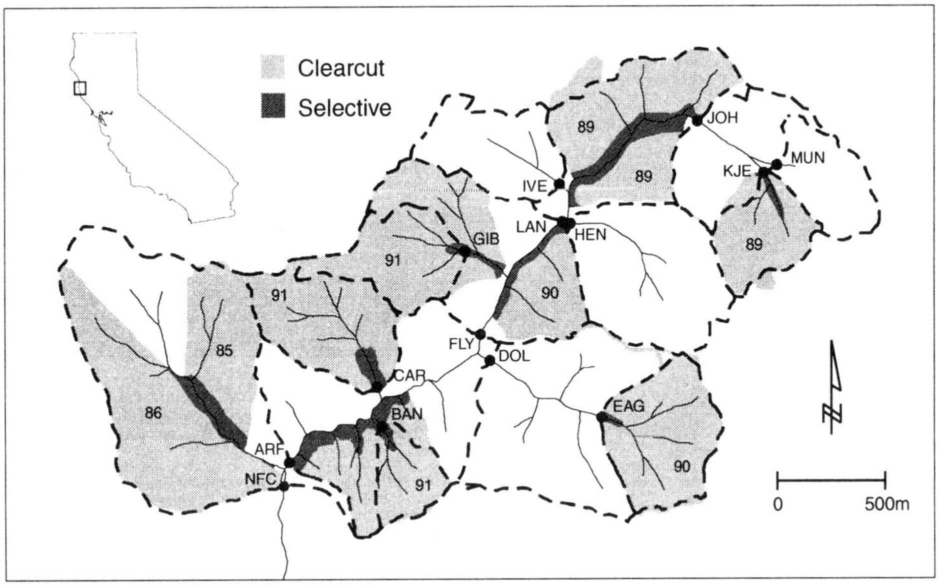

Figure 1. North Fork Caspar Creek. Gaging stations are identified by 3-letter abbreviations and dots, subwatershed boundaries by dashed lines, and logged areas by shading. Inset locates Caspar Creek within California.

timber was selectively removed from 19 ha of stream buffer zones. New roads, landings, skid trails, and firelines occupied from 1.9% to 8.5% of treated subwatersheds. Four cut units, totalling 92 ha, were broadcast burned following harvest.

Three subwatersheds (HEN, IVE, and MUN) within the North Fork were retained in an unlogged condition for use as controls. In addition, the South Fork watershed, unlogged since 1973, was monitored for possible use as a control.

Gaging Stations

A total of 15 gaging stations were monitored: the North and South Fork weirs (NFC and SFC), four stations on the main stem of the North Fork, and nine on tributaries of the North Fork (Figure 1). The channel control structures at the North and South Fork gaging stations are 120° V-notch weirs with concrete upper rectangular sections. The lowest three main-stem stations (ARF, FLY, and LAN) are rectangular plywood sections, rated by discharge measurements. Each rated section has a natural bottom and a stable downstream sill installed to control bed elevation within the rated section. Discharge at the upper main-stem station (JOH) and the nine tributaries is measured with Parshall flumes. Although the rated sections and flume installations were not designed to guarantee complete capture of subsurface intergravel flows, frequent inspections (before, during, and after storms) were made and regular maintenance was performed at these sites to ensure stable discharge estimates throughout the length of the study. Discharge ratings were validated with new measurements each year, and only station ARF required rating equation changes.

Suspended Sediment Data Collection

Selection At List Time (1986-1995). Selection At List Time (SALT) is a variable probability sampling method similar to PPS (probability proportional to size) sampling with replacement [*Hansen* and *Hurwitz*, 1943]. Their estimation formulas are identical. Both methods utilize an auxiliary variable, easily measurable for the entire population, to assign inclusion probabilities to each sampling unit of the population. (We have defined a sampling unit of the sediment population as the suspended sediment load passing a gaged cross-section in 10 min.) The variance is minimized for auxiliary variables that are proportional to the variable of interest. PPS requires enumerating the population and measuring the auxiliary variable on the whole population before sampling. SALT was developed as an alternative to PPS for populations which cannot be enumerated before sampling [*Norick*, 1969]. SALT inclusion probabilities are computed from an estimate of the auxiliary variable total. Immediately upon measuring each unit's auxiliary variable, a decision is made whether or not to select the sampling unit. The auxiliary variable might be a flow-based prediction of unit yield from a sediment rating curve [*Thomas*, 1985]. This results in a sampling rate that is proportional to predicted

TABLE 1. Basic watershed data and percentage in various conditions. Cut area includes portions of stream buffer zones corresponding to the proportion of timber volume removed.

Water-shed	Area (ha)	Cut area	Cable	Trac-tor	Road+ Lndg	Total Bare	Total Burnt	Dates logged
ARF	384	45.5	35.1	7.1	1.8	2.9	24.0	Spr89-Win92
BAN	10	95.0	77.3	13.4	2.6	3.2	0.0	Fall91
CAR	26	95.7	82.1	9.2	2.8	4.4	0.0	Fall91-Win91
DOL	77	36.4	27.4	5.9	2.5	3.7	33.9	Fall90-Fall91
EAG	27	99.9	79.0	15.4	4.9	8.5	97.8	Fall90-Fall91
FLY	217	45.4	34.6	7.6	1.6	3.0	30.4	Spr89-Sum91
GIB	20	99.6	54.9	39.4	4.2	7.9	98.2	Spr91-Sum91
HEN	39	0.0	0.0	0.0	0.0	0.0	0.0	
IVE	21	0.0	0.0	0.0	0.0	0.0	0.0	
JOH	55	30.2	26.4	1.3	2.0	2.1	0.1	Spr89-Fall89
KJE	15	97.1	85.2	3.9	6.5	6.9	0.0	Spr89-Fall89
LAN	156	32.2	27.8	1.9	1.0	1.9	20.3	Spr89-Spr90
MUN	16	0.0	0.0	0.0	0.0	0.0	0.0	
NFC	473	12.7	38.6[a]	7.6[a]	2.0[a]	3.2[a]	19.5[a]	Spr85-Spr86
		+36.9	38.6	7.6	2.0	3.2	19.5	Spr89-Win92

[a] not measured; assumed equal to Spr89-Win92 disturbance proportions

sediment yield. If the discharge and sediment rating curves are power functions of stage (water depth), the sampling rate will also be a power function of stage. In practice, we had to set an upper limit to the sampling rate and modify the parameters of the power function in order to sample small storms as well as large ones [*Thomas*, 1989].

To implement SALT, at each gaging station we interfaced an HP-71 calculator with an automatic pumping sampler and a transducer mounted in a stilling well. The calculator was programmed to "wake up" every 10 min, read the transducer stage height, calculate the auxiliary variable, and, using the SALT algorithm and a set of stored random numbers, decide whether to sample or not. If the decision was to sample, a signal was sent via an interface circuit board to the pumping sampler, which would then collect a specimen (to avoid ambiguity, the word "sample" is reserved to refer to a selected set of "specimens" or "bottles") from a fixed intake nozzle positioned in the center of the channel. Date, time, stage, and other bookkeeping details were recorded on the calculator for subsequent uploading.

Turbidity-controlled sampling (1996). After 10 years of monitoring, the number of gaging sites was reduced to eight: the North and South Fork weirs (NFC and SFC), two controls (HEN and IVE), one main-stem station (ARF), and three tributary stations (CAR, DOL, and EAG). At that time, SALT and the HP-71's were replaced by a turbidity-controlled sampling system utilizing programmable data loggers and *in situ* turbidity probes. Date, time, stage, turbidity, and sampling information are recorded at 10-min intervals. The nephelometric turbidimeters we are using emit infrared light and measure the amount scattered back to the probe. In lab tests, they respond linearly to sediments of a given size distribution. In the field, with mixed-size sediments present, departures from linearity are usually minor. During each storm event, when certain pre-specified turbidity thresholds are reached, the data logger sends a signal to the pumping sampler to collect a concentration specimen. A separate set of thresholds is specified for falling and rising stage conditions. This system reduced sample sizes and field expenses considerably, while still permitting accurate estimation of sediment loads [*Lewis*, 1996].

Data quality control. Field crews typically visited each gaging station one to three times per 24-hour period during storms to check on flumes and equipment, record manual stage observations, measure discharge at rated cross-sections, and collect depth-integrated suspended sediment specimens. Chart recorders provided back-up data. When problems were encountered with the electronic stage record, they were corrected using observer records or digitized data from back-up chart recorders. In a few instances, portions of discharge records were corrected based on correlation with selected alternate gaging stations. All stage data were coded to indicate the quality of the data.

Storms with poor quality or reconstructed peak data were treated as missing data in the peaks analysis. Storms with 25% or more of the flow volume derived from poor quality stage data were treated as missing data in the flow volumes analysis.

In addition to the suspended sediment specimens collected by the SALT algorithm, auxiliary pumped specimens were manually initiated for comparison with simultaneous depth-integrated DH-48 specimens or to augment the sampling algorithm. On occasions when the HP-71/pumping sampler interface failed and could not be immediately repaired, the sampler was set to collect specimens at fixed time intervals. A total of 21,880 bottles were collected: 19,572 under SALT, 378 under the turbidity threshold algorithm, 1048 auxiliary, 686 depth-integrated, and 196 fixed-time specimens.

Suspended sediment concentration was determined in the laboratory using vacuum filtration. Specimens were coded to indicate such conditions as spillage, organic matter content, low volumes, and weighing errors. Those with serious errors were omitted from the analysis. Those with minor errors were re-examined in the context of the whole storm.

Field crews also noted conditions affecting discharge or sediment data including landslides, windthrow, and culvert blockages and diversions. Post-storm surveys of the watershed stream channels and roads were made to document erosion sources potentially affecting sediment loads.

Storm Definition and Feature Identification

A total of 59 storm events occurred during the 11-year study. Storm events were generally included in the study when the peak discharge at SFC exceeded 0.0016 $m^3s^{-1}ha^{-1}$ (recurrence interval about 7 times per year). A few smaller peaks were included in dry years. Multiple peak hydrographs were treated as multiple storms when more than 24 hours separated the peaks and the discharge dropped by at least 50% in the intervening period. When multiple peak hydrographs were treated as a single storm, the discharge for the peaks analysis was identified by selecting the feature corresponding to the highest peak at NFC. Thus the same feature was used at all stations, even if it were not the highest peak of the hydrograph at all stations. However, differences in peak discharge caused by this procedure were very small.

The start of a storm was chosen by seeking a point on the hydrograph, identifiable at all stations, where the discharge began to rise. The start times differed by no more than a few hours at the various stations. At the end of a storm, distinctive hydrograph features are more difficult to identify, unless a new start of rise is encountered. We therefore decided to use the same ending time for a given storm at all stations. The ending time was selected by observing the storm hydrograph for all stations and determining either the time of the next storm, the next significant rainfall, or a stable low-flow recession at all hydrographs, usually within about 3 days after the peak. The end of each storm was always well below the quickflow hydrograph separation point described by Hewlett and Hibbert [1967], except when the recession was interrupted by a new storm.

Dependent Variables

The response variables of interest in this study are storm runoff peak (instantaneous discharge), storm runoff volume (total discharge), and storm suspended sediment load (mass of particles greater than 1 micron in diameter). All are expressed on a unit area (per hectare) basis. The runoff variables were derived from the 10-min electronic record of stage and rating equations relating discharge to stage at each station. The computation of sediment loads is more involved and is described in the next section.

Computation of Suspended Sediment Loads

Correction to obtain cross-sectional average concentration. The pumping sampler intakes were oriented downstream and centered in the inclined throat sections of the Parshall Flumes. In the rated sections (ARF, FLY, and LAN), the intakes were similarly oriented at a fixed position about 9 cm off the bed. To determine whether the specimens were starved or enriched because of sampler efficiency or nozzle orientation or position, simultaneous ISCO and DH-48 depth-integrated (equal transit rate) specimens were collected throughout the study. A log-log regression of depth integrated concentration versus fixed intake concentration was developed for each station. Although only six of thirteen regressions differed significantly from the line $y=x$ (experimentwise $\alpha=0.05$ with Bonferroni method [*Miller*, 1981]), all fixed intake concentrations were adjusted using the back-transformed regression equations and corrected for bias [*Baskerville*, 1972] before storm loads were computed.

Load estimation in 1986-1995. Although sediment sampling followed SALT protocol in hydrologic years 1986-1995, we ultimately applied non-SALT methods of estimation to these samples for two reasons:

1. SALT does not provide a way to estimate sediment loads for periods when the sampling algorithm was inoperative due to equipment problems. Other methods can interpolate over such periods and utilize manually-initiated auxiliary specimens and those collected in fixed-time mode.
2. Using computer simulations on intensively collected storm data, other methods were found to have lower mean squared errors than SALT.

Although unbiased estimates of variance are not available for the alternate methodologies, the simulations strongly suggested that SALT variance estimates could be used as very conservative upper bounds on the variance. Two alternate methods were considered. In both of these methods the total load is computed by summing the products of water discharge and estimated concentration over all 10-min periods in the storm. The concentration, c, between adjacent sampled times t_1 and t_2 is modelled as either:

1. a linear function of time:

$$c = c_1 + (t - t_1)(c_2 - c_1)/(t_2 - t_1)\text{, or}$$

2. a power function of stage: $c = as^b$, where

$$b = \frac{\log c_2 - \log c_1}{\log s_2 - \log s_1},\ a = \frac{c_1}{s_1^b} \qquad (1)$$

in which the subscripts identify concentrations and stages at times t_1 and t_2. These methods will be referred to as "time interpolation" and "stage interpolation" respectively. Stage interpolation has a better physical basis, but computational difficulties frequently arise when s_1 and s_2 are similar or equal, or when c_1 or c_2 is equal to zero. Therefore, time interpolation was substituted for stage interpolation when the power function defined by a pair of stages and sampled concentrations could not be computed or its exponent was not in the range between 1 and 10. If no specimens had been sampled within 10 hours prior to the start of the storm, the starting sediment concentration was assigned a value of zero and time interpolation was applied. An analogous procedure was followed for the end of the storm. The next section describes simulations leading to the decision that stage interpolation be used for estimating the sediment loads in 1986-1995.

Simulations comparing SALT and interpolation estimators. In addition to the usual SALT sampling, in 1994 and 1995 sediment concentration and turbidity at ARF was sampled at 10-min intervals for five storm events. This data, described in greater detail by Lewis [1996], provided realistic populations with known sediment loads that could be used in simulations to evaluate the performance of different load estimation methods. In addition to these five populations, eight storm populations were available from previous studies on the North Fork of the Mad River in northwestern California: three storms from December 1982, January 1983 and December 1983 [*Thomas* and *Lewis*, 1995] and five storms from February 1983 [*Thomas* and *Lewis*, 1993]. The Mad River concentrations were derived from turbidity charts and form a smoother, less realistic, time series than the ARF measurements.

In the simulations, 5000 independent SALT samples were selected from each storm event using SALT sampling parameters that were in use at ARF in 1995 and parameters thought to be optimal at Mad River. The sediment load was estimated for each of the 5000 samples using SALT and time and stage interpolation. The simulation results are strictly applicable only to comparing these estimators under a specific SALT sampling protocol.

The simulation results are summarized in Table 2. While SALT was unbiased as expected, it consistently has much higher root mean square error (RMSE) than the interpolated estimators. This can be attributed to the interpolation methods that take advantage of local trends in concentration that SALT ignores. Because the Mad River storm populations were smoother than those from ARF, they indicate a somewhat greater advantage for the interpolated estimators.

While time interpolation appears to have slightly less bias than stage interpolation, the differences in both bias and RMSE are small relative to the loads. Real data differ from these simulated data in that unexpected time gaps are created during unavoidable equipment malfunctions. Stage interpolation is expected to mimic true concentrations better than time interpolation over large time gaps, so the latter method (with the exceptions noted earlier) was chosen for this study during the SALT years (1986-1995).

Quality control for load estimates (1986-1995). Determining which calculated sediment load data were of high enough quality to include in the analysis was a subjective process and involved an examination of plots showing the storm hydrograph, sediment concentrations, and quality codes. The primary considerations were the number of known concentrations (sample size) and their temporal distribution relative to the hydrograph. Out of 51 storms and 15 stations (765 combinations), 74% of the load estimates were judged acceptable. Because sample sizes were in proportion to the size of storm events, most of the discarded loads were from small events. In those events that were retained, the median sample size was 20 and the median standard error from SALT was 14% of the estimated load. Based on the simulations (Table 2) and the fact that SALT estimates did not utilize all the available concentrations, it is likely that the median error from the interpolated estimates is well under 10% of the estimated load.

Load estimation in 1996. With turbidity-controlled sediment sampling in place in 1996, sediment loads were computed using "turbidity rating curves". Concentration

TABLE 2. Comparison of suspended sediment load estimation by time interpolation, stage interpolation, and SALT algorithms. The load was estimated for 5000 simulated SALT samples from each storm event.

Station	Start of Storm	Load (kg/ha)	$\overline{n}$	Percent RMSE Time Interp	Stage Interp	SALT	Percent Bias Time Interp	Stage Interp	SALT
ARF	950109	178.6	21.2	6.0	6.7	12.2	-2.3	-2.8	0.1
ARF	950113	123.6	22.9	2.8	3.4	8.2	-1.6	-2.0	0.1
ARF	950308	122.4	32.6	4.1	4.1	7.6	-0.3	-0.5	0.0
ARF	950108	99.2	8.6	14.2	14.6	19.8	-6.0	-7.2	-0.0
ARF	940216	33.6	16.5	7.0	6.7	10.0	-3.7	-3.5	-0.2
Mad	821214	846.3	41.8	2.1	1.8	10.0	0.0	-0.3	-0.1
Mad	830209	527.2	36.0	4.2	4.1	13.8	0.4	-1.3	0.1
Mad	830117	198.0	40.8	2.2	2.6	7.2	-0.4	-0.9	0.1
Mad	830225	134.4	22.9	7.8	7.6	19.3	-1.6	-2.6	0.3
Mad	831223	42.8	18.1	5.8	5.4	13.6	-2.7	-2.7	0.0
Mad	830221	33.2	15.7	7.5	8.1	16.1	-4.0	-4.9	-0.3
Mad	830212	27.2	14.0	8.1	7.4	16.2	-3.2	-3.9	0.0
Mad	830218	25.4	14.1	14.7	15.1	22.3	-3.4	-4.2	0.0

was predicted by linear regressions of concentration on turbidity fit to each storm. This method was shown in simulations [*Lewis*, 1996], based on the same five ARF populations as shown in Table 2, to produce load estimates with RMSE of 8% or better while sample sizes were reduced to between 4 and 11, depending on storm size and sampling parameters. The interpolation methods used for 1986-1995 would not be as accurate for the generally smaller sample sizes obtained under turbidity-controlled sampling. However, because of intermittent fouling of the turbidity probes with debris and sediment, valid turbidities were not always available. During such periods, if enough concentration measurements were available (and extras were often triggered by false high turbidities), then time or stage interpolation was used. As a last resort, a sediment rating curve derived from nearby data was used to estimate concentrations. Out of 8 stations and 8 storms in 1996, a total of 46 sediment load estimates were judged to be of acceptable quality. The median sample size was 5 from these events.

Derivation of Independent Variables

The complete data set included both map-derived and field-derived variables. All disturbance variables were coded as proportions of watershed area. The basic watershed descriptors and variables that were useful in the analyses are shown in Table 1.

Topographic contours and streams were digitized from U.S. Geological Survey 7.5 min quadrangle maps. The mapped stream channels in harvest units were then extended to include all channels showing field evidence of annual scour and/or sediment transport before logging. Watershed boundaries were field-mapped using conventional tape-and-compass surveys, respecting diversions of surface runoff where road drainage structures directed flow into or out of the topographic watersheds. During road maintenance, efforts were made to limit changes in drainage due to ruts and berms. Harvest unit boundaries and roads were surveyed using differentially corrected GPS. All these lines were transferred to GIS coverages from which geographic variables were extracted. Burned areas, stratified into two severity classes, and herbicided areas were transferred to the GIS from field maps. For each variable measured, the area within 150 feet of a stream channel, and the length of channel within the affected area were extracted from the GIS.

The areal extent of ground disturbance from roads, landings, skid trails, firelines, and corridors created by dragging logs up the slope by cable were each determined from exhaustive field transects. The areas within 150 feet of a stream channel, and the number of stream crossings were also recorded for these variables.

Cutting age was calculated as the difference in hydrologic year of a given storm and the hydrologic year an area was logged. For watersheds with areas cut at different times, a weighted average cutting age was calculated using the cut unit areas as weights.

An antecedent wetness index intended to reflect seasonal differences in hydrograph response was derived using mean daily discharges from SFC. The daily discharges were accumulated and decayed using a 30-day half-life, i.e.

$$w_i = Aw_{i-1} + q_i \tag{2}$$

where w_i denotes the wetness index on day i, and q_i denotes the daily mean discharge at SFC on day i and the constant $A = 0.97716$ satisfies the relation $A^{30} = 0.5$. The decision to use streamflow rather than precipitation to calculate antecedent conditions was based on the assumption that the history of the streamflow response would be a better predictor of streamflow than would the history of rainfall. The response of streamflow to precipitation is delayed as soil moisture deficit is recharged. A half-life of 30 days was selected to smooth the high frequency variation in streamflow, creating an index that would decline significantly only after lengthy dry periods. No optimization was done on the half-life, but it was found that $\log(w_i)$ made a slightly better predictor. The wetness index time series over the 11-year study period is displayed in Figure 2, with solid circles indicating the wetness level at the start of each storm. The wetness index varied from 13 to 150 at the onset of storms occurring in November and December, but assumed the full range from 13 to 562 at the onset of storms occurring in January, February and March. For two storms that occurred in May, the values of the index were 49 and 84.

Statistical Methods

Initially, simple log-log linear regressions were computed for each dependent watershed against selected control watersheds prior to treatment. The Chow test [*Chow*, 1960; *Wilson*, 1978] was used to test whether the post-treatment data differed in either intercept or slope from the pre-treatment regressions. Following Bonferroni's procedure [*Miller*, 1981] for these tests, an experimentwise error rate of 0.05 for 10 tests required setting the nominal α to 0.005 for each test. Because of their limited sample sizes, these tests, while easy to interpret, are not as powerful as models based on all of the data.

Models incorporating all of the watersheds were initially built up in a stepwise fashion using least squares estimation. At each step, residuals were plotted against candidate predictors to select the next variable and the appropriate transformation or form of interaction. Because a non-standard covariance model was employed, models were ultimately fitted using maximum likelihood estimation and selected using a combination of exploratory and diagnostic techniques.

Models for runoff (storm peaks and flow volumes). Consider the following pretreatment model:

$$\log(y_{ij}) = \beta_{0i} + \beta_{1i}\log(y_{Cj}) + \varepsilon_{ij} \tag{3}$$

where

y_{ij} = unit area response (peak or flow) at treated watershed i, storm j
y_{cj} = unit area response at control watershed in storm j
ε_{ij} = non-independent normally distributed errors (see *Covariance Models* below)

and β_{0i} and β_{1i} are "location" parameters to be estimated for each watershed i. The log transformations are used in order that ε_{ij} appear to be normally distributed. The pretreatment model can be considered as a special case of the following model:

$$\begin{aligned}\log(y_{ij}) = &\left(\beta_{0i} + \beta_4 D_{ij} + \beta_6 D_{ij}\log(w_j) + \beta_7 D_{ij} a_i\right)\\ &+ \left(\beta_{1i} + \beta_5 D_{ij}\right)\log(y_{Cj}) + \varepsilon_{ij}\end{aligned} \tag{4}$$

where

D_{ij} = some measure of disturbance per unit area in watershed i at storm j
w_j = wetness index at start of storm j
a_i = drainage area of watershed i

and β_4, β_5, β_6, and β_7 are parameters to be estimated. The log transformation of w_j is not critical, but was found to improve its explanatory value. Wetness enters the equation only as an interaction with D_{ij} because in the absence of disturbance wetness did not affect the relation between y_{ij} and y_{cj}. As an interaction, it implies that the effect of disturbance on y_{ij} varies linearly with antecedent wetness. The $D_{ij}a_i$ term implies that the disturbance effect also varies linearly with watershed area and it is the key term in this model for detecting a cumulative effect. It describes how watershed impacts propagate downstream and we use it to test the null hypothesis that a unit area disturbance has the same unit area effect in watersheds of all sizes.

The first line of equation (4) permits the intercept of the relation between y_{ij} and y_{cj} to change following disturbance. The second line, via the $D_{ij}\log(y_{cj})$ term, permits the slope of that relation to change following disturbance. Equation (4) can be rearranged as

$$\begin{aligned}\log(y_{ij}) = &\beta_{0i} + \beta_{1i}\log(y_{Cj}) + \varepsilon_{ij}\\ &+ D_{ij}\left[\beta_4 + \beta_5\log(y_{Cj}) + \beta_6\log(w_j) + \beta_7 a_i\right]\end{aligned} \tag{5}$$

We now model the disturbance term using logged area and cutting age to represent loss of transpiration and

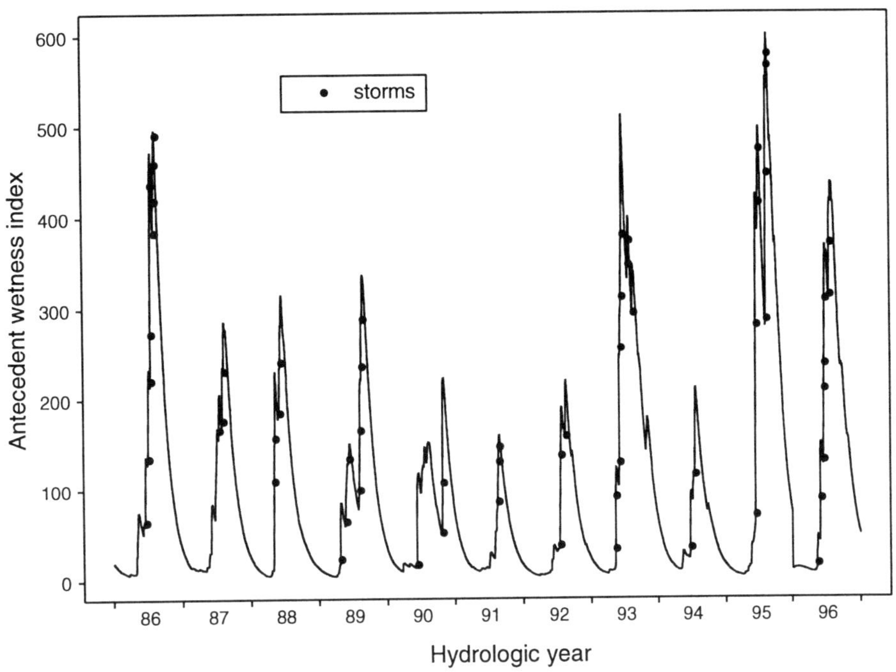

Figure 2. Antecedent wetness index (equation (2)) and temporal distribution of storms for the period of study (1986-1996). Solid circles indicate the wetness level at the start of each storm.

interception following logging. Compacted areas such as roads, landings, skid trails, and firelines were not found to be useful predictors. Since relatively little transpiration occurs at Caspar Creek in the fall and winter, we treat areas logged in the fall or winter prior to the occurrence of a storm as special cases. Let

$$D_{ij} = f(t_{ij})(c_{ij}) + g(c'_{ij}) \tag{6}$$

where

t_{ij} = area-weighted mean cutting age (number of summers passed) in watershed i for areas logged in water years (defined as Aug.1 - July 31) preceding that of storm j

c_{ij} = proportion of watershed i logged in water years prior to that of storm j, and

c'_{ij} = proportion of watershed i logged prior to storm j but in the same water year

We model a linear recovery declining from a maximum of unity the year after cutting:

$$f(t_{ij}) = 1 - \beta_2(t_{ij} - 1) \tag{7}$$

where β_2 is a parameter representing the recovery rate, and we assume the effect of newly cut areas depends only on the season they were cut:

$$g(c'_{ij}) = \beta_3^{(k)} c'_{ij} \tag{8}$$

where $\beta_3^{(k)}$ are parameters for the effect of cutting in the fall (k=1) and winter (k=2) immediately preceding storm j. Equation (6) becomes

$$D_{ij} = \left(1 - \beta_2(t_{ij} - 1)\right)c_{ij} + \beta_3^{(k)} c'_{ij} \tag{9}$$

and the complete model is

$$\begin{aligned} \log(y_{ij}) &= \beta_{0i} + \beta_{1i}\log(y_{C \cdot j}) + \varepsilon_{ij} \\ &+ \left[\left(1 - \beta_2(t_{ij} - 1)\right)c_{ij} + \beta_3^{(k)} c'_{ij}\right] \\ &\times \left[\beta_4 + \beta_5 \log(y_{C \cdot j}) + \beta_6 \log(w_j) + \beta_7 a_i\right] \end{aligned} \tag{10}$$

To investigate whether unit area response increases downstream independently of disturbance, we can look for a relation between β_{0i} and a_i. Alternatively, we can replace β_{0i} with the linear expression $\beta_0^{(1)} + \beta_0^{(2)} a_i$ and test the hypothesis $H_0 : \beta_0^{(2)} = 0$. If unit area responses tend to increase downstream, then cumulative impacts might occur where a response threshold of acceptability is exceeded only below some point in the stream network, even though unit area disturbance is no greater in that point's watershed than in watersheds further upstream.

Model (10) is not a linear model because it involves products of the parameters to be estimated. The non-linearity was introduced as a parsimonious way of modelling recovery with time since logging. It avoids introducing separate recovery parameters for each of the terms in equation (4) that involve D_{ij}.

Models for suspended sediment loads. Suspended sediment load from an untreated control watershed was found to be a much better predictor of sediment load at treated watersheds than water discharge at either location. However, the change in storm flow in the treated watershed, relative to that in the control, was found to be the next best predictor in a model for suspended sediment loads. The change in flow, Δq, was formulated two ways:

1. The residual from the flow model with D_{ij} set to zero

$$\Delta q_{ij}^{(1)} = \log(y_{ij}) - \left(b_{0i} + b_{1i}\log(y_{Cj})\right) \tag{11}$$

where b_{0i} and b_{1i} are estimates of the flow model parameters β_{0i} and β_{1i}.

2. The log of the ratio of the flows between the treated and control watersheds:

$$\Delta q_{ij}^{(2)} = \log\left(y_{ij}/y_{Cj}\right) = \log\left(y_{ij}\right) - \log\left(y_{Cj}\right) \tag{12}$$

The first form makes better sense hydrologically, but treating it as an independent variable may not be statistically legitimate later when estimating precision later on, because it involves parameter estimates from another model. Nevertheless, both forms of Δq were considered. These variables are not useful in a predictive setting because the flows are not known in advance, but the main purpose of these models is explanatory. If prediction is needed, then a third form might be substituted as an approximation to $\Delta q_{ij}^{(1)}$:

3. The predicted change in $\log(y_{ij})$ from equation (10):

$$\Delta q_{ij}^{(3)} = \left[\left(1 - b_2(t_{ij} - 1)\right)c_{ij} + b_3^{(k)}c'_{ij}\right] \times \left[b_4 + b_5\log(y_{Cj}) + b_6\log(w_j) + b_7 a_i\right] \tag{13}$$

where the b's are estimates of the β's in equation (10).

After Δq and one or two disturbance variables were included in the model, no further gains were realized in the sediment models by including factors such as antecedent wetness and cutting age. So, unlike the runoff models, the sediment models remain linear in their parameters:

$$\begin{aligned}\log(y_{ij}) = {} & \beta_{0i} + \beta_{1i}\,\mathrm{bg}(y_{Cj}) + \beta_2\Delta q_{ij} \\ & + \left(\beta_3 + \beta_4\,\mathrm{bg}(y_{Cj}) + \beta_5 a_i\right)x_{ij}^{(1)} \\ & + \left(\beta_6 + \beta_7\,\mathrm{bg}(y_{Cj}) + \beta_8 a_i\right)x_{ij}^{(2)} + \varepsilon_{ij}\end{aligned} \tag{14}$$

where

y_{ij} = unit area sediment load at treated watershed i, storm j

y_{Cj} = unit area sediment load at control watershed in storm j

Δq_{ij} =change in flow as defined by (11) or (12) in watershed i, storm j

a_i = drainage area of watershed i

$x_{ij}^{(1)}$ = a measure of unit area disturbance in watershed i, storm j

$x_{ij}^{(2)}$ = a second measure of unit area disturbance in watershed i, storm j

ε_{ij} = non-independent normally distributed errors (see *Covariance Models* below)

and the β's are parameters to be estimated. The logic behind the interaction terms involving $\log(y_{Cj})\,x_{ij}^{(k)}$ and $a_i x_{ij}^{(k)}$ is the same as in the runoff models. And, as with model (10), we can replace β_{0i} in (14) with the expression $\beta_0^{(1)} + \beta_0^{(2)}a_i$ to investigate whether unit area loads increase downstream independently of disturbance.

Covariance models. The residual covariance was found to depend upon watershed size and location. The correlations decreased with increasing distance between watershed centroids and the variance decreased with increasing watershed size. Serial autocorrelation in the residuals for most watersheds was weak or absent, so responses from different storms were considered independent. The errors were thus assumed to follow a multivariate normal distribution with a covariance matrix for each storm. The dimensions of this square matrix are equal to the number of treated watersheds having good data in that storm. The covariances in the matrix for storm j are modelled as:

$$Cov(\varepsilon_{i_1 j}, \varepsilon_{i_2 j}) = \sigma_{i_1 i_2}^2 = \rho_{i_1 i_2}\sigma_{i_1}\sigma_{i_2} \tag{15}$$

where

$\rho_{i_1 i_2}$ = correlation between $\varepsilon_{i_1 j}$ and $\varepsilon_{i_2 j}$

$\sigma_{i_1}, \sigma_{i_2}$ = standard deviations of $\varepsilon_{i_1 j}$ and $\varepsilon_{i_2 j}$

$\varepsilon_{i_1 j}, \varepsilon_{i_2 j}$ = errors for watersheds i_1 and i_2 in storm j

Subscripts j have been omitted from $\rho_{i_1 i_2}$, σ_{i_1} and σ_{i_2} because these terms are assumed to be independent of storm number and are, in fact, modelled upon the errors from all storms. Two models for the correlation $\rho_{i_1 i_2}$ were found to fit the runoff and sediment data.

1. Exponential decline with distance:

$$\rho_{i_1 i_2} = \frac{\exp(-\theta_1 d_{i_1 i_2}) + \theta_2}{1+\theta_2} \tag{16}$$

where $d_{i_1 i_2}$ is the distance separating watersheds i_1 and i_2, and θ_1 and θ_2 are parameters to be estimated. In this model the correlations decline asymptotically from unity to the value $\theta_2/(1+\theta_2)$.

2. Linear decline with distance:

$$\rho_{i_1 i_2} = \begin{cases} 1, & d_{i_1 i_2} = 0 \\ \theta_1 - \theta_2 d_{i_1 i_2}, & d_{i_1 i_2} > 0 \end{cases} \tag{17}$$

The standard deviations σ_i were modelled as a declining power function of watershed area:

$$\sigma_i = \theta_3 a_i^{-\theta_4} \tag{18}$$

where θ_3 and θ_4 are parameters to be estimated. All peaks models discussed in this paper (other than the least squares fits) employed equations (15), (16), and (18). The flow and sediment models employed equations (15), (17), and (18)

Method of estimation. The parameters of the model were estimated using the method of maximum likelihood [*Mood* et al., 1974]. The likelihood function is assumed to be the multivariate normal density of the ε_{ij} treated as a function of the β and θ parameters. In practice we minimize the negative of the log likelihood. In this problem, the log-likelihood is equal to the sum of the independent storm-wise log-likelihoods. Thus, the dimension of the multivariate density function is the number of watersheds represented in a given storm, a maximum of 10. The log-likelihood functions and their gradients (derivative vectors) are shown in APPENDIX B. They were programmed in S-Plus [*Statistical Sciences*, 1995] and FORTRAN, and solved using the S-Plus function *nlminb* (nonlinear minimization subject to bound-constrained parameters). Least squares estimates of the parameters were used as starting guesses in these iterative numeric calculations.

Model size. The inclusion of up to 31 parameters in these models raises questions about overfitting. These questions were addressed by cross-validation (discussed below) after a model was selected, but the proper model size was selected with the objective of minimizing a variant of Akaike's information criterion [*Burnham* and *Anderson*, 1998],

$$\mathrm{AIC}_c = -2\log(L) + 2K\left(\frac{n}{n-K-1}\right) \tag{19}$$

where L is the maximum likelihood, K is the number of parameters estimated, and n is the sample size. This criterion is recommended over the unmodified AIC when the ratio n/K is small (less than about 40). The inclusion of the 20 location parameters β_{0i} and β_{1i} is strongly supported by AIC_c. Its value increased by 14 to 88 units in the various models when one or two parameters were substituted for either β_{0i} or β_{1i}. Increases of 10 or more AIC units indicate clearly inferior models [*Burnham* and *Anderson*, 1998]. Because of the computational time required to fit each model, it was impractical to obtain the likelihoods of all alternative models. For that reason, parameters other than β_{0i} and β_{1i} were evaluated using hypothesis tests based on the normal distribution, and AIC_c was computed only for the more promising candidate models.

Hypothesis testing. Maximum likelihood parameter estimates are approximately multivariate-normally distributed for large samples [*Rao*, 1973]. The estimated covariance matrix of the estimates was obtained by inverting the observed information matrix, using a finite difference approximation to the Hessian, or matrix of second derivatives of the log-likelihood function [*Bishop* et al., 1975; *McCullagh* and *Nelder*, 1989]. (The observed information matrix is the negative of the Hessian, evaluated at the maximum likelihood estimates.) The standard errors, s_b, of the estimated parameters are the square roots of the diagonal of the covariance matrix. Since the parameter estimates are asymptotically normal, a simple test of the hypothesis H_0: $\beta_i = c$ is provided by observing whether or not the statistic $(b_i - c) / s_b$ is in the rejection zone of the standard normal distribution. The p-values from these hypothesis tests are identified as p_N in this paper. Tests with $p_N < 0.01$ are considered significant in this paper. Tests with $0.01 < p_N < 0.05$ are considered "suggestive" but not conclusive.

Observed change in response. "Observed change" in response was calculated by comparing the observed response, y_{ij}, with an estimate of the expected response, $E(y'_{ij})$, from the same storm and watershed in an undisturbed condition. We define the percentage change in response as

$$p_{ij} = 100\left(\frac{y_{ij} - E(y'_{ij})}{E(y'_{ij})}\right) = 100\left(\frac{y_{ij}}{E(y'_{ij})} - 1\right) \quad (20)$$

The expected undisturbed response, $E(y'_{ij})$, is a function of $E\left(\log(y'_{ij})\right)$:

$$E(y'_{ij}) = \exp\left[E\left(\log(y'_{ij})\right) + \tfrac{1}{2}\sigma_i^2\right] \quad (21)$$

Setting disturbance to zero in either model (10) or (14) above, we have $E\left(\log(y'_{ij})\right) = \beta_{0i} + \beta_{1i}\log(y_{Cj})$. The variances σ_i^2 are a function of θ_3 and θ_4 given by model (18). A nearly unbiased estimator of $E(y'_{ij})$ is given by

$$\hat{y}'_{ij} = \exp\left[b_{0i} + b_{1i}\log(y_{Cj}) + \tfrac{1}{2}\left(\hat{\theta}_3 a_i^{\hat{\theta}_4}\right)^2\right] \quad (22)$$

where b_{0i}, b_{1i}, $\hat{\theta}_3$, and $\hat{\theta}_4$ are the maximum likelihood estimates of β_{0i}, β_{1i}, θ_3, and θ_4, respectively. The term $\tfrac{1}{2}\hat{\sigma}_i^2 = \tfrac{1}{2}\left(\hat{\theta}_3 a_i^{\hat{\theta}_4}\right)^2$ is often called the Baskerville [1972] bias correction. An approximation for p_{ij} that we will call the "observed change in response" is obtained by substituting $\hat{y}'_{ij}$ for $E(y'_{ij})$ in (20):

$$\tilde{p}_{ij} = 100\left(\frac{y_{ij}}{\hat{y}'_{ij}} - 1\right). \quad (23)$$

Of course we are not just interested in the changes in response for the particular values of the explanatory variables encountered during the study. We would like to study the percentage change, p_0, for an arbitrary vector, $\mathbf{x}_0$, of explanatory variables. An unbiased estimator and confidence interval for $E(p_0)$ as well as a prediction interval for p_0 are derived in APPENDIX C. The confidence interval represents the uncertainty of the mean, $E(p_0)$, given $\mathbf{x}_0$. The prediction interval indicates the variability in the individual response p_0, given $\mathbf{x}_0$. Prediction intervals are wider than confidence intervals because they include the variability in the response about its mean value as well as the variability due to uncertainty in the mean itself.

Cross-validation of models. To investigate the possibility that the models were overfitted to the data, ten-fold cross-validation was used [*Efron* and *Tibshirani*, 1993]. The data are split into ten groups. Each observation is predicted from a model based on all of the data except that group to which the observation belongs. The RMSE of these predictions is called the cross-validation prediction error and it may be compared with the RMSE of the models fitted with all the data to assess overfitting.

A regression of the observed responses on the fitted values, known as the *calibration*, should have an intercept near zero and slope near unity. The regression of the observed responses on the cross-validated predictions is expected, in general, to have a slope less than one [*Copas*, 1983]. This phenomenon, known as *shrinkage*, implies that predictions of high or low response values tend to be too extreme. The degree of departure of the calibration slope from unity provides another measure of overfitting.

Because the data were not independent, the cross-validation was repeated using two different methods for splitting the data:

1. Data were randomly divided into groups of equal size.
2. Post-treatment data were omitted systematically, one station at a time.

The latter method does not provide cross-validated predictions for the pre-treatment data, but if all the data from a station, say watershed *i*, are omitted, it becomes impossible to estimate β_{0i} and β_{1i}, which are required to make predictions for that watershed. Nevertheless, the one-station-at-a-time method is probably a more rigorous validation for the inclusion of alternative disturbance variables because it will give higher error rates for models that include variables correlated with the response due to just one or two watersheds.

RESULTS

Storm Peaks

The analysis included 226 pre-treatment and 300 post-treatment observations representing 59 storms on the 10

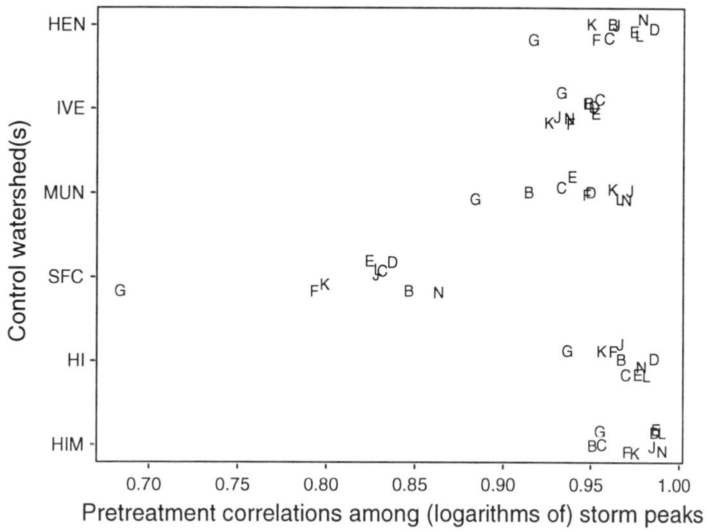

Figure 3. Pretreatment correlations between logarithms of storm peak at treated watersheds and alternative control watersheds. Letters designate watersheds (e.g. G is watershed GIB). Random noise has been added to the vertical plotting positions to improve readability.

treated watersheds. For the 226 pretreatment peaks, the control watersheds correlating best with watersheds to be treated were tributaries HEN and IVE, and MUN (Figure 3). The mean of the peaks at HEN and IVE (designated HI), or at HEN, IVE, and MUN (designated HIM), had higher correlations than did peaks from either HEN, IVE, or MUN individually. Because MUN was not monitored the last year of the study, HI was chosen as the control for the peaks analysis.

The Chow tests [*Chow*, 1960; *Wilson*, 1978], based on the HI control, revealed strong evidence that post-treatment data differed from pre-treatment regressions. Eight of the 10 watersheds departed ($p < 0.005$) from these regressions after logging commenced. The other two, FLY and LAN on the main stem, had p-values less than 0.05. Departures from the pre-logging regression were greatest in the clear-cut tributaries: BAN, CAR, EAG, GIB, and KJE (Figure 4).

Seasonal patterns in the departures from the pre-treatment regressions were evident in most of the treated watersheds. For example, Figure 5 shows the post-logging departures for watershed EAG plotted against storm number. The largest percentage departures occurred early in the season. These were usually, but not always, relatively small storms. Storms 28 and 29 did not show treatment effects, apparently because logging had just taken place the same winter, so insufficient time had elapsed for soil moisture differences to develop between the controls and the logged area. This exemplifies the situation that necessitated modelling of the disturbance term using equation (9).

To develop an overall model, an intercept and slope for each watershed (equation (3)) was initially fit by least squares. The residuals from this model show a strong interaction between proportion of area logged and antecedent wetness (Figure 6). Area logged includes clear-cut areas and a portion of each buffer zone corresponding to the proportion of timber removed (Table 1). The relation of the residuals with area logged is linear, the slope decreasing from strongly positive with increasing wetness (Figure 6, top row). The relation with log(wetness) is linear, the slope becoming strongly negative with increasing logged area (Figure 6, bottom row). These relations imply a product term is an appropriate expression of the interaction, and the coefficient is expected to be negative. The fact that the average residual increases with different categories of area logged but not with wetness shows that a solo logged area term is needed in the model as well as the interaction product, but a solo wetness term is not. No variables related to roads, skid trails, landings, firelines, burning or herbicide application were found to improve the fit of the linear least squares model that includes logged area and its interaction with wetness. Adding logged area and the wetness interaction to the model, a plot of post-treatment residuals against time after logging (Figure 7) indicates an approximately linear recovery trend in the first 7 years.

When model (10) was fit to the data, the coefficient b_7 on the cumulative effect term did not differ significantly from zero (Table 3, $p_N = 0.21$). The coefficient b_5 was negative but not highly significant ($p_N=0.047$), weakly suggesting that the effect of logged area on peak flows

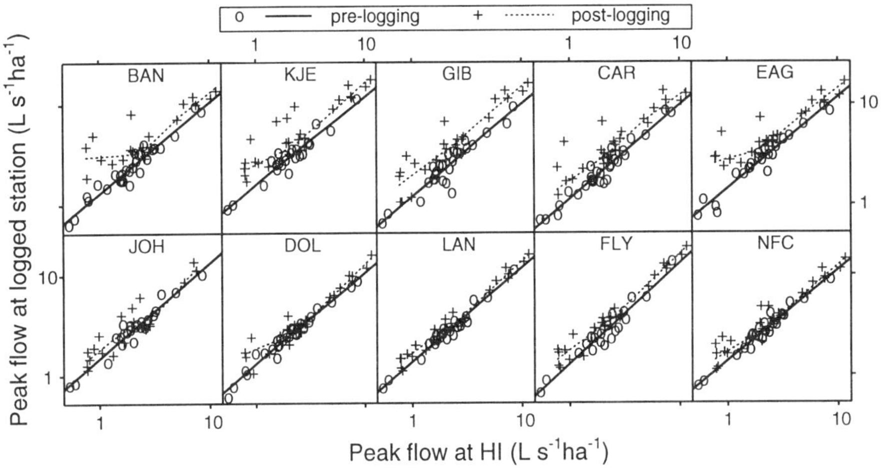

Figure 4. Relation between peak streamflow in the 10 treated tributaries in the North Fork of Caspar Creek, and that of the HI control. Post-logging relations were fitted by loess method [*Cleveland*, 1979]. The top row represents 95-100% clear-cut watersheds.

tends to diminish in larger storms. The coefficient b_4 on logged area was positive as expected and its interaction with wetness, b_6, was negative as expected. The recovery coefficient, b_2, indicates an average recovery rate of about 8% per year. The null hypothesis for each of the parameters $\beta_3^{(k)}$ is $H_0 : \beta_3^{(k)} = 1$, because the recovery model assumes a value of unity the year after logging. The coefficient $b_3^{(1)} = 0.59$ (standard error 0.10) indicates a reduced effect from fall logging on peaks in the following winter and $b_3^{(2)} = 0.00$ suggests that the effects of winter logging on peak flow are delayed until a growing season has passed.

There was no indication of a dependency on watershed area in either the coefficients b_{0i} or b_{1i} from model (10). When we replaced β_{0i} in model (10) with the expression $\beta_0^{(1)} + \beta_0^{(2)} a_i$, the coefficient $b_0^{(2)}$ was not significantly

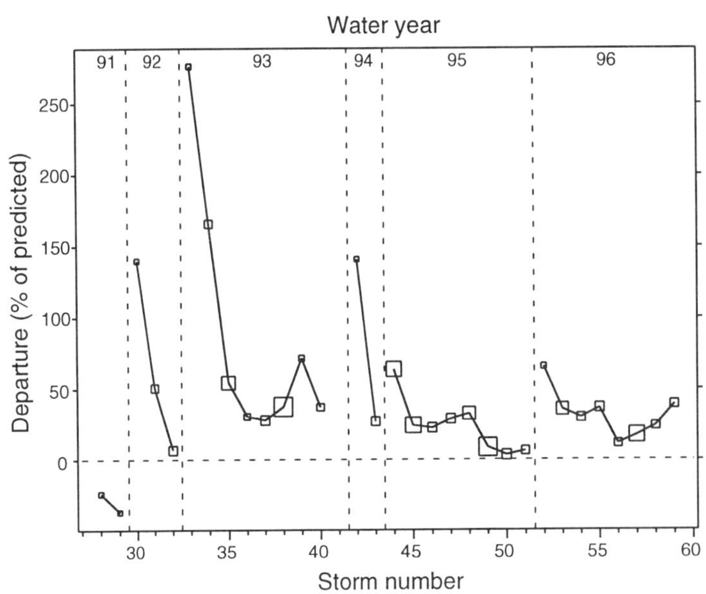

Figure 5. Post-logging departures of storm peaks (as percentage of predicted) at watershed EAG from those predicted from pretreatment regression on HI control. Axes are logarithmic. Symbol sizes indicate relative size of storm peak at HI control. Vertical dotted lines separate water years. About half the watershed was winter-logged before storm 28 and logging was completed by storm 30.

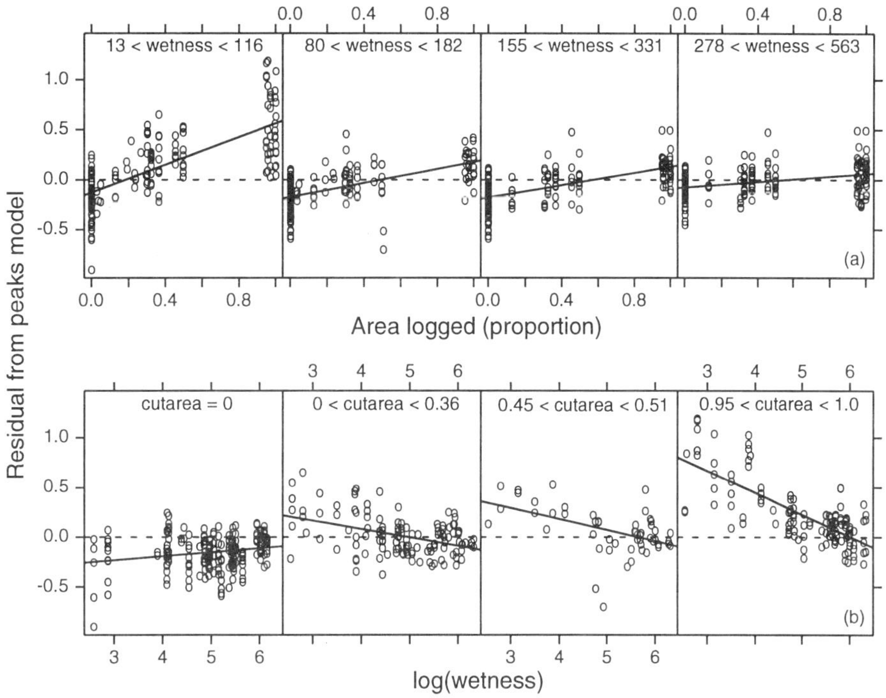

Figure 6. Conditioning plots of residual from storm peaks model (3) and interaction between area logged and antecedent wetness index with (a) wetness index fixed in each frame, and (b) proportion of area logged fixed in each frame.

different from zero (p_N=0.58), indicating no trend of unit area storm peak with watershed area.

The exponentially declining correlation model (18) was used when solving model (10) for peak flows (with β_7 fixed at zero), and it can be seen to be a reasonable fit (Figure 8). The variance model (18) also seems reasonable (Figure 9). The Box-Pierce test [*Shumway*, 1988] did not indicate the presence of serial autocorrelation at any of the stations (minimum p=0.089). The residuals conform very well to the normal distribution (Figure 10), as do plots for individual stations (not shown), validating our choice of likelihood function. The lone outlier is from a storm at GIB that produced 2 peaks at all stations except GIB. (The first peak was selected for the storm but was identifiable only as a shoulder of the hydrograph at GIB.) The model fits the data very well (Figure 11). For the regression between observed and fitted values, $r^2 = 0.946$. This compares with $r^2 = 0.848$ for a model with no disturbance variables and $r^2 = 0.937$ for model (3) fit to only the pre-treatment data, so the model fits the post-treatment data as well as the pre-treatment data.

Magnitude of observed changes. Maximum peak flow increases based on equations (22) and (23) were about 300%, but most were less than 100% (Figure 12). The mean percentage increase declined with wetness but was still positive even under the wettest conditions of the study ($w_i > 500$), when it was 23% for clear-cuts but only 3% in partially cut watersheds. Increases more than 100% generally only occurred in clear-cuts under relatively dry conditions ($w_i < 50$) and when peaks in the control were less than 0.0025 $m^3s^{-1}ha^{-1}$ (return period 3-4 times per year). Large increases occurred less frequently as the winters progressed, but increases over 100% did occur in January and February. The mean percentage increase in peak flow declined with storm size and then leveled at an average increase of 35% in clear-cuts and 16% in partially cut watersheds for peaks greater than 0.004 $m^3s^{-1}ha^{-1}$ (return periods longer than 0.5 years) (Figure 13). For a storm size having a 2-year return period, the average peak-flow increase in 100% clear-cuts was 27% [*Ziemer*, 1998].

Figure 14 shows 95% confidence intervals for the modelled mean response in a 20-ha watershed that has been 50% clear-cut, for two wetness conditions and two cutting ages within the range of our data. The effect of antecedent wetness is a greater influence on the response than time since cutting, although the recovery data only span 7 years. Prediction intervals are much wider than confidence intervals, revealing post-treatment variability that is greater than the treatment effect itself.

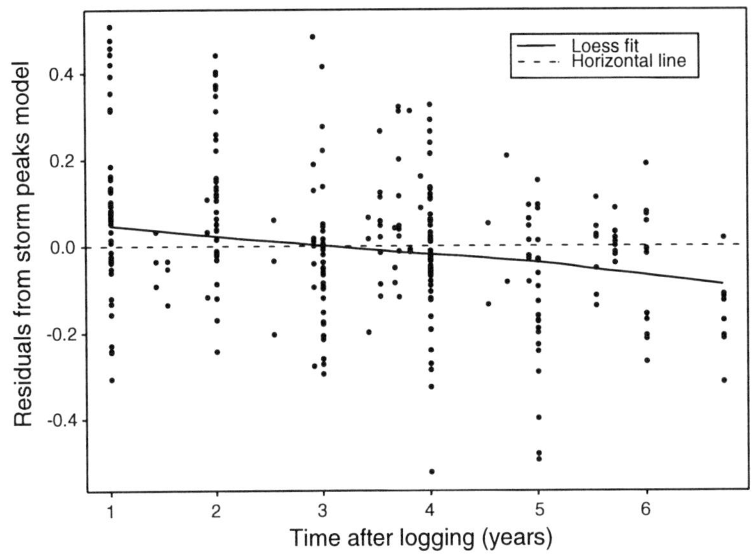

Figure 7. Relation between storm peak residuals and time after logging. Curve is fit by loess method [*Cleveland*, 1979]. Residuals are from least squares fit to the model $\log(y_{ij}) = \beta_{0i} + \beta_{1i}\log(y_{C\cdot j}) + \beta_4 D_{ij} + \beta_6 D_{ij}\log(w_j) + \varepsilon_{ij}$.

Storm Runoff Volume

The analysis included 527 observations representing 59 storms. For the same reasons as in the peaks analysis, HI (the mean of HEN and IVE) was chosen as the control. The modeling results are similar to the peaks analysis results, except that the watershed area interaction b_7 was marginally significant (Table 4, p_N=0.012) and watershed correlations were found to decline linearly with distance, so model (17) was used instead of (16) in the covariance model. For the sake of brevity, the modeling results for storm runoff volume are omitted, and we report only the coefficients (Table 4) and the magnitude of observed changes.

Magnitude of observed changes. The maximum storm runoff volume increase from equations (22) and (23) was 400%, but most were less than 100% . The mean percentage increase declined with wetness but was still positive even under the wettest conditions of the study ($w_i > 500$), when it was 27% for clear-cuts and 16% in partially cut watersheds. Increases more than 100% generally only occurred in clear-cuts under relatively dry conditions ($w_i < 100$) and when runoff volume in the control was less than 250 m^3ha^{-1}. Large increases occurred less frequently as the winters progressed, but in-creases over 100% did occur in January and February. The mean percentage increase in storm runoff volume declined with storm size and then leveled at an average increase of 30%

TABLE 3. Maximum likelihood parameter estimates for storm peaks model {(10),(16),(18)}, excluding β_{0i} and β_{1i}. p_N is normal probability value for H_0: $\beta = 0$.

Parameter	Effect	Estimate	Standard Error	p_N
β_2	Recovery	0.0771	0.0183	<0.0001
$\beta_3^{(1)}$	Fall logging	0.5939	0.0996	<0.0001
$\beta_3^{(2)}$	Winter logging	0.0000	0.2843	1.0000
β_4	Amount logged	1.1030	0.3409	0.0012
β_5	Storm size interaction	-0.0963	0.0484	0.0468
β_6	Wetness interaction	-0.2343	0.0251	<0.0001
β_7	Watershed area interaction	3.553E-4	2.861E-4	0.2142
θ_1	Correlation shape parameter	2.809E-3	6.188E-4	<0.0001
θ_2	Correlation limit parameter	0.4698	0.1564	0.0027
θ_3	Variance magnitude	0.2285	0.0242	<0.0001
θ_4	Variance shape	-0.0937	0.0238	0.0001

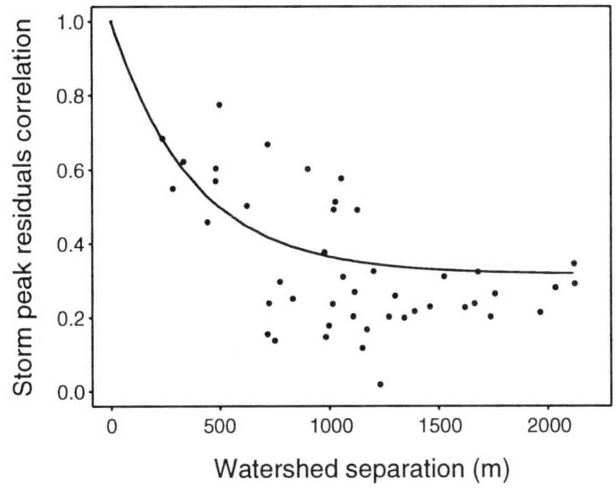

Figure 8. Relation between storm peak residuals correlation and distance between watershed centroids. Residuals are from maximum likelihood fit to storm peak model {(10),(16),(18)}. Curve depicts equation (16), with estimated parameters $\hat{\theta}_1$ and $\hat{\theta}_2$.

in clear-cuts and 13% in partially cut watersheds for storm runoff greater than 250 m^3ha^{-1}.

Annual storm runoff volume (sum of storms) increased an average of 58% (1119 m^3ha^{-1}) in clear-cut watersheds and 23% (415 m^3ha^{-1}) in partly clear-cut watersheds (Table 5). Based on the complete discharge record at NFC, the runoff volume for the storms included in this analysis represents 41 to 49% of the total annual runoff volume in individual tributaries.

Figure 15 shows confidence intervals and prediction intervals for storm runoff volume in a 20-ha watershed that has been 50% clear-cut, under two wetness conditions and two cutting ages within the range of our data.

Suspended Sediment Loads

The relatively large number of missing observations resulting from quality control screening complicated the selection of controls for the sediment analysis. The use of synthetic controls such as HI and HIM permitted larger sample sizes because these means could be computed from any combination of non-missing controls. Thus the sample size was 376 with the HIM control, but only 333 with the HI control, and less than 300 with HEN or IVE alone. Although HIM control permitted the largest sample size, its correlations tended to be lower than those of HI (Figure 16). We therefore present the analysis twice, once with the HIM control and once with the HI control.

Chow tests [*Chow*, 1960; *Wilson*, 1978] for treatment effects at individual stations gave mixed results (Table 6). Only 2 of the tests were significant when HIM was used as the control and 3 were significant with the HI control. The tributaries all had more significant changes than the main-stem stations. Figure 17 (top row) indicates that suspended sediment loads increased in all the clear-cut tributaries except KJE, where loads appear to have decreased after logging. The only partly clear-cut watershed on a tributary (DOL) also showed highly significant increases in sediment loads. The upper main-stem stations (JOH and LAN) showed no effect after logging, and the lower main-stem stations (FLY and ARF) experienced increases only in smaller storms. Summing suspended sediment over *all* storms, the four main-stem stations all showed little or no change (Table 7). Sediment loads at the North Fork weir, below ARF, increased by about 89% per year, mainly as a result of a large landslide in the ungaged subwatershed that enters between ARF and NFC.

Models with HI control. The analysis included 333 observations representing 43 storms. In these models (14), the change in storm flow volume $\Delta q_{ij}^{(1)}$ was found to be the best explanatory variable after sediment load from the HI control, y_{HI}. Figure 18 shows the relation between the post-treatment sediment departures from pretreatment model (3)) and $\Delta q_{ij}^{(1)}$. Since both variables are differences in logarithms, it is convenient to express them as ratios of observed to predicted response, obtained by exponentiating the differences. The linear correlation between the sediment and flow departures is 0.54.

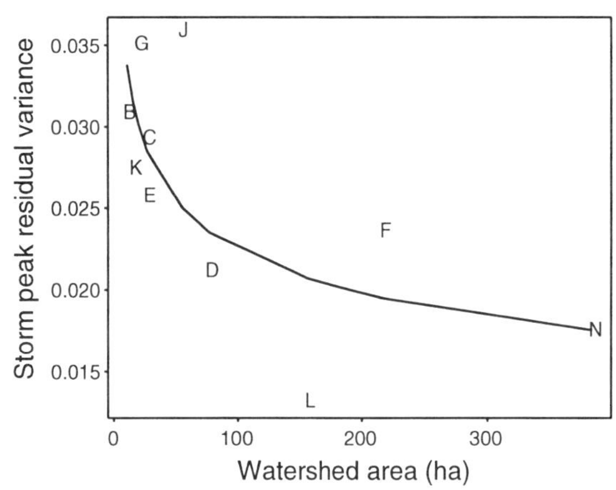

Figure 9. Relation between variance of storm peak residuals and watershed area. Residuals are from maximum likelihood fit to storm peak model {(10),(16),(18)}. Curve depicts equation (18) with estimated parameters $\hat{\theta}_3$ and $\hat{\theta}_4$. Letters designate watersheds (e.g. G is watershed GIB).

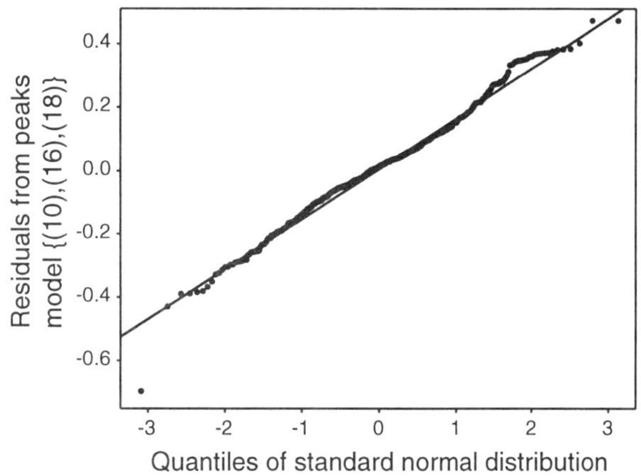

Figure 10. Normal quantile plot of residuals from storm peak model {(10),(16),(18)}. Line is least squares fit.

After $\Delta q_{ij}^{(1)}$ is in the model, disturbance variables explain only a very small part of the remaining variation (Figure 19). The length of unbuffered stream channel in clear-cut areas was one of the more useful disturbance variables in the sediment models. Under California Forest Practice Rules in effect during the North Fork logging, vegetation buffers were not required for stream channels that do not include aquatic habitat. The best models were found when this variable was separated into channels in burned clear-cuts and channels in unburned clear-cuts. The variable did not need to be separated, however, in the interaction terms. Thus the model (14) was modified to:

$$\begin{aligned}\log(y_{ij}) = \beta_{0i} + \beta_{1i}\log(y_{(\mathrm{HI})j}) + \beta_2 \Delta q_{ij}^{(1)} \\ + \beta_3 x_{ij}^{(1)} + \beta_4 x_{ij}^{(2)} + \beta_5 (x_{ij}^{(1)} + x_{ij}^{(2)}) \log(y_{(\mathrm{HI})j}) \quad (24) \\ + \beta_6 (x_{ij}^{(1)} + x_{ij}^{(2)}) a_i + \varepsilon_{ij}\end{aligned}$$

where

$x_{ij}^{(1)}$ = length of stream channel in burned clear-cuts, and

$x_{ij}^{(2)}$ = length of stream channel in unburned clear-cuts

To indicate the relative contribution of the various terms in model (24), the increase in residual sum of squares is shown for least squares models after dropping each explanatory variable (Table 8).

The maximum likelihood estimates for model (24) are shown in Table 9. The coefficient estimate b_3 is about 1.8 times b_4, suggesting that streams in burned clear-cuts contribute more sediment than those in unburned clear-cuts. The estimate, b_5, of the storm size interaction is negative, suggesting that the ratio between post-treatment and pre-treatment sediment loads diminishes for larger events. The estimate, b_6, of the cumulative effect coefficient in this model was negative and was found marginally significant ($p_N = 0.044$). This interaction in the sediment model only partly offsets the small positive interaction that was noted in the runoff model and is hidden in the term $\Delta q_{ij}^{(1)}$. Other variables being equal, the model still predicts larger unit area sediment loads from larger watersheds (Figure 20). Because of its marginal significance, the β_6 term was dropped from the model for the remainder of this section.

The fitted intercepts b_{0i} from model (24), with β_6 fixed at zero, tend to increase with watershed area (Figure 21), with the exceptions of KJE (K) and JOH (J). This pattern in the intercepts is confirmed by substituting $\beta_0^{(1)} + \beta_0^{(2)} a_i$ for the term β_{0i}. The fitted coefficient $b_0^{(2)}$ is positive and differs significantly from zero (p_N=0.0031). The slope coefficients b_{1i}, are all between 0.8 and 1, except BAN (0.73) and EAG (1.06), and show no trend with area. Thus, ignoring the anomalous KJE and JOH for the moment, the unit area sediment loads from the watersheds prior to disturbance (Figure 22) tend to be highest in the four largest watersheds (ARF, FLY, LAN, and DOL), followed by the tributaries CAR, GIB, and EAG, and are lowest in the smallest watershed BAN.

Although there are signs of positive or negative trends in some individual watersheds, the residuals from model (24) display little if any trend with time (Figure 23). If the anomalous JOH and KJE, which did not show treatment effects, are omitted, hints of a recovery trend disappear entirely.

The covariance model fit rather well for the sediment models based on HI. Correlations declined linearly with watershed separation (Figure 24) and variance declined as

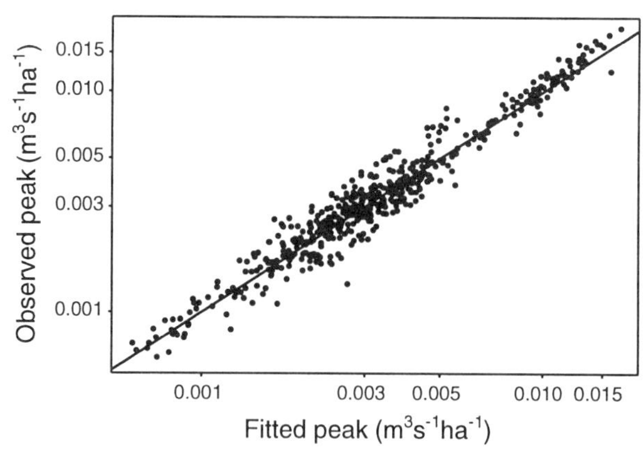

Figure 11. Observed storm peaks versus fitted values from model {(10),(16),(18)}. Line is $y = x$.

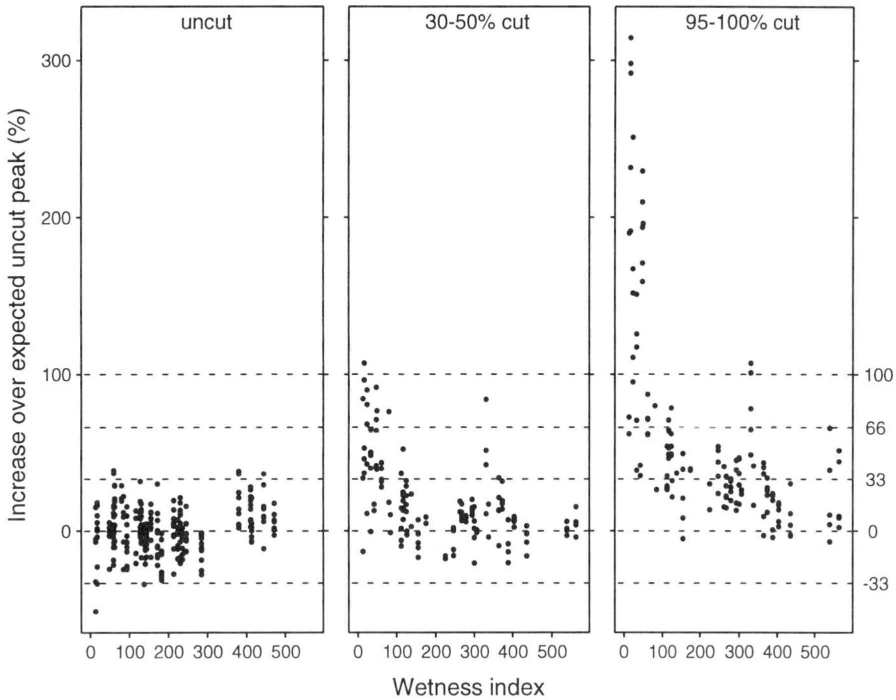

Figure 12. Percentage increase over expected uncut storm peak as related to antecedent wetness index for uncut (before treatment), partly (30-50%) clear-cut, and (95-100%) clear-cut watersheds. Bias-corrected predictions are from model {(10),(16),(18)} with disturbance set to zero.

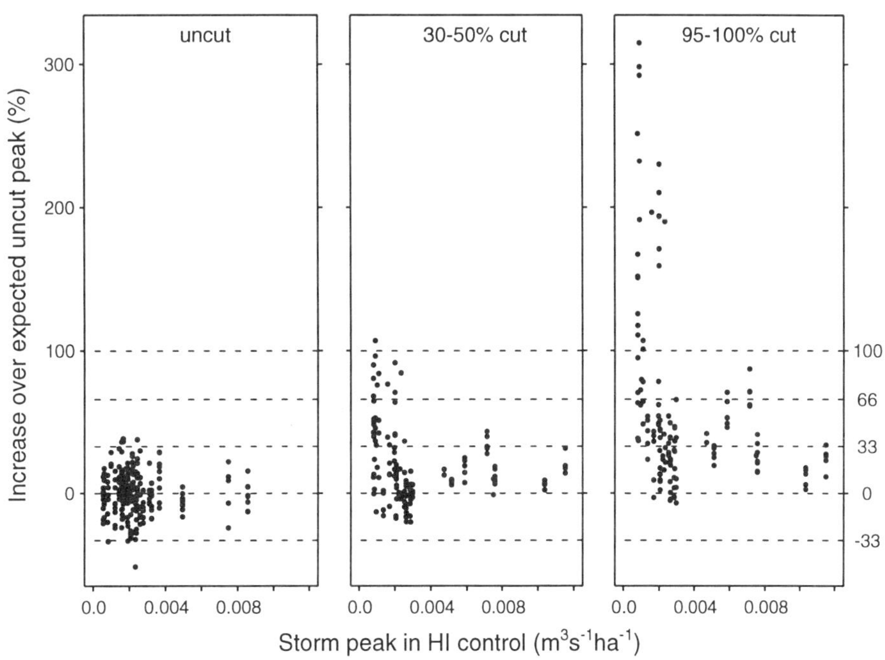

Figure 13. Percentage increase over expected uncut storm peak as related to peak size in the HI control for uncut (before treatment), partly (30-50%) clear-cut, and (95-100%) clear-cut watersheds. Bias-corrected predictions are from model {(10),(16),(18)} with disturbance set to zero.

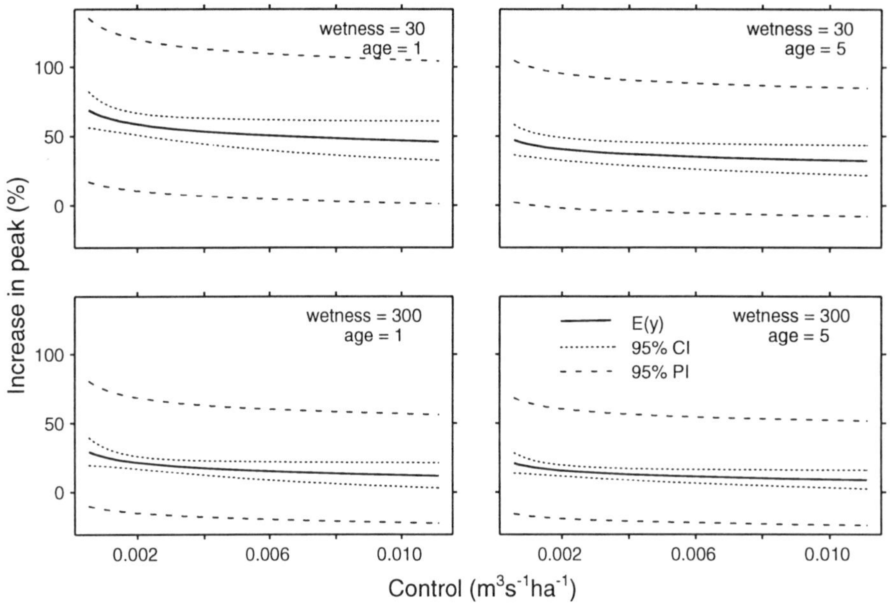

Figure 14. The effect of wetness and age after cutting on predictions from storm peak model {(10), (16),(18)} after clear-cutting 50% of a 20 ha watershed. Expected increases and 95% confidence (CI) and prediction (PI) intervals are shown for two levels of antecedent wetness 1 and 5 years after cutting.

a power function of watershed area (Figure 25). The Box-Pierce test [*Shumway*, 1988] indicated (using an experimentwise error rate of 0.05) the presence of serial autocorrelation at four stations (ARF, BAN, GIB, and KJE) and suggests that we conservatively assess marginally significant terms in the model. The residuals again conform very well to the normal distribution and there is only one outlier (associated with stream bank collapses in EAG). The regression between observed and fitted values has $r^2 = 0.915$. This compares with $r^2 = 0.828$ for a model with no disturbance variables and $r^2 = 0.948$ for model (3) fit to only the pre-treatment data. So the complete model (without the cumulative effects term) explains $(0.915 - 0.828) / (0.948 - 0.828) = 72\%$ of the variation introduced by the post-treatment data.

Models with HIM control. This analysis included 376 observations representing 51 storms. In models developed with the HIM control, the log-ratio flow variable $\Delta q_{ij}^{(2)}$ was found to be a better explanatory variable than the flow model residual $\Delta q_{ij}^{(1)}$. The most important disturbance

TABLE 4. Maximum likelihood parameter estimates for storm runoff model {(10),(17),(18)}, excluding β_{0i} and β_{1i}. p_N is normal probability value for H_0: $\beta = 0$.

Parameter	Effect	Estimate	Standard Error	p_N
β_2	Recovery	0.0912	0.0143	<0.0001
$\beta_3^{(1)}$	Fall logging	0.8117	0.0910	<0.0001
$\beta_3^{(2)}$	Winter logging	-0.196	0.225	0.3843
β_4	Amount logged	2.3054	0.2646	<0.0001
β_5	Storm size interaction	-0.1103	0.0467	0.0181
β_6	Wetness interaction	-0.2362	0.0236	<0.0001
β_7	Watershed area interaction	6.481E-4	2.578E-4	0.0119
θ_1	Correlation intercept	0.6697	0.0587	<0.0001
θ_2	Correlation slope	-1.898E-4	4.962E-5	0.0001
θ_3	Variance magnitude	0.1987	0.0190	<0.0001
θ_4	Variance shape	-0.0873	0.0209	<0.0001

TABLE 5. Percentage and absolute departures from predicted annual storm runoff volume (sum of storms).

	Uncut	30-50% Clearcut	95-100% Clearcut
Mean (%)	2	23	58
Median (%)	2	19	51
Mean ($m^3\ ha^{-1}\ yr^{-1}$)	54	415	1119
Median ($m^3\ ha^{-1}\ yr^{-1}$)	29	387	1050

variable in these models is proportion of the watershed occupied by road cuts and fills. The length of stream channel in clear-cuts and the interaction terms in model (24) were not significant when tested in maximum likelihood models with the HIM control. This is partly explained by a high correlation (0.80) between road cut/fills and stream length in burned areas. A negative interaction between road cut/fills and watershed area was marginally significant (p_N=0.037). The maximum likelihood estimates for the model

$$\log(y_{ij}) = \beta_{0i} + \beta_{1i}\log(y_{(\mathrm{HIM})j}) + \beta_2\Delta q_{ij}^{(2)} + \beta_3 x_{ij} + \beta_4 x_{ij} a_i + \varepsilon_{ij} \quad (25)$$

where x_{ij} is the proportion of the watershed occupied by road cuts and fills, are shown in Table 10. As with model (24), the interaction only serves to partly offset the positive interaction hidden in the $\Delta q_{ij}^{(2)}$ term, and we do not consider it significant. The trend in intercepts that was seen for model (24) is also present in model (25). Setting β_4 to zero, and substituting $\beta_0^{(1)} + \beta_0^{(2)} a_i$ for β_{0i}, we test $\beta_0^{(2)}$ and again find that it is positive and differs significantly from zero (p_N=0.0023). The residuals from model (25), with β_4 fixed at zero, do not display a significant trend with time since logging.

Magnitude of observed changes. Sediment load increases were calculated using equations (22) and (23) with the coefficients estimated from model (25). Median increases were 64% in partly clear-cut watersheds and 107% in clear-cut watersheds (Figure 26). Absolute increases were similar in clear-cut and partly clear-cut watersheds (Figure 27). Most of the larger percentage increases in clear-cuts were from small events and equated to relatively minor absolute increases in load. As one would expect, there is a tendency for percentage increases to decrease with storm

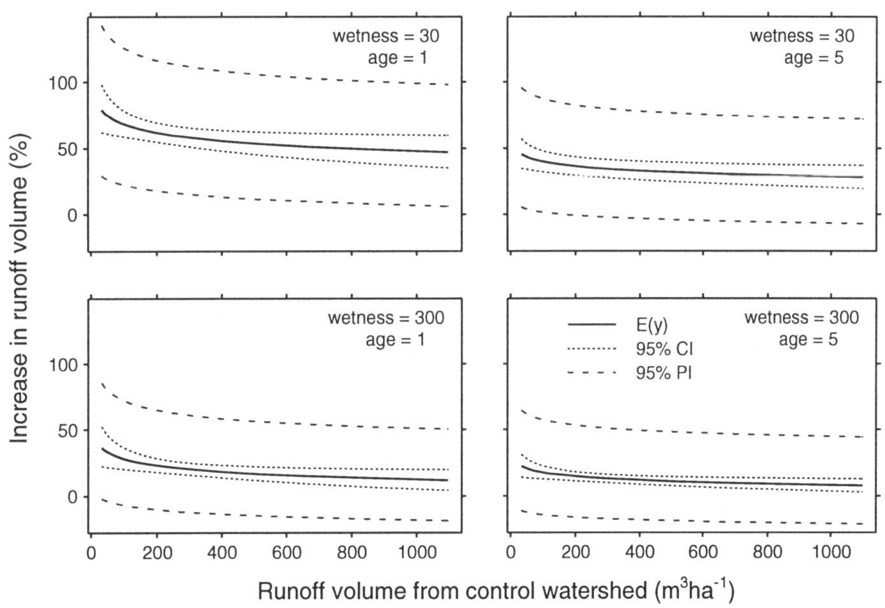

Figure 15. The effect of wetness and age after cutting on predictions from storm runoff volume model {(10),(17),(18)}, after clear-cutting 50% of a 20 ha watershed. Expected increases and 95% confidence (CI) and prediction (PI) intervals are shown for two levels of antecedent wetness 1 and 5 years after cutting.

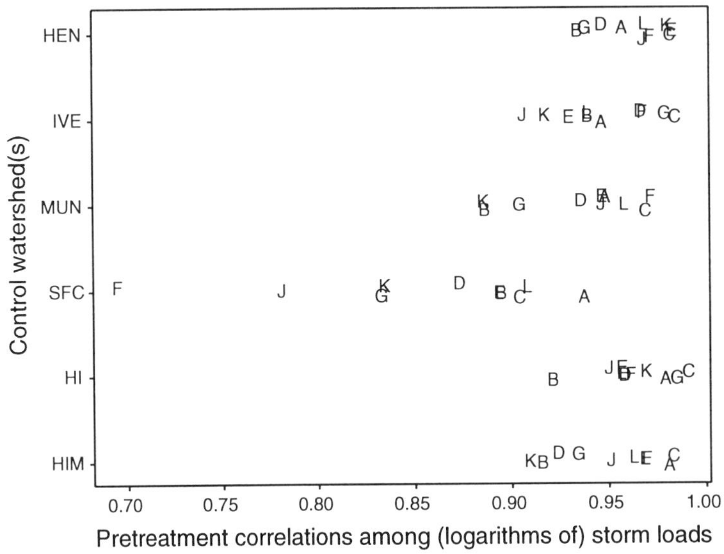

Figure 16. Pretreatment correlations between logarithms of storm sediment load at treated watersheds and alternative control watersheds. Letters designate watersheds (e.g. G is watershed GIB). Random noise has been added to the vertical plotting positions to improve readability.

TABLE 6. Chow test [Wilson, 1978] p-values for hypothesis of no change in suspended sediment load after logging.

Watershed	HI control	HIM control
ARF	0.1649	0.0215
BAN	0.0128	0.0292
CAR	0.0000[a]	0.0001[a]
DOL	0.0198	0.0093
EAG	0.0056	0.0013[a]
FLY	0.3528	0.0955
GIB	0.0002[a]	0.0096
JOH	0.0983	0.0476
KJE	0.0026[a]	0.0384
LAN	0.8018	0.2453

[a] significant at $\alpha = 0.005$ (experimentwise error rate = 0.05)

size, and for absolute increases to increase with storm size. Figure 28 shows 95% confidence intervals and prediction intervals for the sediment model (25), with the area×disturbance interaction, β_4, set to zero. The watersheds are ranked by increasing proportion of road cuts and fills (x_{ij}). The uncertainty in the model and the variability in suspended sediment loads is much greater than for peak flow or storm runoff volume.

Summing storms by year, annual suspended sediment loads increased an average of 212% (262 kg·ha^{-1}yr^{-1}) in clear-cut watersheds and 73% (263 kg·ha^{-1}yr^{-1}) in partly clear-cut watersheds (Table 11). The absolute increases are heavily influenced by outlying data points that tend to occur in wet years (1993 and 1995), while the percentage increases weight all years approximately equally. If the extreme outlier in the partly clear-cut population (Figure 27) is omitted, the mean increase in that category drops to 67% (180 kg·ha^{-1}yr^{-1}). Because of the highly skewed distribution of sediment loads, median increases were much smaller: 109% (59 kg·ha^{-1}yr^{-1}) in clear-cut watersheds and 52% (46 kg·ha^{-1}yr^{-1}) in partly clear-cut watersheds. Based on the complete discharge record at NFC, the storms included in this analysis represent 36 to 43% of the total annual runoff in individual tributaries. However, these storms include roughly 90% of the annual suspended sediment load [*Rice* et al., 1979].

Cross-Validation of Models for Runoff Peaks, Volumes, and Sediment Loads

Predictions of storm runoff from random 10-fold cross-validation had RMSE only 2 to 3% (peaks) and 4% (volumes) higher than those from the original fitted models, for both pre-treatment and post-treatment responses (Table 12). The systematic cross-validation, omitting the post-treatment data one station at a time, gave RMSE 5% and 7% higher than the apparent post-treatment RMSE from the original runoff peaks and volume models, respectively. The systematically cross-validated RMSE values of 0.1739 and 0.1676 for logarithms of peaks and volumes correspond to prediction errors of about 20% for the untransformed responses. Calibration slopes (for regression of the observed versus predicted runoff) are

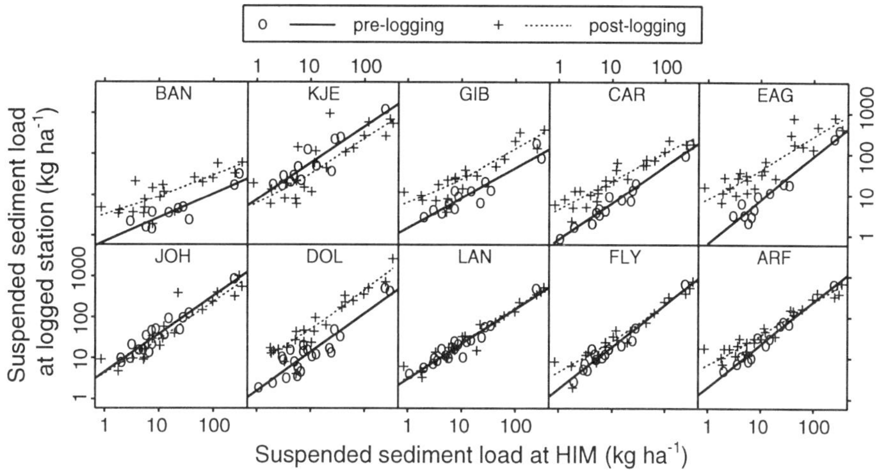

Figure 17. Relations between storm suspended sediment loads at logged subwatersheds in the North Fork and the the HIM control from 1986 to 1995. Post-logging relations were fitted by loess method [*Cleveland*, 1979]. The top row represents 95-100% clear-cut watersheds.

very close to unity (Table 13) for both peaks and volumes. Both the random and systematic cross-validation calibrations are nearly indistinguishable from $y = x$ on 600 dpi letter-size plots. Both the RMSE and calibration results indicate the models for runoff peaks and volumes are not overfit. Remarkably, they appear to predict independent data nearly as well as the data to which the models were fit.

Predictions of suspended sediment loads from random cross-validation had RMSE 7% (HI control) and 4% (HIM control) higher than those from the original fitted models, for both pre-treatment and post-treatment responses (Table 12). On the other hand, the systematic cross-validation gave RMSE 32% (HIM control) and 50% (HI control) higher than the apparent post-treatment RMSE from the original sediment models. The systematically cross-validated RMSE values of 0.6724 and 0.6966 for logarithms of sediment loads correspond to prediction errors of about 100% for the untransformed responses. Calibration slopes for the sediment models are similar to the original models for the random cross-validation, but the systematic cross-validation has calibration slopes significantly smaller (Table 13), indicating substantial shrinkage in prediction of data from subwatersheds not

TABLE 7. Summary of changes in suspended sediment load (summed over storms) after logging in North Fork subwatersheds. Predicted loads are computed from pre-treatment lincar regressions between the logarithms of the storm sediment load in the treated watershed and the mean of the storm sediment loads at the control watersheds HEN, IVE, and MUN. Predictions were corrected for bias when back-transforming from logarithmic units. The number of years in the post-logging period varies from four to six, depending upon when the watershed was logged and whether or not monitoring was discontinued in water year 1996.

Treated watershed	Number of years	Observed (kg ha^{-1}yr^{-1})	Predicted (kg ha^{-1}yr^{-1})	Change (kg ha^{-1}yr^{-1})	Change (%)
ARF	6	505	591	-86	-15
BAN	4	85	28	57	203
CAR	5	240	108	132	123
DOL	5	1130	306	824	269
EAG	5	710	210	500	238
FLY	5	536	555	-19	-3
GIB	4	358	119	239	200
JOH	5	667	865	-198	-23
KJE	5	821	1371	-551	-40
LAN	5	420	400	20	5
NFC	6	465	246	219	89

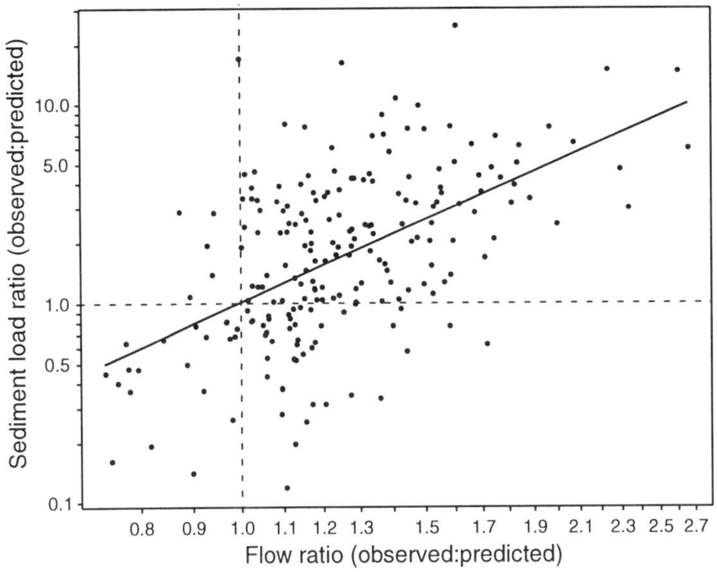

Figure 18. Relation between post-treatment sediment load departures from pretreatment relationship (3) and flow departures $\Delta q_{ij}^{(1)}$. Departures are expressed as the ratio of observed to predicted response.

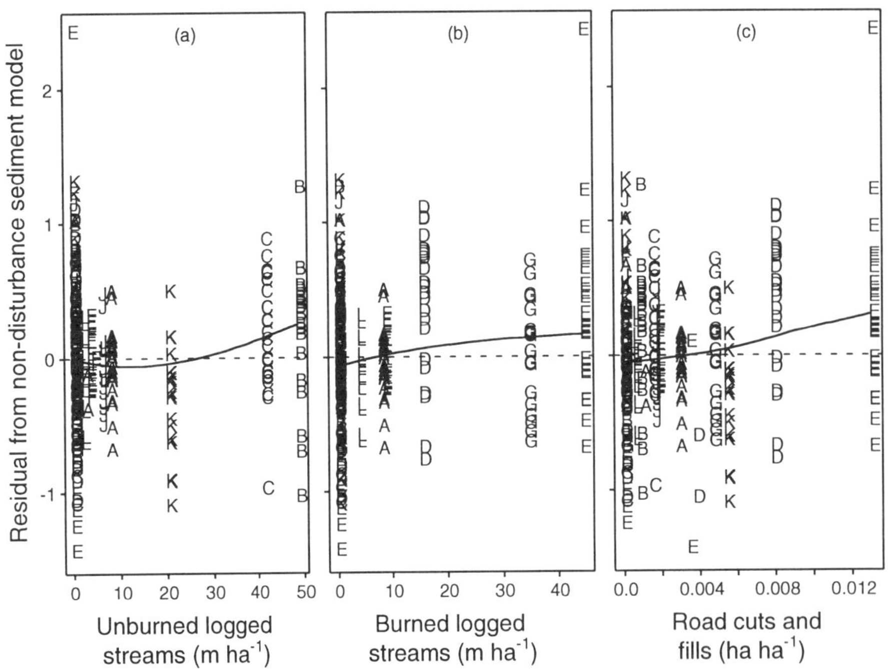

Figure 19. Relation between sediment load residuals and disturbance per unit watershed area. Curves are fit by loess method [*Cleveland*, 1979] to least squares residuals from the model: $\log(y_{ij}) = \beta_{0i} + \beta_{1i}\log(y_{(\mathrm{HI})j}) + \beta_2 \Delta q_{ij}^{(1)} + \varepsilon_{ij}$. Disturbance variables shown are (a) length of stream in burned clear-cut areas, (b) length of stream in unburned clear-cut areas, and (c) road cut and fill area. Letters designate watersheds (e.g. G is watershed GIB).

TABLE 8. Increase in residual sum of squares after dropping variables from least squares fit to model (24).

Coefficient	Variable	SS Reduction
β_2	Change in flow	25.33
β_3	Burned stream channel	10.21
β_4	Unburned stream channel	3.51
β_5	Storm size interaction	1.62
β_6	Watershed area interaction	0.62

used in model-fitting. The cross-validations indicate that the sediment models are not likely to predict future sediment loads well, and the associations identified between sediment loads and the disturbance variables in these models may be coincidental.

DISCUSSION

Storm Peaks

The effect of logging second-growth forests on streamflow peaks in Caspar Creek is consistent with the results from studies conducted over the past several decades throughout the Pacific Northwest. That is, the greatest effect of logging on streamflow peaks is to increase the size of the smallest peaks occurring during the driest antecedent conditions, with that effect declining as storm size and watershed wetness increases. However, increases were still apparent even in the largest storm of this study, which had a recurrence interval of 7 years at NFC.

Although the relative increases in peak flows tend to decline as storm size increases, the effects on large storms may still be important when recurrence intervals of a given size peak are considered. The curve for m=2, for example, in Figure 29 shows the increase in peak needed to reach a size that formerly had twice the recurrence interval, based on a curve fitted to the 28-year pre-logging partial duration series at NFC. Equivalently these are the increases necessary to halve the recurrence interval of the peaks that would result from the increased flow regime. Under such a flow regime, the frequency of large peaks of a given size would double, roughly doubling the geomorphic work performed on the channel. For comparison, the increased peak flows observed in this study (Figure 13) have been included in Figure 29, assuming unit-area flow frequencies in the tributaries are the same as at NFC. Although the variability is very great, it appears that the average observed increases in clear-cuts are great enough to roughly halve the recurrence intervals for storm sizes greater than 0.004 $m^3s^{-1}ha^{-1}$ (return periods longer than 0.5 years). Average observed increases in partly cut watersheds were smaller.

Accounting for the amount of watershed disturbance, there was no evidence that either storm peaks or the logging effect on peaks was related to watershed size. Peaks in the smallest drainages tended to have greater responses to logging than in larger watersheds, but this was because the smaller watersheds had greater proportions disturbed. That is the typical pattern because Forest Practice Rules and economics usually limit the amount of intense activity occurring within any given watershed in any year. Therefore, it is possible for entire small first-order watersheds to be logged within a single year. However, as the size of the watershed increases, a smaller proportion of the watershed is likely to be logged in any given year. In the largest watersheds, harvesting may be

TABLE 9. Maximum likelihood parameter estimates for suspended sediment load model {(24),(17),(18)}, excluding β_{0i} and β_{1i}. p_N is normal probability value for H_0: $\beta = 0$. Control is HI, the mean sediment load from watersheds HEN and IVE.

Parameter	Effect	Estimate	Standard Error	p_N
β_2	Change in flow	1.3276	0.1609	<0.0001
β_3	Stream length, burned	0.0376	0.0057	<0.0001
β_4	Stream length, unburned	0.0204	0.0053	0.0001
β_5	Storm size interaction	-0.0051	0.0017	0.0031
β_6	Watershed area interaction	-3.316E-5	1.649E-5	0.0443
θ_1	Correlation intercept	0.6222	0.0846	<0.0001
θ_2	Correlation slope	-3.802E-4	9.218E-5	<0.0001
θ_3	Variance magnitude	1.0841	0.1565	<0.0001
θ_4	Variance shape	-0.2286	0.0338	<0.0001

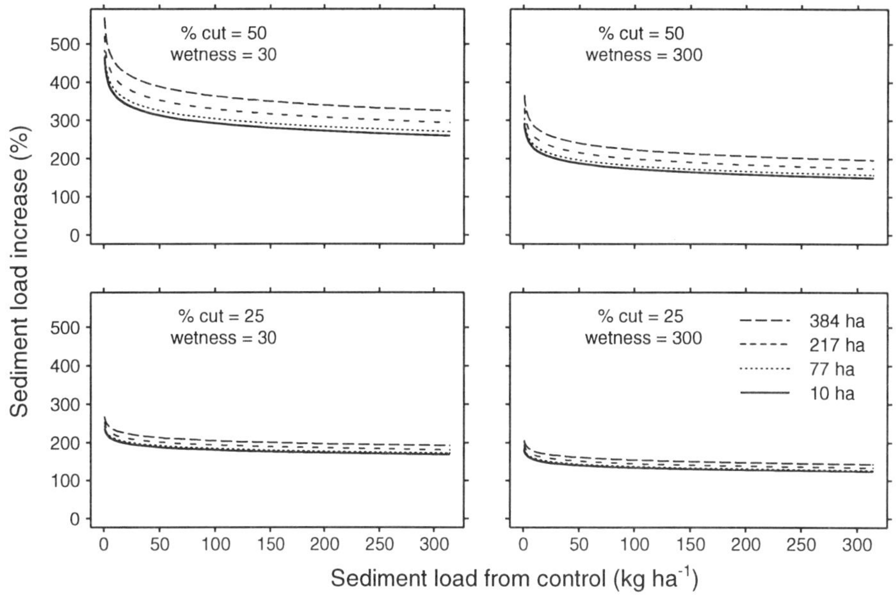

Figure 20. Effect of watershed area on predictions from sediment model {(24),(17),(18)} for two levels of cutting and two levels of antecedent wetness. Watershed areas are those of ARF, FLY, DOL, and BAN (Table 1). Predictions are for first year after cutting with $x_{ij}^{(1)} = x_{ij}^{(2)} = 12 \text{ m ha}^{-1}$.

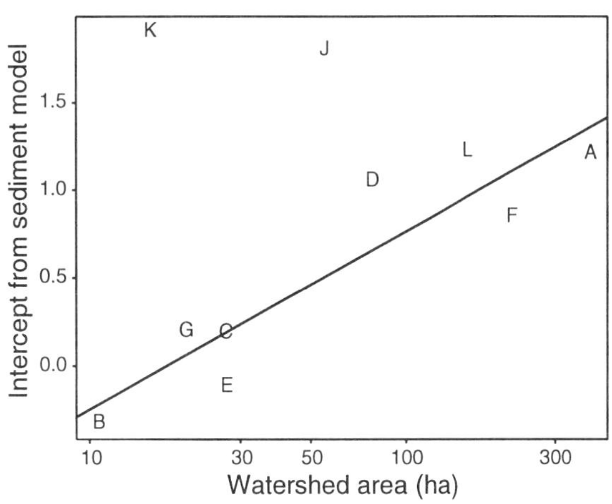

Figure 21. Relation between watershed area and fitted intercepts b_{0i} from model {(24),(17),(18)}, with β_6 fixed at zero. Watersheds JOH (J) and KJE (K) are omitted from regression. Letters designate watersheds (e.g. G is watershed GIB).

spread over decades, within which time the earliest harvested areas will have revegetated.

The data from the streamflow, pipeflow [*Ziemer*, 1992; *Keppeler* and *Brown*, 1998], and soil moisture studies [*Keppeler* et al., 1994] at Caspar Creek all suggest that the peak flow response to logging is related to a reduction in vegetative cover. Reducing vegetative cover, in turn, reduces transpiration and rainfall interception. Since little soil moisture recharge occurs during the spring and summer growing season at Caspar Creek, large differences in soil moisture can develop between logged and unlogged watersheds by late summer because of differences in evapotranspiration. For example, by late summer, a single mature pine tree in the northern Sierra Nevada depleted soil moisture to a depth of about 6 m and to a distance of 12 m from the trunk [*Ziemer*, 1968]. This single tree transpired about 88 m^3 more water than the surrounding logged area, equivalent to about 180 mm of rainfall over the affected area. In the South Fork of Caspar Creek, the largest changes in peak streamflow after logging were found to be for the first storms after lengthy dry periods [*Ziemer*, 1981]. Similarly, after logging the North Fork, there was a strong interaction between the proportion of the area logged and watershed wetness that explained differences in streamflow peaks.

Evaporation of rainfall intercepted by the forest canopy can result in a substantial reduction in the amount of water that reaches the ground. Preliminary measurements at Caspar Creek suggest that average rainfall interception is about 20% of gross winter rainfall. Studies elsewhere have also reported that a large portion of annual rainfall is intercepted and evaporated from the forest canopy. For example, Rothacher [1963] reported that under dense Douglas-fir stands in the Oregon Cascades, canopy

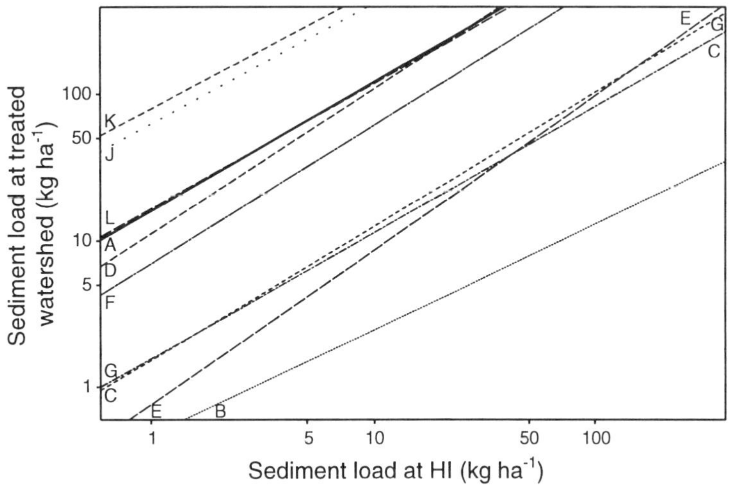

Figure 22. Regression lines for each watershed based on intercepts b_{0i} and slopes b_{1i} of sediment model {(24),(17),(18)}, with β_6 fixed at zero. Letters designate watersheds (e.g. G is watershed GIB).

interception loss averaged 24% of gross summer precipitation and 14% gross winter precipitation. Percentage interception losses are greatest during low-intensity rainfall interspersed with periods of no rain. As with transpiration, rainfall interception can contribute to important differences in antecedent conditions between logged and unlogged watersheds. And during the large high-intensity storms that result in large streamflow peaks, rainfall interception is still important; about 18% of the rainfall from a 96-mm 24-hour storm was intercepted by the forest canopy at Caspar Creek. Differences in interception loss between logged and unlogged areas probably explain most of the observed increases in the larger winter peaks, when transpiration is at its annual minimum.

Road construction and logging were not applied as separate treatments in this study. And, because they are correlated, it is difficult to distinguish their effects statistically. However, soil compaction from roads and timber harvest represents only 3.2% of the North Fork watershed and ranges from 1.9% to 8.5% for the tributary watersheds. Further, roads, landings, and skid-trails in the

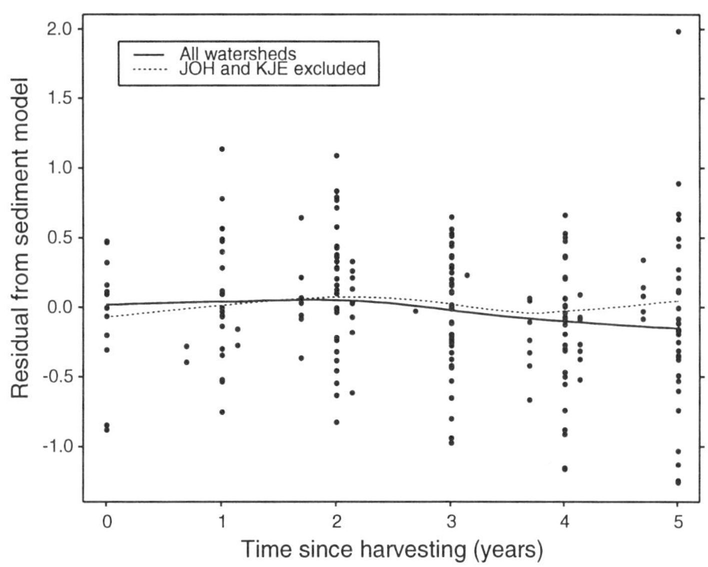

Figure 23. Relation between residuals from sediment model {(24),(17),(18)} and time after logging. Curves are fit by loess method [*Cleveland*, 1979], with and without the anomalous watersheds JOH and KJE.

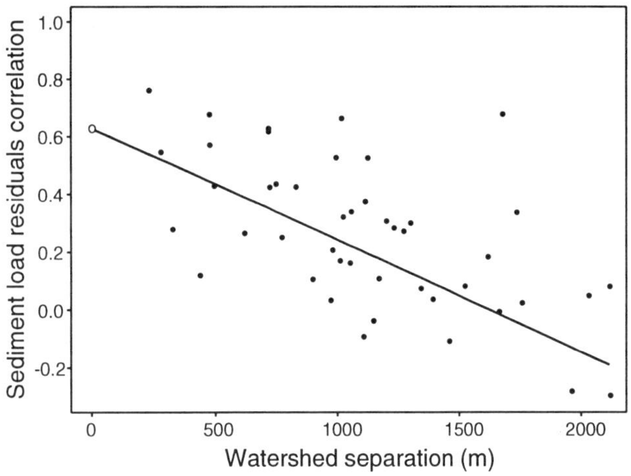

Figure 24. Relation between sediment residuals correlation and distance between watershed centroids. Residuals are from maximum likelihood fit to sediment model {(24),(17),(18)}. Curve depicts equation (17), with estimated parameters $\hat{\theta}_1$ and $\hat{\theta}_2$.

North Fork are all located near the ridges and well away from any streams. Consequently, roads, soil compaction, and overland flow probably did not produce important changes in peak flow response of the North Fork watersheds. The recovery rate of about 8% per year for storm peaks supports the hypothesis that changes in peak flows are largely controlled by changes in vegetation.

Storm Runoff Volume

Analogous to the storm peaks model, the model for storm flow volumes showed that flow increases could be largely explained by the proportion of a watershed logged, an antecedent wetness index, and time since logging. Logging probably impacted both storm peaks and flow volumes via the same mechanisms: reduction of rainfall interception and transpiration.

Suspended Sediment Loads

The most important explanatory variable identified by the sediment models was in-creased volume of streamflow during storms after logging. This result is not unexpected because, after logging, increased storm flows in the treated watersheds provide additional energy to deliver and transport available sediment and perhaps to generate additional sediment through channel and bank erosion.

Whereas individual watersheds show trends indicating increasing or decreasing sediment loads, there is no overall pattern of recovery apparent in a trend analysis of the residuals from the model (Figure 23). This is in contrast with the parallel model for storm flow volume, and suggests that some of the sediment increases are unrelated to flow increases.

Other variables found to be significant, depending on the control watersheds used, were road cut and fill area and length of unbuffered stream channel, particularly in burned areas. One must be cautious about drawing conclusions about cause and effect when treatments are not randomly assigned to experimental units and replication is limited. Increases in sediment load in one or two watersheds can create associations with any variable that happens to have higher values in those watersheds, whether or not those variables are physically related to the increases. In this study, the contrast in response was primarily between watershed KJE, where sediment loads decreased, versus watersheds BAN, CAR, DOL, EAG, and GIB. Watershed KJE was unburned and also had the smallest amount of unbuffered stream of all the cut units. Watersheds EAG and GIB were burned and had the greatest amount of unbuffered stream in burned areas. Watershed EAG experienced the largest sediment increases and also had the greatest proportion of road cut and fill area. EAG was not unusually high in road surface area, and the larger road cut and fill area in EAG reflects roads that are on steeper terrain than in the other cut units.

Road systems would typically be expected to account for much of the sediment. During storm events frequent

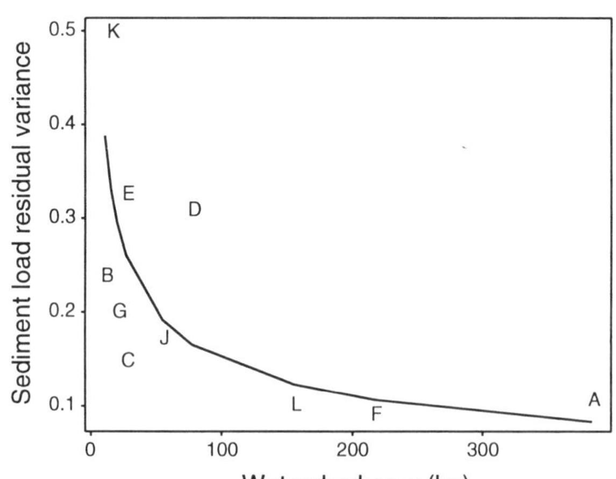

Figure 25. Relation between variance of sediment residuals and watershed area. Residuals are from maximum likelihood fit to model {(24),(17),(18)}. Curve depicts equation (18) with estimated parameters $\hat{\theta}_3$ and $\hat{\theta}_4$. Letters designate watersheds (e.g. G is watershed GIB).

TABLE 10. Maximum likelihood parameter estimates for suspended sediment load model {(25),(17),(18)}, excluding β_{0i} and β_{1i}. p_N is normal probability value for H_0: $\beta = 0$. Control is HIM, the mean sediment load from watersheds HEN, IVE, and MUN.

Parameter	Effect	Estimate	Standard Error	p_N
β_2	Flow increase (log ratio)	1.3564	0.1414	0.0000
β_3	Road cut and fill area	107.11	13.071	0.0000
β_4	Watershed area interaction	-0.1822	0.0872	0.0367
θ_1	Correlation intercept	0.6848	0.0643	0.0000
θ_2	Correlation slope	-3.949E-4	7.618E-5	0.0000
θ_3	Variance magnitude	1.1839	0.1473	0.0000
θ_4	Variance shape	-0.2330	0.0290	0.0000

cutbank failures and culvert blockages along the pre-existing North Fork perimeter all-season road (dating back more than half a century) resulted in drainage diversions and sediment input to North Fork tributaries both before and after logging. But there is little field evidence of sediment delivery from the *new* spur roads in the North Fork watershed. In an inventory of failures greater than 7.6 m^3, only 8 of 96 failures, and 1,686 of 7,343 m^3 of erosion were related to roads and none were associated with the new roads. Based on 129 random erosion plots [*Rice*, 1996; *Lewis*, 1998] in the North Fork, the road erosion in EAG was 9.3 m^3ha^{-1}, compared to 34.5 m^3ha^{-1} for KJE and 16.6 m^3ha^{-1} for all roads in the North Fork. Thus it seems that the appearance of road cuts and fills in the model resulted from a spurious correlation. The *new* roads were relatively unimportant as a sediment source in the North Fork, probably because of their generally stable locations on upper hillslopes far from stream channels, the use of outsloping and frequent rolling-dips (drains), and negligible rainy season use.

Field evidence suggesting that unbuffered stream channels contributed to suspended sediment loads is more consistent. Channel reaches subjected to intense broadcast burns showed increased erosion from the loss of woody debris that stores sediment and enhances channel roughness. Annual surveys evaluating bank stability,

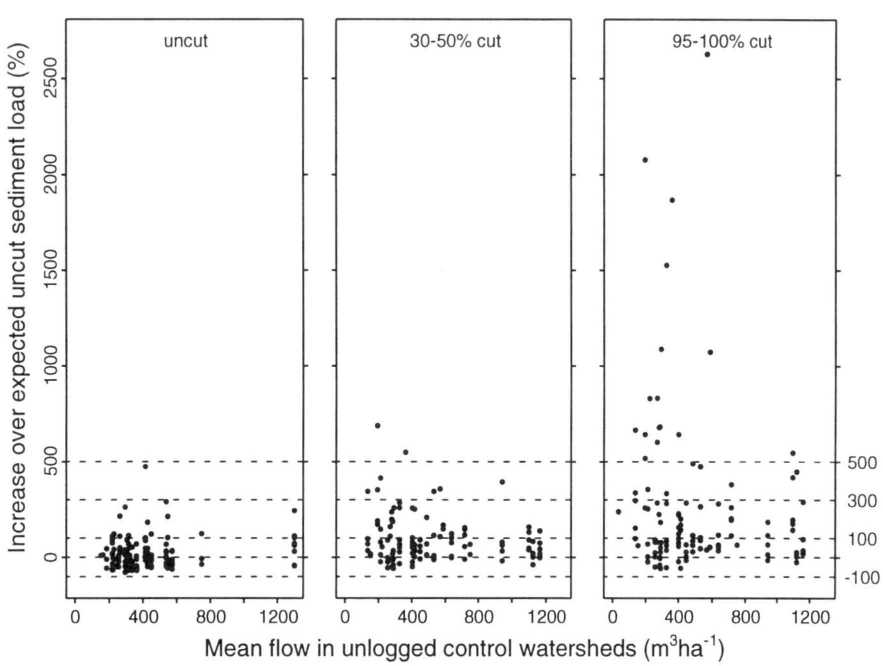

Figure 26. Percentage increase over expected uncut storm sediment load as related to mean of storm runoff volume in HIM control watersheds for uncut (before treatment), partly (30-50%) clear-cut, and (95-100%) clear-cut watersheds. Bias-corrected predictions are from model {(25),(17),(18)} with disturbance set to zero.

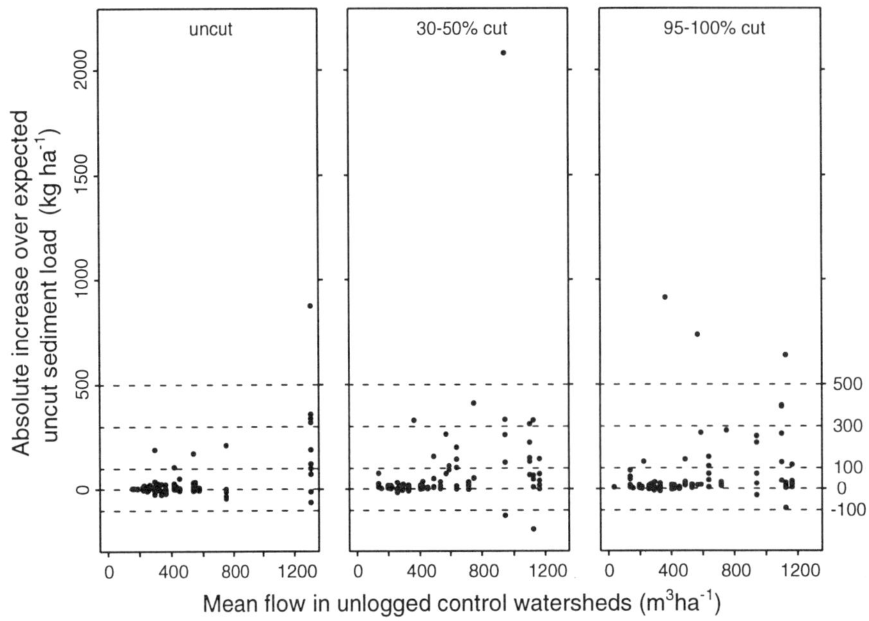

Figure 27. Absolute increase over expected uncut storm sediment load as related to mean of storm runoff volume in HIM control watersheds for uncut (before treatment), partly (30-50%) clear-cut, and (95-100%) clear-cut watersheds. Bias-corrected predictions are from model {(25),(17),(18)} with disturbance set to zero.

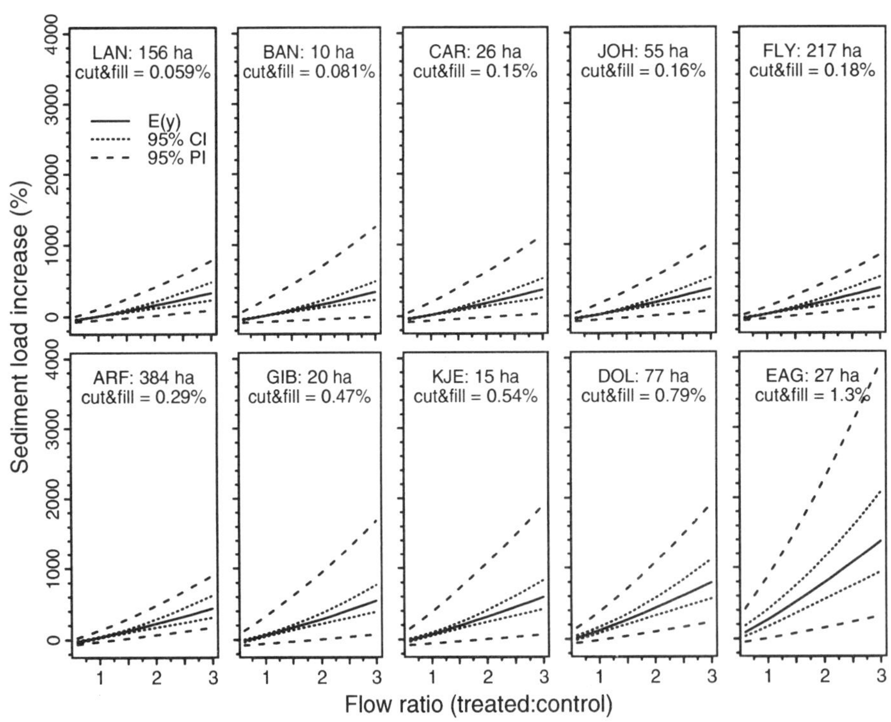

Figure 28. Predictions of sediment load as a function of flow ratio ($\Delta q_{ij}^{(2)}$) based on sediment model {(25),(17),(18)}, with area interaction term for cumulative impacts (β_4) fixed at zero. Expected increases and 95% confidence (CI) and prediction (PI) intervals are shown for each treated watershed, ordered by proportion of the watershed occupied by road cuts and fills.

TABLE 11. Percentage and absolute departures from annual (sum of storms) sediment load predicted from HIM control. Parenthesized values omit outlier in middle frame of Figure 27.

	Uncut	30-50% Clearcut	95-100% Clearcut
Mean (%)	35	73 (67)	212
Median (%)	15	52	109
Mean (kg ha^{-1} yr^{-1})	65	263 (180)	262
Median (kg ha^{-1} yr^{-1})	1	46	59

vegetative cover, and sediment storage potential suggest the greatest sediment production and transport potential existed in the burned channel reaches. Bank disturbances from timber falling and yarding were evident in the unburned channels, but slash and residual woody debris provided both potential energy dissipation and sediment storage sites for moderating sediment transport. Increased flows, accompanied by soil disruption and burning in headwater swales, may have accelerated channel headward expansion and soil pipe enlargements and collapses observed in watershed KJE [*Ziemer*, 1992] and in EAG, DOL, and LAN.

Based on 175 random 0.08-ha erosion plots in harvest areas [*Rice*, 1996; *Lewis*, 1998] in the North Fork, total erosion after logging in the burned watersheds EAG and GIB was 153 m^3ha^{-1} and 77 m^3ha^{-1}, respectively, higher than all other watersheds. Total erosion for the unburned clear-cut watersheds BAN, CAR, and KJE averaged 37 m^3ha^{-1}. These figures include estimates of sheet erosion, which is difficult to measure and may be biased towards burned areas because it was easier to see the ground where the slash had been burned. About 72% of EAG and 82% of GIB were judged to be thoroughly or intensely burned, and the remainder was burned lightly or incompletely. It is unknown how much of this hillslope erosion was delivered to stream channels, but the proportion of watershed burned was not a useful explanatory variable for suspended sediment transport. A plausible conclusion is that only burned areas in or adjacent to stream channels contributed appreciable amount of sediment to the streams.

The inventory of failures greater than 7.6 m^3 identified windthrow as another fairly important source of sediment. Of failures greater than 7.6 m^3, 68% were from windthrow. While these amounted to only 18% of the failure volume measured, 91% of them were within 15 m of a stream, and 49% were in or adjacent to a stream channel. Because of the proximity of windthrows to streams, sediment delivery from windthrow would be expected to be high. Windthrows are also important as contributors of woody debris to these channels, and play a key role in pool formation. Because woody debris traps sediment in transport, the net effect of windthrow on sediment transport can be either positive or negative. Woody debris inputs into the channel have been unusually high in the years since logging, partly because of a number of severe windstorms and partly because of the buffer strip design [*Reid* and *Hilton*, 1998]. While this has led to substantial bank cutting and channel reworking, the bulk of the increased sediment loads after logging watersheds BAN, CAR, EAG, and GIB has not yet reached the main stem stations FLY and ARF, much of it having been stored in reaches affected by blowdown [*Lisle* and *Napoletano*, 1998].

Cumulative effects. We have considered three types of information that the sediment models provide about the cumulative effects of logging activity on (unit area) suspended sediment loads. Keep in mind that the response being considered in all these questions is the suspended sediment load per unit watershed area for a given storm

TABLE 12. Apparent and cross-validated RMSE for model predictions.

Data Omitted	Data Predicted	Model: Peaks[a]	Model: Volume[b]	Model: Sed (HI)[c]	Model: Sed (HIM)[d]
None	All	0.1589	0.1426	0.4584	0.5046
10% at random	All	0.1633	0.1483	0.4900	0.5238
None	Post-treatment	0.1654	0.1560	0.4644	0.5094
10% at random	Post-treatment	0.1692	0.1623	0.4948	0.5291
Systematic by station	Post-treatment	0.1739	0.1676	0.6966	0.6724

[a] model {(10),(16),(18)}, HI control, $\beta_7 = 0$

[b] model {(10),(17),(18)}, HI control, $\beta_3^{(2)} = 0$

[c] model {(24),(17),(18)}, HI control

[d] model {(25),(17),(18)}, HIM control

TABLE 13. Regression slope of observed versus predicted response.

Data Omitted	Data Predicted	Model			
		Peaks[a]	Volume[b]	Sed (HI)[c]	Sed (HIM)[d]
None	All	1.0039	1.0103	1.0012	0.9986
10% at random	All	1.0028	1.0047	0.9920	0.9947
None	Post-treatment	1.0077	1.0103	0.9921	0.9651[e]
10% at random	Post-treatment	1.0085	1.0020	0.9825	0.9611[e]
Systematic by station	Post-treatment	1.0014	0.9998	0.8601[f]	0.8775[f]

[a] model {(10),(16),(18)}, HI control, $\beta_7 = 0$

[b] model {(10),(17),(18)}, HI control, $\beta_3^{(2)} = 0$

[c] model {(24),(17),(18)}, HI control

[d] model {(25),(17),(18)}, HIM control

[e] $0.01 < p < 0.05$ for one-sided test of H_0: slope=1 (H_A: slope < 1)

[f] $p < 10^{-6}$ for one-sided test of H_0: slope=1 (H_A: slope < 1)

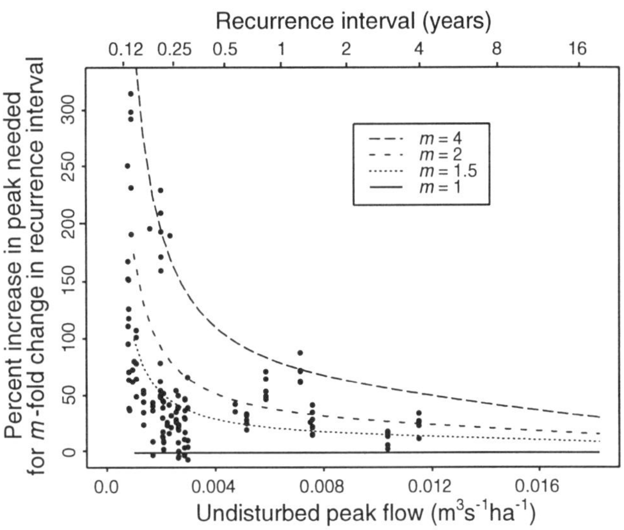

Figure 29. The curve shows the percentage increase in peak flow necessary to reach a size that formerly had 1 to 4 times the recurrence interval. The data points are from Figure 13c, representing the observed percentage increases in storm peak flow (based on the HI control, plotted on the abscissa) in 95-100% clear-cut watersheds.

event and that watershed area was used in the model to represent distance downstream.

Question 1. Were the effects of multiple disturbances additive in a given watershed? This question may be answered partly by looking at the forms of the storm flow and sediment models. Analyses of residuals and covariance structures provide good evidence that the models are appropriate for the data, including the use of a logarithmic response variable. A logarithmic response implies a multiplicative effect for predictors that enter linearly and a power function for predictors that enter as logarithms. The flow response to logged area in model (10) is multiplicative, and the sediment response to flow increases in models (24) and (25) is a power function because Δq (equations (11), (12)) is equivalent to the log of a ratio. We next examine how much these relations differ from an additive relationship in the range of data we observed.

Consider $E(r_{ij})$, the expected value of the ratio between an observation and its expectation in an unlogged condition. From equations (9) and APPENDIX C, equations (35) and (36),

$$E(r_{ij}) = \exp\left[D_{ij} T_{ij}\right] \qquad (26)$$

where $T_{ij} = \beta_4 + \beta_5 \log(y_{C\cdot j}) + \beta_6 \log(w_j) + \beta_7 a_i$. The expected effect of combining two simultaneous disturbances D_1 and D_2 is

$$E(r_{1+2}) = \exp\left[\left(D_1 + D_2\right) T_{ij}\right] = E(r_1)E(r_2) \qquad (27)$$

where $E(r_1) = \exp[D_1 T_{ij}]$ and $E(r_2) = \exp[D_2 T_{ij}]$ are the expected effects of the individual disturbances. The combined effect departs most from additive when $E(r_1) = E(r_2)$. For example, disturbances that individually would result in 10% and 30% increases in the response produce a combined increase of 43% ($1.10 \times 1.30 = 1.43$), while disturbances that individually would result in 20% increases, produce a greater combined increase of 44% ($1.20 \times 1.20 = 1.44$). If the disturbances were additive the combined increase would be a 40% increase in either case. For more than two disturbances, the departures from additivity can be somewhat greater. In general, multiple disturbances that have a combined effect of r on the response under a multiplicative model will result in a minimum increase of $\log(r)$ in the response under an additive model, where r is defined in the sense of r_{ij} above.

(This results from a mathematical limit as the number of equal-magnitude disturbances contributing to the effect r becomes large.)

In the storm flow data, only the main-stem gaging stations received waters from multiple disturbances. The maximum observed increase in storm flow on any main stem gaging station was 118%, but 8 out of 10 increases were under 40% and the median increase was just 16%. Taking the logarithms of 2.18, 1.40, and 1.16, we find that multiple disturbances that could produce these increases in a multiplicative model would produce minimum increases of 78%, 34%, and 15%, respectively, under an additive model. Therefore, in the range of most of the data (increases less than 40%) the disturbance effect on storm flow is approximately additive.

Now we can evaluate the additivity of the disturbance effect on sediment load, since this is expressed mainly through Δq. For this evaluation we fit model {(25),(17),(18)}, but fixing the parameters involving road cuts and fills at zero. Under this model, analogously to equation (26) for the flow model, the expected value of the ratio between an observation and its expectation in an unlogged condition is given by

$$E(r_{ij}) = \exp\left[\beta_2 \Delta q_{ij}^{(2)}\right] = \exp\left[\beta_2 \log\left(\frac{y_{ij}}{y_{Cj}}\right)\right] = \left(\frac{y_{ij}}{y_{Cj}}\right)^{\beta_2} \tag{28}$$

The ratio of y_{ij} and y_{Cj}, the unit area flow volumes in storm j from the treated and control watersheds, is an expression of the increased flow related to tree removal. A plot of equation (28) using the maximum likelihood estimate of 1.514 for β_2 passes through (1,1) and is very nearly linear in the range $0.82 \le y_{ij}/y_{Cj} \le 1.92$, which includes 95% of the observations on the main-stem stations. It follows that the effect of flow on suspended sediment is approximately additive for stations which receive waters from multiple logging units. For example, a flow ratio of 1.40 corresponds to a 66% increase in sediment load, while a flow ratio of 1.80 corresponds to a 143% increase in sediment load. An additive flow effect would produce an increase of 66 + 66 = 132% in sediment load, not much less than 143%. Examples of smaller flow ratios deviate from additivity even less than this example.

So, in the range of data we observed, the effect of disturbance on flow is approximately additive, and the effect of flow on sediment loads is approximately additive. In summary, the mathematical approach indicates that the combined effect of multiple disturbances on sediment loads is very similar to the sum of the effects of the individual disturbances.

Question 2. Were downstream changes greater than would be expected from the proportion of area disturbed? This question was addressed by testing the coefficients of terms formed from the product of disturbance and watershed area. If the coefficient of this term were positive, it would imply that the effect of a given disturbance proportion increases with watershed size. The interactions of those disturbance measures that had explanatory utility in the sediment models were considered, including road cut and fill area and length of unbuffered stream channels. None of the product terms were found to have coefficients significantly greater than zero, indicating that suspended load increases were not disproportionately large in larger watersheds. To the contrary, the sum of the observed sediment loads at the four main-stem stations were all within 25% of the sum of the loads predicted for undisturbed watersheds (Table 7). Channel cross-section measurements indicate 1040 metric tons of net filling in the main stem during the post-logging period [*Lisle* and *Napolitano*, 1998]. Much of the logging-related sediment from the tributaries has apparently been deposited in the main stem, especially in reaches affected by blowdowns and in alluvial bars near tributary confluences, and therefore has not reached downstream gages.

There is, however, one subwatershed where this second type of cumulative effect may be occurring. Watershed DOL, only 36% cut, includes the 100% cut watershed EAG, yet the percentage sediment increases have been similar (269% at DOL versus 238% at EAG). Several mechanisms appear to be responsible for the unexpectedly high loads at DOL. In the incised lower reach, bank failures and channel widening have occurred. In addition, a major stream diversion caused by a windthrow resulted in the formation of a major gully eroding 87 m^3 directly into the stream. Sediment is also being released from behind decaying logs that were placed in the channel for skidding by oxen during historic logging. Finally, all these processes would have been augmented by the increased storm flows that followed modern logging.

Question 3. Were sediment loads in the lower watershed elevated to higher levels than in the tributaries? Regardless of the control watersheds used, suspended sediment transport per unit watershed area tended to increase downstream before logging (Figure 21). This tendency may reflect a greater availability of fine sediment downstream in lower gradient channels. If unit area sediment loads increase downstream and result in water quality levels of concern with a smaller proportion of watershed disturbance than upstream locations, then cumulative effects may be said to have occurred, in the sense that activities producing acceptable local impacts resulted in impacts that are unacceptable by the same standard downstream.

To the extent that larger watersheds reflect average disturbance rates and therefore have smaller proportions of disturbance than the smallest disturbed watersheds upstream, one might expect sediment loads downstream to increase by less than those in the logged tributaries. In addition, as mentioned before, some of the sediment may be temporarily stored before reaching the lower stations. Indeed, in this study the post-treatment regression lines were much more similar among watersheds than the pretreatment lines, and the main-stem stations no longer transported the highest unit area sediment loads. However, larger watersheds will not necessarily behave the same way. For example, in geographically similar Redwood Creek in northwestern California, two main-stem gaging stations (175 km^2 and 720 km^2) yield higher sediment loads per unit area than three intensively logged tributaries [*Lewis*, 1998].

Cumulative effects considered in this paper were limited to a few hypotheses about water quality that could be statistically evaluated. But cumulative effects can occur in many ways. For example, resources at risk are often quite different in downstream areas, so an activity that has acceptable local impacts might have unacceptable offsite impacts if critical or sensitive habitat is found downstream. Different physical processes also tend to dominate upstream and downstream reaches. Channel aggradation may be the biggest problem downstream, while channel scour may be of concern upstream.

Subwatersheds and KJE anomaly. Analyses of the 5 clear-cut tributaries in the North Fork drainage show suspended load increases at all gaging stations located immediately below clear-cut units except at KJE, where loads have decreased. KJE had the highest pre-logging (1986-1989) unit area sediment loads of any of the tributaries (Figure 22), but, after logging, loads were similar to the other logged tributaries (Figure 17).

Prior to logging, the stream channel above KJE was unique. The KJE channel was an active gully with an abundant supply of sediment and the lowest gradient of any of the tributaries. After logging, the number of small debris jams doubled in the buffered channel above KJE, and further upstream the channel contained a large amount of logging debris and dense vegetative regrowth. Thus, opportunities for temporary sediment storage increased, and net energy available for sediment transport may have decreased, despite moderately increased flows, because of the increased channel roughness. The other tributaries were stable, vegetated, steep channels with limited sediment supplies and relatively low unit area sediment loads prior to logging. In these tributaries the increased sediment introduced by logging was readily transported. While this explanation is speculative, response in sediment transport to a disturbance certainly will vary with channel morphology and the relative availability of sediment and energy.

CONCLUSIONS

The main conclusions from these analyses are:

- Models based upon the proportion of watershed area logged, an antecedent wetness index, time since logging, and the responses in unlogged control watersheds explained 95% of the variation in the logarithms of both storm discharge peaks and volumes. Goodness-of-fit is similar for pre-logging and post-logging data, and cross-validation indicates that the models were not overfit to the data.
- Storm discharge peaks and volumes after extended periods with little or no precipitation increased up to 300% and 400% respectively, but most increases were below 100%.
- The effect of logging on storm discharge peaks and volumes declines with increasing regional antecedent wetness, as indexed by a decay function of prior runoff at a control watershed. However, even under the wettest conditions of the study, increases in storm runoff from clear-cut watersheds averaged 23% for peaks and 27% for volumes.
- Relative increases in storm discharge peaks and volumes decline with storm size but were positive even in the largest storms of the study period.
- Average increases in annual storm runoff were 58% from 95-100% clear-cut watersheds and 23% from 30-50% clear-cut watersheds.
- Recovery rates in the first 4-7 years after logging are estimated to be 8% per year for peak flows and 9% per year for storm flow volumes.
- Effects of multiple disturbances on storm discharge peaks and volumes are approximately additive, and there is little evidence for magnification of effects downstream.
- Reduction in rainfall interception and transpiration by forest vegetation is the probable cause of increased storm discharge peaks and volumes following logging.
- Annual sediment loads increased 123-269% in the tributaries, but, at main-stem stations, increased loads were detected only in small storms and had little effect on annual sediment loads. At the North Fork weir, an

increase of 89% was caused mainly by a landslide in an ungaged tributary that enters just above the weir.

- Much of the increased sediment load in North Fork tributaries was related to increased storm flow volumes. With flow volumes recovering as the forest grows back, flow-related increases in sediment load are expected to be short-lived.
- The effects of multiple disturbances on suspended loads in a watershed were approximately additive.
- In general, downstream suspended load increases were no greater than would be expected from the proportion of area disturbed. In one tributary, increased flows evidently impacted the channel in an uncut area downstream by mobilizing stored sediment and aggravating bank instabilities, but most of the increased sediment produced in the tributaries was apparently stored in the main stem and has not yet reached the main-stem stations.
- Before logging, sediment loads on the main stem were higher than on most tributaries. This was no longer the case after logging, apparently because sediment exported from tributaries was deposited at temporary storage sites, and smaller proportions of downstream watersheds were disturbed.
- Sediment increases in North Fork tributaries probably could have been reduced by avoiding activities that denude or reshape the banks of small drainage channels.
- Sediment loads are affected as much by channel conditions (e.g. organic debris, sediment storage sites, channel gradient, and width-to-depth ratio) as by sediment delivery from hillslopes.

APPENDIX A. Notation Used in the Text

a_i	Drainage area of watershed i
b_i	Estimate of parameter β_i
c_{ij}	Proportion of watershed i logged in water years prior to that of storm j, and
c'_{ij}	Proportion of watershed i logged prior to storm j but in the same water year
D_{ij}	Some measure of disturbance per unit area in watershed i at storm j
$d_{i_1i_2}$	Distance between centroids of watersheds i_1 and i_2
K	Number of parameters estimated in a model
n	Number of observations used in an analysis
p_{ij}	True (unknown) percentage change in response of watershed i in storm j as a result of treatment
$\tilde{p}_{ij}$	"Observed" percentage change in response of watershed i in storm j based on a comparison of y_{ij} and $\hat{y}'_{ij}$
p_0	Percentage change in response, given an arbitrary vector $\mathbf{x}_0$ of explanatory variables
p_N	Significance level of a hypothesis test based on the normal distribution
$\Delta q_{ij}^{(1)}$	Residual from the flow model (3) containing only β_{0i} and β_{1i}
$\Delta q_{ij}^{(2)}$	Difference between the logarithms of flow in the treated and control watersheds
$\Delta q_{ij}^{(3)}$	Predicted change after logging in the logarithm of storm flow from eqn (10)
t_{ij}	Area-weighted mean cutting age (number of summers passed) in watershed i for areas logged in water years preceding that of storm j
w_j	Wetness index at start of storm j
$x_{ij}^{(1)}, x_{ij}^{(2)}$	Generic measures of unit area disturbance in watershed i at storm j
$\mathbf{x}_0$	Arbitrary vector of explanatory variables
y_{ij}	Unit area response at treated watershed i in storm j
y_{cj}	Unit area response at control watershed in storm j
y'_{ij}	Unknown response at watershed i, if it had been left untreated, in storm j
$\hat{y}'_{ij}$	Estimate of y'_{ij}
β_{0i}, β_{1i}	Location parameters (slope and intercept) to be estimated for each watershed i
$\beta_0^{(1)}, \beta_0^{(2)}$	Parameters used to model β_{0i} as a function of a_i
$\rho_{i_1i_2}$	Correlation between ε_{i_1j} and ε_{i_2j}
$\sigma_{i_1}, \sigma_{i_2}$	Standard deviations of ε_{i_1j} and ε_{i_2j}
ε_{ij}	Error or deviation of y_{ij} from model at treated watershed i in storm j
$\varepsilon_{i_1j}, \varepsilon_{i_2j}$	Errors for watersheds i_1 and i_2 in storm j
θ_i	Parameter in covariance model
$\hat{\theta}_i$	Estimate of parameter θ_i

APPENDIX B. Likelihood Function and Gradient

The model for the mean response can be written

$$\mathbf{u} = E(\mathbf{y}) = f(\beta) \tag{29}$$

where $\mathbf{y}$ is an $n\times1$ response vector and β is a $p\times1$ vector of unknown parameters. The error, $\mathbf{e} = \mathbf{y} - \mathbf{u}$, is modelled as a multivariate normal variable depending on q parameters:

$$\begin{aligned} \mathbf{e} &\sim \mathbf{N}(0,\Sigma) \\ \Sigma &= \mathbf{G}(\theta) \end{aligned} \tag{30}$$

where Σ is the $n \times n$ covariance matrix of **e** depending on θ, a $q \times 1$ vector of unknown parameters. The elements of Σ are paramaterized by equations (15)-(18). The likelihood function and its logarithm are

$$\begin{aligned} L &= (2\pi)^{-n/2} |\Sigma|^{-1/2} \exp\left[-\tfrac{1}{2}(\mathbf{y}-\mathbf{u})^T \Sigma^{-1}(\mathbf{y}-\mathbf{u})\right] \text{ and} \\ \ell &= \log(L) \\ &= -\tfrac{n}{2}\log(2\pi) - \tfrac{1}{2}\log|\Sigma| - \tfrac{1}{2}(\mathbf{y}-\mathbf{u})^T \Sigma^{-1}(\mathbf{y}-\mathbf{u}) \end{aligned} \quad (31)$$

respectively, where $|\Sigma|$ is the determinant of Σ. The gradient consists of the partial derivatives of ℓ with respect to β and θ:

$$\begin{aligned} \mathbf{grad} &= \left(\frac{\partial \ell}{\partial \beta_1}, \cdots, \frac{\partial \ell}{\partial \beta_p}, \frac{\partial \ell}{\partial \theta_1}, \cdots, \frac{\partial \ell}{\partial \theta_q} \right) \\ \frac{\partial \ell}{\partial \beta_i} &= \frac{\partial \mathbf{u}^T}{\partial \beta_i} \Sigma^{-1}(\mathbf{y}-\mathbf{u}), \qquad i = 1, \cdots, p \\ \frac{\partial \ell}{\partial \theta_j} &= -\frac{1}{2} tr\left(\Sigma^{-1} \frac{\partial \Sigma}{\partial \theta_j} \right) \qquad j = 1, \cdots, q \\ &\quad + \frac{1}{2}(\mathbf{y}-\mathbf{u})^T \Sigma^{-1} \frac{\partial \Sigma}{\partial \theta_j} \Sigma^{-1}(\mathbf{y}-\mathbf{u}) \end{aligned} \quad (32)$$

in which $tr(\cdot)$ refers to the trace (sum of the diagonal elements) of the matrix. The partial derivatives, $\partial \mathbf{u}^T/\partial \beta_i$ and $\partial \Sigma / \partial \theta_j$, are model-specific and can be derived from equations (10) and (14)-(18).

APPENDIX C. An Unbiased Estimator, and Confidence and Prediction Intervals for Percentage Change in Response

Let y_0 be the response given an arbitrary predictor vector $\mathbf{x}_0$ and let y_0' be the unknown response for the same storm assuming the watershed were undisturbed. A prediction interval is sought for $p_0 = 100[y_0/E(y_0') - 1]$, the percentage change in response, and an unbiased estimator and confidence interval are sought for its expectation, $E(p_0)$. It will be convenient to obtain the unbiased estimator and confidence interval first. Since $\log(y_0)$ and $\log(y_0')$ are assumed to be normally distributed,

$$\begin{aligned} E(y_0) &= \exp\left[E(\log(y_0)) + \tfrac{1}{2}\sigma^2\right] \quad \text{and} \\ E(y_0') &= \exp\left[E(\log(y_0')) + \tfrac{1}{2}\sigma^2\right] \end{aligned} \quad (33)$$

Let us denote the ratio of the actual response to its expected undisturbed value by

$$r_0 = \frac{y_0}{E(y_0')} \quad (34)$$

Its expectation is

$$\begin{aligned} E(r_0) &= \frac{E(y_0)}{E(y_0')} \\ &= \frac{\exp\left[E(\log(y_0)) + \tfrac{1}{2}\sigma^2\right]}{\exp\left[E(\log(y_0')) + \tfrac{1}{2}\sigma^2\right]} \\ &= \exp\left[f_0(\beta)\right] \end{aligned} \quad (35)$$

where, for the runoff models (10),

$$\begin{aligned} f_0(\beta) = &\left[\left(1 - \beta_2(t_0 - 1)\right)c_0 + \beta_3^{(k)} c_0'\right] \times \\ &\left[\ \beta_4 + \beta_5 \log(y_{c'0}) + \beta_6 \log(w_0) + \beta_7 a_0 \right] \end{aligned} \quad (36)$$

Since $\mathbf{b}$, the vector of estimates for β, is asymptotically distributed normal, we have that $f_0(\mathbf{b})$ is asymptotically distributed normal with $E[f_0(\mathbf{b})] = f_0(\beta)$ and unknown variance σ_*^2 [*Bishop* et al., 1975]. In shorthand, $f_0(\mathbf{b}) \sim N\left(f_0(\beta), \sigma_*^2\right)$ for large samples. The variance σ_*^2 may be approximated using the delta method [*Bishop* et al., 1975]:

$$\tilde{\sigma}_*^2 = \sum_{i=1}^{p} \sum_{j=1}^{p} \frac{\partial f_0}{\partial b_i} \frac{\partial f_0}{\partial b_j} Cov\left[b_i, b_j\right] \quad (37)$$

The covariances are estimated by the elements of the inverted information matrix [*McCullagh* and *Nelder*, 1989]. The information matrix is the negative of the matrix of second derivatives (Hessian) of ℓ with respect to the parameters, β and θ.

Let us introduce an estimator $\hat{r}_0 = \exp\left[f_0(\mathbf{b}) - \tfrac{1}{2}\sigma_*^2\right]$. Its expected value is

$$\begin{aligned} E(\hat{r}_0) &= \exp\left(-\tfrac{1}{2}\sigma_*^2\right) E\left\{\exp\left[f_0(\mathbf{b})\right]\right\} \\ &= \exp\left(-\tfrac{1}{2}\sigma_*^2\right) \exp\left\{E\left[f_0(\mathbf{b})\right] + \tfrac{1}{2}\sigma_*^2\right\} \\ &= \exp\left\{E\left[f_0(\mathbf{b})\right]\right\} \\ &= \exp\left\{f_0(\beta)\right\} \\ &= E(r_0) \end{aligned} \quad (38)$$

Hence $\hat{r}_0$ is an asymptotically unbiased estimator for $E(r_0)$, and $100(\hat{r}_0 - 1)$ is an asymptotically unbiased estimator for $100(E(r_0) - 1) = E(p_0)$. In practice, because σ_* is unknown, we replace it with $\tilde{\sigma}_*$ in the expression for $\hat{r}_0$.

Next we will compute a confidence interval for $E(r_0)$, and convert it to a confidence interval for $E(p_0)$. A 100(1-α)% confidence interval for $f_0(\beta)$ is defined by the joint probability

$$\Pr\left\{\begin{matrix} f_0(\mathbf{b}) - z_{\alpha/2}\sigma_* \le f_0(\beta) \\ f_0(\mathbf{b}) + z_{\alpha/2}\sigma_* \ge f_0(\beta) \end{matrix}\right\} = 1-\alpha \qquad (39)$$

where $z_{\alpha/2}$ is the $\alpha/2$ cutoff point of the standard normal distribution. Applying the monotone transformation **exp** to both inequalities yields a confidence interval for $E(r_0)$:

$$\Pr\left\{\begin{matrix} \exp\left[f_0(\mathbf{b}) - z_{\alpha/2}\sigma_*\right] \le E(r_0) \\ \exp\left[f_0(\mathbf{b}) + z_{\alpha/2}\sigma_*\right] \ge E(r_0) \end{matrix}\right\} = 1-\alpha \qquad (40)$$

Noting that $E(p_0) = 100(E(r_0) - 1)$, the above confidence interval is readily transformed into a confidence interval for $E(p_0)$.

$$\Pr\left\{\begin{matrix} 100\left[\exp\left(f_0(\mathbf{b}) - z_{\alpha/2}\sigma_*\right) - 1\right] \le E(p_0) \\ 100\left[\exp\left(f_0(\mathbf{b}) + z_{\alpha/2}\sigma_*\right) - 1\right] \ge E(p_0) \end{matrix}\right\} = 1-\alpha \qquad (41)$$

Since σ_* is unknown, we replace it with $\tilde{\sigma}_*$.

Finally, we will compute a prediction interval for r_0, and convert it to a prediction interval for p_0. Using model (10) and (33), we find

$$\begin{aligned} r_0 &= \frac{y_0}{E(y_0')} \\ &= \frac{\exp\left[\beta_0 + \beta_1 \log(y_{C0}) + f_0(\beta) + \varepsilon_0\right]}{\exp\left[\beta_0 + \beta_1 \log(y_{C0}) + \frac{1}{2}\sigma^2\right]} \qquad (42) \\ &= \exp\left[f_0(\beta) + \varepsilon_0 - \tfrac{1}{2}\sigma^2\right] \end{aligned}$$

Since $\varepsilon_0 \sim N(0, \sigma^2)$ and, asymptotically, $f_0(\mathbf{b}) \sim N\left(f_0(\beta), \sigma_*^2\right)$, and they are independent random variables, it follows that $f_0(\mathbf{b}) - \varepsilon_0 \sim N\left(f_0(\beta), \sigma_*^2 + \sigma^2\right)$. Thus

$$\Pr\left\{\begin{matrix} f_0(\mathbf{b}) - z_{\alpha/2}\left(\sigma_*^2 + \sigma^2\right)^{\frac{1}{2}} \le f_0(\beta) + \varepsilon_0 \\ f_0(\mathbf{b}) + z_{\alpha/2}\left(\sigma_*^2 + \sigma^2\right)^{\frac{1}{2}} \ge f_0(\beta) + \varepsilon_0 \end{matrix}\right\} = 1-\alpha \qquad (43)$$

Subtracting $0.5\sigma^2$ in both inequalities and applying the monotone transformation, **exp**, converts the right-hand terms to r_0, yielding the following $100(1-\alpha)$ prediction interval for r_0:

$$\exp\left(f_0(\mathbf{b}) - \tfrac{1}{2}\sigma^2 \pm z_{\alpha/2}\left(\sigma_*^2 + \sigma^2\right)^{\frac{1}{2}}\right) \qquad (44)$$

which is readily transformed to a $100(1-\alpha)$ prediction interval for p_0:

$$100\left[\exp\left(f_0(\mathbf{b}) - \tfrac{1}{2}\sigma^2 \pm z_{\alpha/2}\left(\sigma_*^2 + \sigma^2\right)^{\frac{1}{2}}\right) - 1\right] \qquad (45)$$

Since σ_* and σ are unknown, we replace them with $\tilde{\sigma}_*$ and $\hat{\sigma} = \hat{\theta}_3 a_0^{\hat{\theta}_4}$, where a_0 is the watershed area.

Confidence and prediction intervals for sediment models (24) and (25) are similar, but $f_0(\mathbf{b})$ is replaced by the linear functions $g_0(\mathbf{b})$ and $h_0(\mathbf{b})$, respectively, where

$$\begin{aligned} g_0(\mathbf{b}) = {} & \beta_2 \Delta q_0^{(1)} + \beta_3 x_0^{(1)} + \beta_4 x_0^{(2)} \\ & + \beta_5 (x_0^{(1)} + x_0^{(2)}) \log(y_{(\mathrm{HI})0}) \qquad (46) \\ & + \beta_6 (x_0^{(1)} + x_0^{(2)}) a_0 \end{aligned}$$

$$\text{and} \quad h_0(\mathbf{b}) = \beta_2 \Delta q_0^{(2)} + \beta_3 x_0 + \beta_4 x_0 a_0 \qquad (47)$$

Since these functions are linear, the delta method yields the exact variance, but, as before, the covariance matrix of **b** must be estimated from the observed information matrix, so σ_*^2 is still only known approximately.

Acknowledgments. This research was a result of a cooperative effort by the California Department of Forestry and Fire Protection and the USDA Forest Service, Pacific Southwest Research Station. The study design was a result primarily of the efforts of Raymond Rice, who was the Principal Investigator until his retirement in 1989. Robert Thomas designed the SALT sampling algorithm and Rand Eads designed the hardware/software interface used in the field implementation. In addition, the authors are grateful to Dave Thornton and the many individuals, too numerous to mention here, who spent thousands of hours in the field (often during hazardous storm conditions), in the laboratory, and in the office to ensure the best possible data quality.

Detailed hydrologic and climatic data collected at Caspar Creek between 1963 and 1997 are available on compact disk from the authors.

REFERENCES

Bailey, E. H., W. P. Irwin, and D. L. Jones, Franciscan and related rocks, and their significance in the geology of western California, *Calif. Div. Mines Geol. Bull., 183*, 177 pp., 1964.

Baskerville, G. L., Use of logarithmic regression in the estimation of plant biomass, *Can. J. Forest Res., 2*, 49-53, 1972.

Beschta, R. L., M. R. Pyles, A. E. Skaugset, and C. G. Surfleet, Peak flow responses to clear-cutting and roads in small and large basins, western Cascades, Oregon: An alternative analysis, unpublished report supplied by author, Department of Forest Engineering, Oregon State University, Corvallis, 1997.

Bishop, Y. M., S. E. Fienberg, and P. W. Holland, *Discrete Multivariate Analysis: Theory and Practice*, MIT Press, Cambridge, Mass., 557 pp., 1975.

Brown, G. W. and J. T. Krygier, Clear-cut logging and sediment production in the Oregon Coast Range, *Water Resour. Res.* 7(5):1189-1198, 1971.

Burnham, K. P. and D. R. Anderson, *Model Selection and Inference: A Practical Information-Theoretic Approach*, Springer, New York, 353 pp., 1998.

Chow, G. C., A test of equality between sets of observations in two linear regressions, *Econometrica, 28*, 591-605, 1960.

Cleveland, W. S., Robust locally weighted regression and smoothing scatterplots, *J. Amer. Stat. Assoc., 74*, 829-836, 1979.

Copas, J. B., Regression, prediction and shrinkage, *J. R. Statist. Soc. B, 45*(3): 311-354, 1983.

Efron, B. and R. J. Tibshirani, *An Introduction to the Bootstrap*, Chapman and Hall, New York, 436 pp., 1993.

Fredricksen, R. L., Sedimentation after logging road construction in a small western Oregon watershed, in *Proceedings of the Federal Inter-Agency Sedimentation Conference, 1963*, Agricultural Research Service Misc. Pub. No. 970, pp. 56-59, 1963.

Grant, G. E. and A. L. Wolff, Long-term patterns of sediment transport after timber harvest, western Cascade Mountains, Oregon, USA, *IAHS Publ. 203*, 31-40, 1991.

Hansen, M. H. and W. N. Hurwitz, On the theory of sampling from finite populations, *Ann. Math. Stat., 14*, 333-362, 1943.

Harr, R. D., Forest practices and streamflow in western Oregon, *Gen. Tech. Rep. PNW-49*, 18 pp., For. Serv., U.S. Dep. Of Agric., Portland, Oreg., 1976.

Harr, R. D., W. C. Harper, J. T. Krygier, and F. S. Hsieh, Changes in storm hydrographs after road building and clear-cutting in the Oregon Coast Range, *Water Resour. Res., 11*, 436-444, 1975.

Harr, R. D., R L. Fredriksen, and J. Rothacher, Changes in streamflow following timber harvest in southwestern Oregon, *Res. Paper PNW-249*, 22 pp., For. Serv., U.S. Dep. Of Agric., Portland, Oreg., 1979.

Harris, D. D., Hydrologic changes after logging in two small Oregon coastal watersheds, *U.S. Geol. Surv. Water Supply Paper 2037*, 31 pp., 1977.

Henry, N., Overview of the Caspar Creek watershed study, in Proceedings, Conference on Coastal Watersheds: The Caspar Creek Story, *Gen. Tech. Rep. PSW-168*, pp. 1-9, For. Serv., U.S. Dep. Of Agric., Albany, Calif., 1998.

Hewlett, J. D. and A. R. Hibbert, Factors affecting the response of small watersheds to precipitation in humid areas, in *Forest Hydrology*, edited by W. E. Sopper and H. W. Lull, pp. 275-290, Pergamon, New York, 1967.

Jones, J. A. and G. E. Grant, Peak flow responses to clear-cutting and roads in small and large basins, western Cascades, Oregon, *Water Resour. Res., 32*, 959-974, 1996.

Keppeler, E. T., R. R. Ziemer, and P. H. Cafferata, Changes in soil moisture and pore pressure after harvesting a forested hillslope in northern California, in *Effects of Human-induced Changes on Hydrologic Systems*, edited by R. A. Marston, and V. R. Hasfurther, pp. 205-214, American Water Resources Association, Herndon, Virginia, 1994.

Keppeler, E. T. and D. Brown, Subsurface drainage processes and management impacts, in Proceedings, Conference on Coastal Watersheds: The Caspar Creek Story, *Gen. Tech. Rep. PSW-168*, pp. 25-34, For. Serv., U.S. Dep. Of Agric., Albany, Calif., 1998.

Keppeler, E. T. and R. R. Ziemer, Logging effects on streamflow: water yields and summer flows at Caspar Creek in northwestern California. *Water Resour. Res., 26*(7), 1669-1679, 1990.

Leaf, C. F., Sediment yields from central Colorado snow zone, *J. Hydraulics Division, ASCE Proc. 96*(HY1): 87-93, 1970.

Lewis, J., Turbidity-controlled suspended sediment sampling for runoff-event load estimation, *Water Resour. Res., 32*(7), 2299-2310, 1996.

Lewis, J., Evaluating the impacts of logging activities on erosion and suspended sediment transport in the Caspar Creek watersheds, in Proceedings, Conference on Coastal Watersheds: The Caspar Creek Story, *Gen. Tech. Rep. PSW-168*, pp. 55-69, For. Serv., U.S. Dep. Of Agric., Albany, Calif., 1998.

Lisle, T. E. and M. B. Napolitano, Effects of recent logging on the main channel of North Fork Caspar Creek, in Proceedings, Conference on Coastal Watersheds: The Caspar Creek Story, *Gen. Tech. Rep. PSW-168*, pp. 81-85, For. Serv., U.S. Dep. Of Agric., Albany, Calif., 1998.

McCullagh, P. and J. A. Nelder, *Generalized Linear Models*, 2nd edn., Chapman & Hall, London, 1989.

Megahan, W. F., J. G. King, and K. A. Seyedbagheri, Hydrologic and erosional responses of a granitic watershed to helicopter logging and broadcast burning, *Forest Science, 41*(4), 777-795, 1995.

Miller, R. G., *Simultaneous Statistical Inference*, 2nd edn., Springer-Verlag, New York, 299 pp., 1981.

Miller, E. L., Sediment yield and storm flow response to clear-cut harvest and site preparation in the Ouachita Mountains, *Water Resour. Res. 20*(4): 471-475, 1984.

Mood, A. M., F. A. Graybill, and D. C. Boes, *Introduction to the Theory of Statistics*, 3rd edn., McGraw-Hill, New York, 564 pp., 1974.

Napolitano, M. B., Persistence of historical logging impacts on channel form in mainstem North Fork Caspar Creek, in

Proceedings, Conference on Coastal Watersheds: The Caspar Creek Story, *Gen. Tech. Rep. PSW-168,* pp. 97-101, For. Serv., U.S. Dep. Of Agric., Albany, Calif., 1998.

Norick, N. X., SALT (selection at list time) sampling: a new method and its application in forestry. M.A. Thesis, 50 pp., Univ. of California, Berkeley, 1969.

Olive, L. J. and W. A. Rieger, Assessing the impact of land use change on stream sediment trans-port in a variable environment, *IAHS Publ. 203,* 73-81, 1991.

O'Loughlin, C. L., L. K. Rowe, and A. J. Pearce, Sediment yield and water quality responses to clearfelling of evergreen mixed forests in western New Zealand, *IAHS Publ. 130,* 285-292, 1980.

Plamondon, A. P., Logging and suspended sediments input in small streams: concentration, origin, and duration, presented at *XVII IUFRO World Congress, Kyoto, Japan,* Sept. 1981.

Rao, C. R. *Linear Statistical Inference and its Applications,* 2nd edn., Wiley and Sons, New York, 625 pp., 1973.

Reid, L. M. and S. Hilton, Buffering the buffer, in Proceedings, Conference on Coastal Watersheds: The Caspar Creek Story, *Gen. Tech. Rep. PSW-168,* pp. 71-80, For. Serv., U.S. Dep. Of Agric., Albany, Calif., 1998.

Rice, R. M., F. B. Tilley, and P. A. Datzman, A watershed's response to logging and roads: South Fork of Caspar Creek, California, 1967-1976. *Res. Pap. PSW-146,* 12 pp., For. Serv., U.S. Dep. Of Agric., Berkeley, Calif.,1979.

Rice, R. M., Sediment delivery in the North Fork of Caspar Creek, unpublished final report, Agreement CA94077, Calif. Dep. of For. and Fire Prot., Sacramento, 1996.

Rothacher, J., Net precipitation under a Douglas-fir forest. *Forest Science, 4,* 423-429, 1963.

Rothacher, J., Regimes of streamflow and their modification by logging, in *Proceedings of the Symposium on Forest Land Use and Stream Environment,* edited by J. T. Krygier and J. D. Hall, pp. 55-63, Oregon State Univ., Corvallis, 1971.

Rothacher, J., Does harvest in west slope Douglas-fir increase peak flow in small streams? Pacific Northwest Forest and Range Experiment Station, *Res. Pap. PNW-163,* 13 pp., For. Serv., U.S. Dep. of Agric., Portland, Oreg., 1973.

Shumway, R. H., *Applied Statistical Time Series Analysis,* Prentice Hall, Englewood Cliffs, N. J., 379 pp., 1988.

Statistical Sciences, S-Plus Version 3.3 for Unix, StatSci, a division of MathSoft, Inc., Seattle, 1995.

Thomas, R. B., Estimating total suspended sediment yield with probability sampling, *Water Resour. Res., 21*(9), 1381-1388, 1985.

Thomas, R. B., Piecewise SALT sampling for estimating suspended sediment yields, *Gen. Tech. Rep., PSW-114,* 11 pp., For. Serv., U.S. Dep. Of Agric., Berkeley, Calif., 1989.

Thomas, R. B. and J. Lewis, A comparison of selection at list time and time-stratified sampling for estimating suspended sediment load, *Water Resour. Res. 29*(4), 1247-1256, 1993.

Thomas, R. B. and J. Lewis, An evaluation of flow-stratified sampling for estimating suspended sediment loads, *J. Hydrol., 170,* 27-45, 1995.

Thomas, R. B. and W. F. Megahan., Peak flow responses to clear-cutting and roads in small and large basins, western Cascades, Oregon: A second opinion, *Water Resour. Res., 34*(12):3393-3403, 1998.

Wilson, A. L., When is the chow test UMP? *The American Statistician, 32*(2), 66-68, 1978.

Wright, K. A., K. H. Sendek, R. M. Rice, and R. B. Thomas, Logging effects on streamflow: storm runoff at Caspar Creek in northwestern California, *Water Resour. Res., 26*(7), 1657-1667, 1990.

Ziemer, R. R., Soil moisture depletion patterns around scattered trees, *Res. Note PSW-166,* 13 pp., For. Serv., U.S. Dep. of Agric., Berkeley, Calif., 1968.

Ziemer, R. R., Stormflow response to roadbuilding and partial cutting in small streams of northern California, *Water Resour. Res., 17*(4), 907-917, 1981.

Ziemer, R. R., Effect of logging on subsurface pipeflow and erosion: coastal northern California, USA., *IAHS Publ. 209,* 187-197, 1992.

Ziemer, R. R., Flooding and stormflows, in Proceedings, Conference on Coastal Watersheds: The Caspar Creek Story, *Gen. Tech. Rep. PSW-168,* pp. 15-24, For. Serv., U.S. Dep. Of Agric., Albany, Calif., 1998.

Jack Lewis, Pacific Southwest Research Station, Redwood Sciences Laboratory, 1700 Bayview Drive, Arcata, California, U.S. 95521

Sylvia R. Mori, Pacific Southwest Research Station, P.O. Box 245, Berkeley, California, U.S. 94701

Elizabeth Keppeler, Pacific Southwest Research Station, Caspar Creek Watershed Laboratory, 802 N. Main St., Ft. Bragg, California, U.S. 95437

Robert Ziemer, Pacific Southwest Research Station, Redwood Sciences Laboratory, 1700 Bayview Drive, Arcata, California, U.S. 95521

Simulating the Effects of Forest Roads on Watershed Hydrology

Mark S. Wigmosta and William A. Perkins

Pacific Northwest National Laboratory, Richland, Washington

The Distributed Hydrology-Soil-Vegetation Model (DHSVM) was enhanced to represent the interception and redirection of surface and subsurface flow by road cuts and stream channels. Two channel networks are imposed in vector form on the DHSVM grid: one representing roads, the other representing streams. Impacts of the road network on channel flows are modeled explicitly; road location and geometry, along with soil moisture, determine the timing and volume of intercepted water, while road drainage characteristics and culvert placement determine the flow path and travel time to the channel system. The active road drainage/channel network may expand and contract as grid cell water tables rise and fall below their channel beds. The model was applied to a 10-km^2 experimental watershed to study the influence of road network design on streamflow characteristics. It was demonstrated that even where the road network produces minor changes in streamflow at the basin outlet, subsurface flow interception by roads could alter significantly streamflow in tributary channels. The impact of the road network on channel flows was shown to vary during a storm based on the road design, storm characteristics, topography, local geology, and soil moisture conditions. The impact occurs earlier in a storm when the road network diverts flow into the drainage rather than out of it. When flows are redirected into a channel, the road contributing area quickly increases the current total contributing area for the channel, causing a proportional increase in discharge. If the road design allows some water to pass under the road, there may be only minor impacts early in the storm with increasing impact as watertables rise and subsurface flow begins to be intercepted. When flows are redirected out of the natural drainage the impact is not felt until the channel contributing area extends upslope to the road system.

INTRODUCTION

The impact of timber harvesting and road construction on stream flow is a key concern in the management of forested lands. Timber harvesting may result in greater snow accumulation and more rapid melt compared to unharvested stands. Evapotranspiration may be reduced as trees are replaced by shrubs. The increased snowmelt volumes and suppressed evapotranspiration can result in locally higher soil moisture, raised water tables, and local changes in runoff production. These local changes may be propagated down-gradient through surface and subsurface water movement. Runoff production on road surfaces and the interception of subsurface flow by road ditches and cut-banks may also alter flow production and routing within a watershed. To alter streamflow substantially, subsurface

Land Use and Watersheds: Human Influence on Hydrology and Geomorphology in Urban and Forest Areas
Water Science and Application Volume 2, pages 127-143

flow must first be intercepted by the logging road, and then routed rapidly to the stream channel. Intercepted water that discharges from mid-slope culverts and reinfiltrates before reaching a stream channel will have less impact on peak flows than intercepted water routed directly to the channel network. The cumulative impact on receiving water bodies from changes in hillslope water balances and flow routing depends on local hydrology, the spatial distribution of timber harvesting, the road network, and the channel system.

Numerous studies have attempted to evaluate the effects of road construction and vegetation removal on basin hydrologic response, including annual water yield [Rothacher, 1970], peak stream flows [King and Tennyson, 1984; Harr et al., 1975; Harr and McCorison, 1979; Jones and Grant, 1996; Thomas and Megahan, 1998], and summer low flows [Rothacher, 1965]. The impacts of logging roads have been evaluated through a number of field studies [Megahan, 1972, 1983; Wemple et al., 1996; Wemple, 1998]. While these studies provide insight into the effects of forest treatments in the experimental basins, the findings can not be readily extrapolated beyond the study sites. Spatially distributed hydrologic modeling provides another means for exploring the influence of forest practices on streamflow, and potentially extending these findings to new areas.

The Distributed Hydrology Soil Vegetation Model (DHSVM) is a physically-based distributed parameter hydrologic model that allows the simulation of runoff processes in forested, mountainous environments [Wigmosta et al., 1994]. Recent applications of the model have been used to evaluate the impacts of forest harvesting and road construction on watersheds in western Washington [Bowling and Lettenmaier, 1997; Lamarche and Lettenmaier, 1998; Bowling and Lettenmaier, this issue] and western British Columbia [Wigmosta and Perkins, 1997]. Here we provide an overview of DHSVM, describe model representation of road/channel networks, and present results from an application to the Carnation Creek watershed, British Columbia.

THE DISTRIBUTED HYDROLOGY-SOIL-VEGETATION MODEL

DHSVM is described in detail elsewhere [Wigmosta et al., 1994; Wigmosta and Perkins, 1997; Storck et al. 1998; Wigmosta and Lettenmaier 1994, 1999]. The essential features of DHSVM are discussed before describing the enhancements that permit modeling the influence of constructed roads on hillslope hydrology. DHSVM provides an integrated representation of watershed processes at the spatial scale described by digital elevation model (DEM) data. It includes a multi-layer canopy model for evapotranspiration, energy balance models for canopy snow interception and ground snowpack processes, a multi-layer rooting zone model, and subsurface, surface, and channel flow modules. The modeled landscape is divided into computational grid cells centered on DEM elevation nodes (Figure 1). Digital elevation data are used to model topographic controls on absorbed shortwave radiation, precipitation, air temperature, and down-slope water movement. At each time step, the model provides a simultaneous solution to the energy and water balance equations for every grid cell in the watershed. Each cell exchanges water with its adjacent neighbors, resulting in a three-dimensional redistribution of surface and subsurface water across the landscape.

Each grid cell is assigned a surface cover type and soil properties. These properties may vary spatially throughout the basin. Local meteorological conditions (precipitation, air temperature, etc.) are specified for each cell at a specified distance above the canopy. Solar radiation and wind speed are attenuated through each canopy based on cover density and Leaf Area Index (LAI), providing separate values for each canopy layer, and the soil. Evapotranspiration from vegetation is modeled using a multi-canopy representation with each canopy partitioned into wet and dry areas. Canopy snow interception and release are modeled as described by Storck et al. [1995]. Snow interception is governed by the current rate of snowfall, air temperature, the amount of intercepted snow, and the canopy LAI. Snow can be removed from the canopy by wind, melt, or mass release to the ground. Wind removal increases linearly with wind speed and the current amount of intercepted snow. Snowmelt can cause either liquid water to drip from the canopy or initiate a mass release. Snow accumulation and melt below the canopy or in the open are simulated using a two-layer energy- and mass-balance model [Storck et al., 1995] that explicitly incorporates the effects of topography and vegetation cover on the energy exchange at the snow surface.

The vertical movement of unsaturated soil moisture is simulated using a multi-layer soil rooting zone model. Discharge is calculated via Darcy's Law assuming a unit hydraulic gradient using the soil vertical unsaturated hydraulic conductivity [Wigmosta et al., 1994]. In each grid cell, percolation from the lowest rooting zone recharges the local water table (Figure 2). Every grid cell is allowed to exchange water with its adjacent neighbors as a function of local hydraulic conditions; thus, a given grid cell will receive water from up-gradient neighbors and discharge to adjacent down-gradient neighbors (Figure 2). If the computed water table in a grid cell is above the

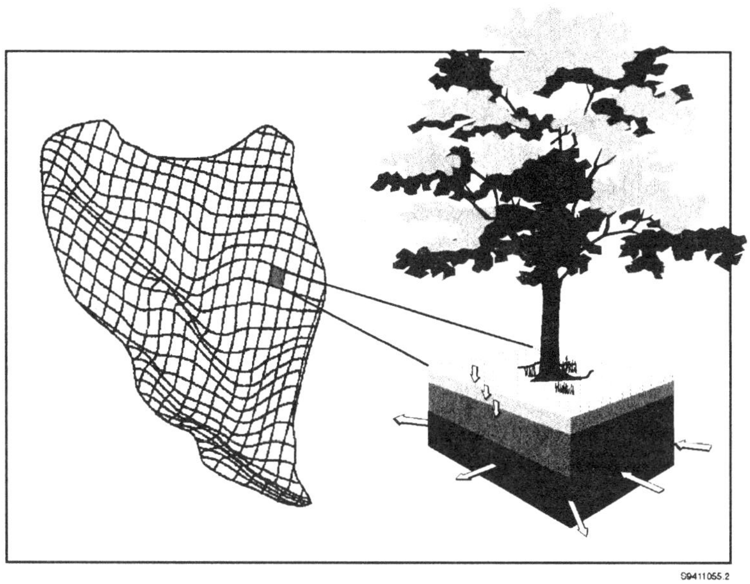

Figure 1. Model representation of a watershed. DEM data are used to model topographic controls on absorbed solar radiation, precipitation, air temperature, and downslope water movement. Grid cells are allowed to exchange water with their adjacent neighbors, resulting in a three-dimensional redistribution of surface and subsurface water across the landscape.

ground surface, the excess volume represents groundwater exfiltration to the surface, which contributes to surface runoff. Rainfall or snowmelt occurring on saturated areas contributes directly to surface runoff.

The drainage network is represented as a series of connected reaches with each reach passing through one or more DEM grid cells (Figure 3). A road reach begins to intercept subsurface flow when grid cell water tables rise above the elevation of the associated road drainage ditches. The road drainage network also receives runoff directly from the surface of inwardly sloped roads. Surface water in roadside ditches is routed through the road drainage network until it reaches a culvert or stream channel. If the road intersects a stream channel, the water is input to the appropriate channel reach and routed through the channel system. The discharge from a culvert without a defined channel is allowed to reinfiltrate as it moves downslope below the culvert. Impacts of the road network on channel flows are modeled explicitly, road location and geometry along with soil moisture determine the volume of water intercepted; road drainage characteristics and culvert placement determine the path and travel time to the channel system. The active road drainage/channel network may expand and contract as grid cell water tables rise and fall below their channel beds. Flow in road drainage ditches and stream channels is routed using either a cascade of linear reservoirs or the Muskingum-Cunge method [Garbrecht and Brunner, 1991].

INTERCEPTION OF SATURATED SUBSURFACE FLOW

Saturated Flow Routing

DHSVM employs a cell-by-cell approach to route saturated subsurface flow [Wigmosta et al., 1994; Wigmosta and Lettenmaier, 1999] using either a kinematic or diffusion approximation. The conceptual model is developed for elevation data on a rectangular grid with cell dimensions Δx and Δy, with the x- and y-axis pointing east and north, respectively. Model grid cells are centered on each elevation point. Directions between a node and its neighbors are assigned the index k and numbered from 0 to 7 in a clockwise direction from north. On steep slopes with thin, permeable soils, hydraulic gradients may be approximated by local ground surface slopes (kinematic assumption). In areas of low relief, hydraulic gradients must be approximated by local water table slopes (diffusion assumption).

The rate of saturated subsurface flow from cell i, j in the k-direction $q_{Si,j,k}$ is given by:

$$q_{Si,j,k} = w_{i,j,k}\beta_{i,j,k}T_{i,j}(z,D) \quad (1)$$

where $w_{i,j,k}$ is the grid cell flow width in the k-direction, $\beta_{i,j,k}$ is the water table slope in the k direction, and

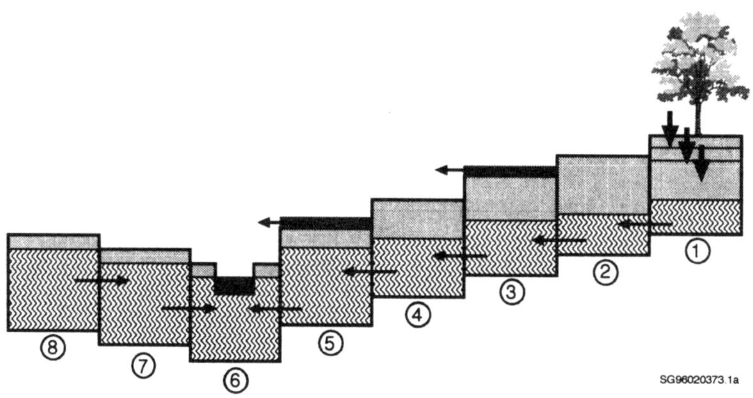

Figure 2. Model representation of downslope water movement to a stream channel. Dark shading represents surface water in the form of overland flow (model cells 3 and 5) or as channel flow (model cell 6). Light shading represents unsaturated soil, while the wave pattern corresponds to saturated soil below the water table. In each grid cell, percolation from the lower rooting zone recharges local grid cell water tables (shown by the downward arrows in cell 1). Grid cells exchange water with adjacent neighbors resulting in the downslope movement of water (horizontal arrows) to stream channels. The stream channel receives subsurface flow when grid cell water tables rise above the elevation of the channel bed (model cell 6). Surface flow may be reinfiltrated in downslope grid cells where the zone of soil saturation is below the ground surface (surface flow from cells 3 and 5 reinfiltrate in cells 4 and 6, respectively).

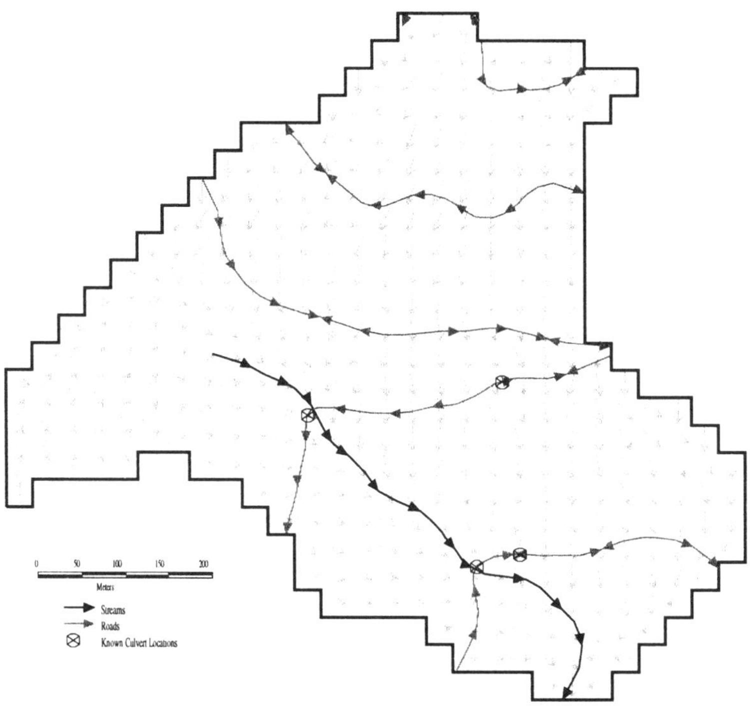

Figure 3. Model representation of hydrologic features showing flow directions for DEM grid cells, road locations and drainage directions, culvert locations, and the stream channel. Water moving downslope coincident with the grid cell flow directions may be intercepted and diverted by the road network. The road and stream networks are represented in vector form as a series of connected reaches; each associated with one or more grid cells. Road drainage flow directions are calculated in ARC/INFO using the road coverage and the DEM. Areas with convergent flow directions are possible low spots in the road drainage network, where water may reinfiltrate or over-top the road. These locations are subject to field verification.

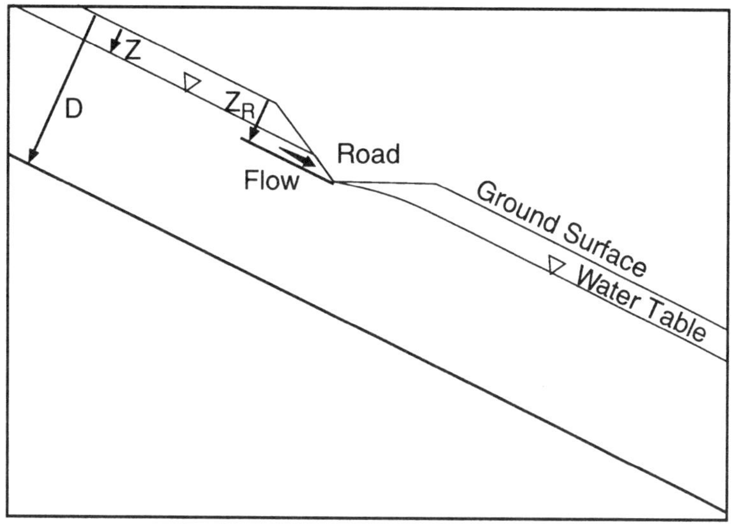

Figure 4. Subsurface flow interception by a logging road in a model grid cell. Flow interception occurs when the grid cell water table rises above the elevation of the road drainage ditch. Subsurface flow below the road-cut continues downslope to adjacent grid cells.

$T_{i,j}(z,D)$ is the grid cell transmissivity. Soil saturated hydraulic conductivity is assumed to decrease exponentially with depth, allowing soil transmissivity in (1) to be calculated as:

$$T_{i,j}(z,D) = \frac{K_{i,j}}{f_{i,j}}\left[\exp\left(-f_{i,j}z_{i,j}\right) - \exp\left(-f_{i,j}D_{i,j}\right)\right] \tag{2}$$

where $K_{i,j}$ is the grid cell saturated hydraulic conductivity at the soil surface, $f_{i,j}$ is a decay coefficient, and $D_{i,j}$ is the grid cell soil thickness.

The total subsurface outflow from a grid cell $Q_{S_{i,j}}$ is equal to the sum of the component flows by (1). The change in saturated zone water volume ($\Delta S_{S_{i,j}}$) over the time step Δt is given by:

$$\Delta S_{Si,j} = \left(\left(Q_{Sin_{i,j}} + R_{ei,j}\right) - \left(Q_{S_{i,j}} + ET_{i,j}\right)\right)\Delta t \tag{3}$$

where $Q_{Sin_{i,j}}$ is the total inflow to cell i, j from up-gradient cells, $R_{ei,j}$ is the current rate of recharge from the unsaturated zone to the water table and $ET_{i,j}$ is the rate of evapotranspiration calculated by the multi-layer canopy model. The change in $z_{i,j}$ over the time step is given by:

$$\Delta z_{i,j} = -\frac{\Delta S_{S_{i,j}}}{\phi_{i,j}A} \tag{4}$$

where $\phi_{i,j}$ is the local effective porosity, and A is the grid cell area (horizontal projection). Negative values of $z_{i,j}$ represent "exfiltration" of subsurface water to the surface, available for overland flow routing.

Flow Interception by Roads

When the depth to the water table is less than the depth of the road cut ($z < z_R$) (Figure 4), saturated subsurface flow will be intercepted by the road at a rate given by:

$$Q_{R_{i,j}} = w_{R_{i,j}}\beta_{i,j}T_{i,j}(z, z_R) \tag{5}$$

where $w_{R_{i,j}}$ is the length of road in the grid cell orthogonal to the cell aspect $\alpha_{i,j}$, $\beta_{i,j}$ is the grid cell slope corresponding to $\alpha_{i,j}$, and $T_{i,j}(z,z_R)$ is the transmissivity of the saturated zone above the road-cut, obtained by substituting z_R for D in (2). $w_{R_{i,j}}$ is given by:

$$w_{R_{i,j}} = \left|\sin\left(\varepsilon - \alpha_{i,j}\right)\right|L_R \tag{6}$$

where L_R is the straight-line road length crossing the grid cell and ε is the aspect of the road. The total flux of subsurface flow passing beneath the road ($Q_{S_{i,j}}$) is obtained using z_R in place of z in (1) when calculating flux components. The change in saturated zone water volume ($\Delta S_{S_{i,j}}$) over the time step Δt is given by:

$$\Delta S_{Si,j} = \left(\left(Q_{Sin_{i,j}} + R_{ei,j}\right) - \left(Q_{S_{i,j}} + Q_{R_{i,j}} + ET_{i,j}\right)\right)\Delta t \tag{7}$$

The change in $z_{i,j}$ over the time step is given by (4) with $A_{i,j}$ equal to the grid cell plan view area less the plan view area of the road. When the water table remains below the road cut ($z \geq z_R$), $A_{i,j}$ is equal to the full grid cell plan view area and $Q_{R_{i,j}} = 0$.

Flow Interception by Stream Channels

A grid cell will contribute water to a stream reach when the grid cell water table rises above the streambed (Figure 2). Subsurface flow will be intercepted by the channel at a rate given by:

$$Q_{C_{i,j}} = 2L_{C_{i,j}}\beta_{C_{i,j}}T_{i,j}(z,z_C) \tag{8}$$

where is z_C is the depth to the channel bed and $L_{C_{i,j}}$ is the length of channel crossing the grid cell. $T_{i,j}(z,z_C)$ is the transmissivity of the saturated zone above the streambed, obtained by substituting z_C for D in (2). The hydraulic gradient is approximated by:

$$\beta_{C_{i,j}} = \frac{z_{C_{i,j}} - z_{i,j}}{0.5w_{C_{i,j}}} \tag{9}$$

where $w_{C_{i,j}}$ is the width of the stream reach. If surface water is available within the cell, it is contributed to the stream reach in the same time interval.

Comparision to Analytic Solutions

A numerical simulation was conducted to evaluate the DHSVM representation of hillslope hydraulics with a road under steady-state and transient conditions. The simulation was designed to test the model using conditions where topography dominates the downslope movement of subsurface flow, i.e. steep hillslopes with thin, permeable soils. Under such conditions the slope of the water table is often assumed parallel to the ground surface and saturated subsurface flow can be approximated using a kinematic wave equation. Beven [1982] presents an analytic solution to the kinematic wave equation for subsurface flow down a plane of constant slope. This solution was modified to account for a road-cut.

DHSVM simulated discharge and water table profiles were compared with the analytic solution to the kinematic wave equation for saturated subsurface flow down a 100-m plane of constant slope ($\beta = 0.4$) containing a mid-slope road cut to bedrock ($z_R = D$ in (5)). The hydraulic gradient in each DHSVM grid cell was set equal to the slope of the plane to allow direct comparison with the kinematic wave solution. Soil properties were taken uniform in the downslope direction with transmissivity expressed as an exponential function of water table depth. Soil properties used in the model comparison were: $K = 2$ m/h, $f = 1.5$, $D = 1.0$ m, and $\phi = 0.25$. A constant rainfall of 0.004 m/h was applied uniformly over the slope for 48 h. DHSVM simulated water table profiles show close agreement with the analytic solution at 12-, 24-, 36-, and 48-h (Figure 5). DHSVM does equally well replicating the subsurface discharge hydrograph.

OVERLAND FLOW ROUTING

Surface runoff is generated in a grid cell when: 1) the input of throughfall and snowmelt, exceeds the user defined infiltration capacity (infiltration excess runoff); 2) throughfall or snowmelt occurs on a saturated cell (saturation excess runoff); or 3) the water table rises above the ground surface (return flow). The downslope movement of surface runoff is done on a cell-by-cell basis similar to the method used for subsurface flow. Outflow in the k-direction $(q_{O_{i,j,k}})$ is given by:

$$q_{o_{i,j,k}} = w_{i,j,k}\, v(n,y)_{i,j}\, y_{i,j} \tag{10}$$

where $v_{i,j}$ is the grid cell flow velocity, $y_{i,j}$ is the grid cell flow depth, and $w_{i,j,k}$ remains are defined in (1). The total outflow ($Q_{O_{i,j}}$) is the sum of component flows from (10). The change in the surface water volume (ΔS_O) over the time step Δt is given by:

$$\Delta S_{O\,i,j} = \left(\left(Q_{O\,in_{i,j}} + P_{i,j} + M_{i,j}\right) - \left(Q_{O_{i,j}} + Q_{I_{i,j}}\right)\right)\Delta t \tag{11}$$

where $Q_{o_{in_{i,j}}}$ is the total amount of overland flow into cell i, j from up-gradient cells, $P_{i,j}$ is the rate of throughfall, $M_{i,j}$ is the rate of snowmelt, and $Q_{I_{i,j}}$ is the soil infiltration rate (a user defined constant for each soil type). The model currently uses a constant velocity of $v = \Delta x/\Delta t$, meaning the volume of overland flow leaving a grid cell over the time step is equal to the surface water storage at the start of the time step. The influence of this approach on simulation results will be discussed later in the text.

CHANNEL AND ROAD DRAINAGE NETWORKS

Flow in road drainage ditches and stream channels is routed using either the Muskingum-Cunge method [Garbrecht and Brunner, 1991] or a cascade of linear reservoirs. The stream and road channel networks comprise of any number of individual segments, each of which will have its own hydraulic parameters. Lateral inflow to a channel segment, from the watershed cells through which it passes, consists of overland flow (Q_O via (10)) and subsurface flow intercepted by roads (Q_R via (5)) or channels (Q_C via (8)). Outflow from a segment may drain to another segment or exit the watershed. Outflow from a road channel segment may also be input back into a watershed cell. This happens only if the road channel

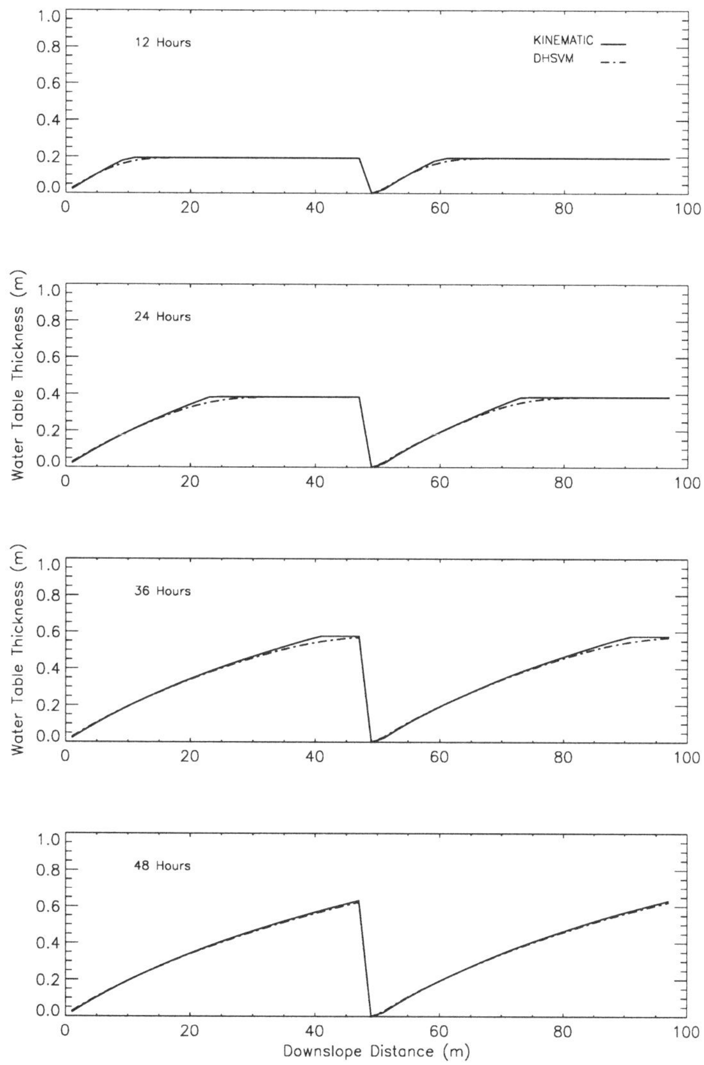

Figure 5. Simulated water table profiles for a plane of constant slope with a mid-slope road cut to bedrock and a uniform recharge rate of 0.004 m/h applied for 48 h. Water tables are show at (a) 12 h; (b) 24 h; (c) 36 h, and (d) 48 h.

segment outflow point has been designated a "sink", or culvert. In which case, the outflow from the channel segment is added to the surface water of the cell, making it available for reinfiltration and overland flow routing through (10).

Muskingum-Cunge Routing

The Muskingum-Cunge method expressed in finite difference notation [Chow et al., 1988], represents outflow from channel segment $s+1$ at time $t+1$ as:

$$Q_{s+1}^{t+1} = C_1 Q_s^{t+1} + C_2 Q_s^t + C_3 Q_{s+1}^t \tag{12}$$

where

$$C_1 = \frac{\Delta t - 2KX}{2K(1-X)+\Delta t},\; C_2 = \frac{\Delta t + 2KX}{2K(1-X)+\Delta t},$$

$$\text{and } C_3 = \frac{2K(1-X)-\Delta t}{2K(1-X)+\Delta t} \tag{13}$$

Equation (12) is applied to each segment individually, using the inflow, Q_s^t and Q_s^{t+1}, from all segments from which it receives flow. In a given time step, the lateral inflow received by all of the cells it passes through is divided equally between the segment's inflow (Q_s^{t+1}) and outflow (Q_{s+1}^{t+1}) in equation (12).

The routing parameters K and X are computed as (Chow et al., 1988)

$$K = \frac{\Delta L}{c_k} = \frac{\Delta L}{\mathrm{d}Q_r/\mathrm{d}A_r} \tag{14}$$

and

$$X = \frac{1}{2}\left(1 - \frac{Q_r}{c_k w_r S_o \Delta L}\right) \tag{15}$$

where ΔL is the length of the channel segment, S_o is the channel slope, and Q_r is a reference discharge. The remaining terms are taken at Q_r where c_k is the kinematic wave celerity, w_r is the water surface width, and A_r is the channel cross-sectional area. Assuming that at Q_r the channel friction slope is approximately equal to S_o, Manning's equation can be used to represent Q_r

$$Q_r = \frac{1}{n} A_r R_r^{2/3} \sqrt{S_o} \tag{16}$$

where n is the Manning roughness coefficient and R_r is the hydraulic radius. For a rectangular channel cross-section

$$c_k = \frac{Q_r}{w_r}\left(\frac{5w_r + 6y_r}{3y_r(w_r + 2y_r)}\right) \tag{17}$$

where y_r is the reference flow depth. The values of K and X are computed in advance for all segments and remain constant over the simulation. For these computations, y_r is assumed to be the normal flow for the segment when the depth is 75% of the bank height. Some checks are used to ensure that the values computed for K and X will not cause instability or gross errors, during the simulation. The stability of (12) requires that $0 \le X \le \frac{1}{2}$. If this condition is not met, X is set to 0 and c_k in (14) taken equal to $Q_r/w_r S_o \Delta L$, which satisfies (15) with X set to 0. Equation (12) can produce negative outflows, especially on a steep falling limb of a hydrograph, if $K < \Delta t/2X$. If the computed K does not satisfy this criterion, K is set equal to $\Delta t/2X$.

Linear Storage Routing

In some situations the Muskingum-Cunge method may exhibit minor instability over small portions of the hydrograph, despite the checks on X and K described above. A relatively simple linear storage routing algorithm is therefore available as an alternative to the Muskingum-Cunge method. The linear storage algorithm has provided satisfactory results when applied to a large range of basin sizes and topographic characteristics. This is the method of routing used in most current applications of DHSVM, including the Carnation Creek application described later.

Each channel reach is treated as a reservoir of constant width with outflow linearly related to storage (V_c)

$$Q = kV_c = kA\Delta L \tag{18}$$

where k is the reach storage parameter (equal to the inverse of the average residence time). The linear storage-discharge relationship in (18) implies a constant flow velocity that can be calculated from Manning's equation using the reference flow depth and hydraulic radius

Equating (16) and (18) yields

$$k = \frac{R_r^{2/3}\sqrt{S_o}}{n\Delta L} \tag{19}$$

Mass balance for the reach in given by:

$$\frac{\mathrm{d}V_c}{\mathrm{d}t} = Q_{in} - kV_c \tag{20}$$

Solving (20) for storage at time $t+1$ yields

$$V_c^{\,t+1} = \frac{Q_{in}}{k} + \left(V_c^{\,t} - \frac{Q_{in}}{k}\right)\exp(-k\Delta t) \tag{21}$$

with the average outflow from the reach given by

$$Q_{out} = Q_{in} - \left(V_c^{\,t+1} - V_c^{\,t}\right)/\Delta t \tag{22}$$

where Q_{in} is the total lateral and upstream inflow to the reach over the time step.

The Muskingum-Cunge and the linear storage routing algorithms both use a reference depth to precalculate routing coefficients. To evaluate the implications of this approach, consider a 200-m long channel segment, 1 m wide, on a slope of 0.1 with a roughness coefficient of 0.03. Changing the reference depth an order of magnitude, from 0.1-m to 1.0-m reduces X less than 4% and K by about 90%. Under the same conditions the linear storage parameter k is increased approximately 60%. It is obvious that Q_r must be selected with care. This is particularly true in large watersheds where the channel properties play a more important role in determining hydrograph characteristics and flood frequency (Wolff and Burges, 1994). Small watersheds where hillslope processes dominate the hydrograph will be less sensitive to selection of the reference discharge.

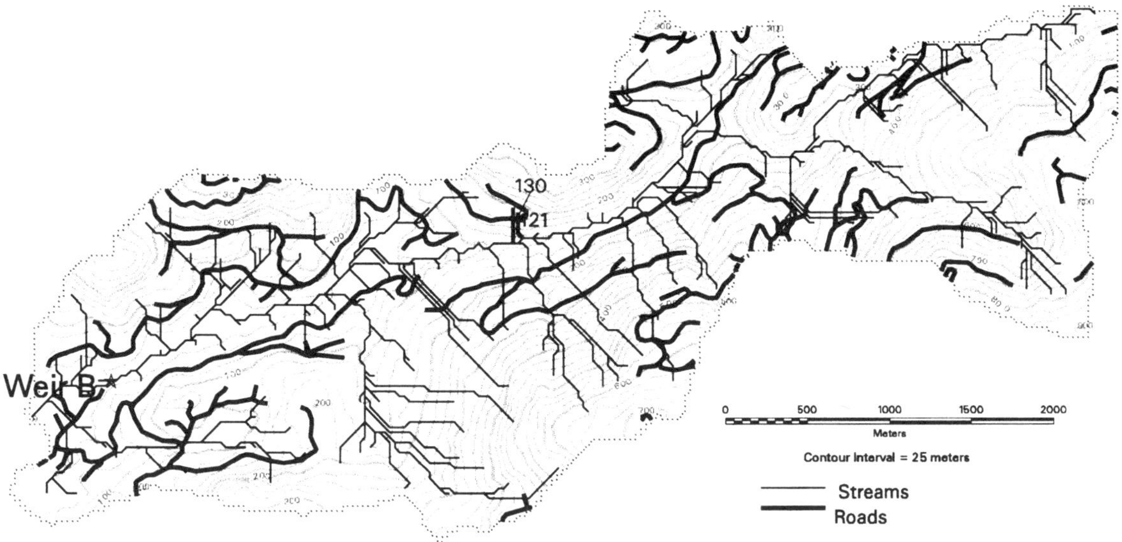

Figure 6. Contours from thirty-meter digital elevation model of the Carnation Creek watershed, showing the stream and road channel networks, and the locations of Weir B and stream channel segments 121 and 130. The two segments have similar natural drainage areas. The road bisecting these segments captures water that would have flowed into segment 130 and redirects it to segment 121.

APPLICATION TO CARNATION CREEK

The 10-km^2 Carnation Creek watershed (Figure 6) is located on the southeastern side of Barkley Sound on the west coast of Vancouver Island, British Columbia. The watershed is subject to frequent winter storms with annual precipitation ranging from 210 to over 500 cm. Approximately 90 percent of the precipitation occurs as rain, and greater than 90 percent of the annual precipitation leaves the watershed as runoff [Hartman and Scrivener, 1990]. Topography in the basin is steep and irregular. Valley walls have gradients up to 80 percent that are interrupted by bluffs and rock outcrops. Slope soils of 80-cm mean depth overlie bedrock. Soils are composed of coarse colluvial material, gravelly loam and loamy sand with an organic surface layer [Hartman and Scrivener, 1990]. Primary forest trees in the watershed are western hemlock (Tsuga heterophylla), western red cedar (Thuja plicata), amabilis fir (Abies amabilis), Douglas fir (Pseudotsuga menziesii), Sitka spruce (Picea sitchensis), and red alder (Alnus rubra). The watershed has been used to study the effects of forest harvest activities on hydrology, vegetation, soils, and fisheries since 1970 [Hartman and Scrivener, 1990]. There was a prelogging monitoring stage 1970-1975, a logging stage encompassing the winter of 1975-76 through winter 1981, and a post-logging stage from spring 1981 to 1986.

MacMillan Bloedel, Limited provided Arc/INFO coverages containing the road network, 212 culvert locations, locations of major channels, and 75-m resolution point ground surface elevations. 75-m was too coarse a spatial resolution for modeling this small basin. As a result, the 75-m elevations and channel network were used to construct a 30-m DEM consistent with the main channel network. A higher resolution stream network was then generated from the 30-m DEM using Arc/INFO (Figure 6). The resulting network has a significant number of straight channel segments. This partially the result of the flow direction algorithm employed by Arc/INFO (Costa-Cabral and Burges, 1994). The 30-m DEM may also be rather coarse for defining smaller scale channel features.

The prelogging stage was used for model calibration and testing. Meteorological data and streamflow measurements were available beginning October 1, 1972. The model was calibrated using data from Water Year 1973 (October 1, 1972 – September 30, 1973), and tested using data from Water Year's 1974 and 1975. The model was driven using hourly precipitation, air temperature, relative humidity, wind speed, and solar radiation measurements at Weir B, located near the basin outlet at an elevation of 40 m (Figure 9). Temperature was adjusted for elevation using a consistent environmental lapse rate of –6.5 °C per km increase in elevation. Precipitation was adjusted using a linear regression of elevation versus precipitation

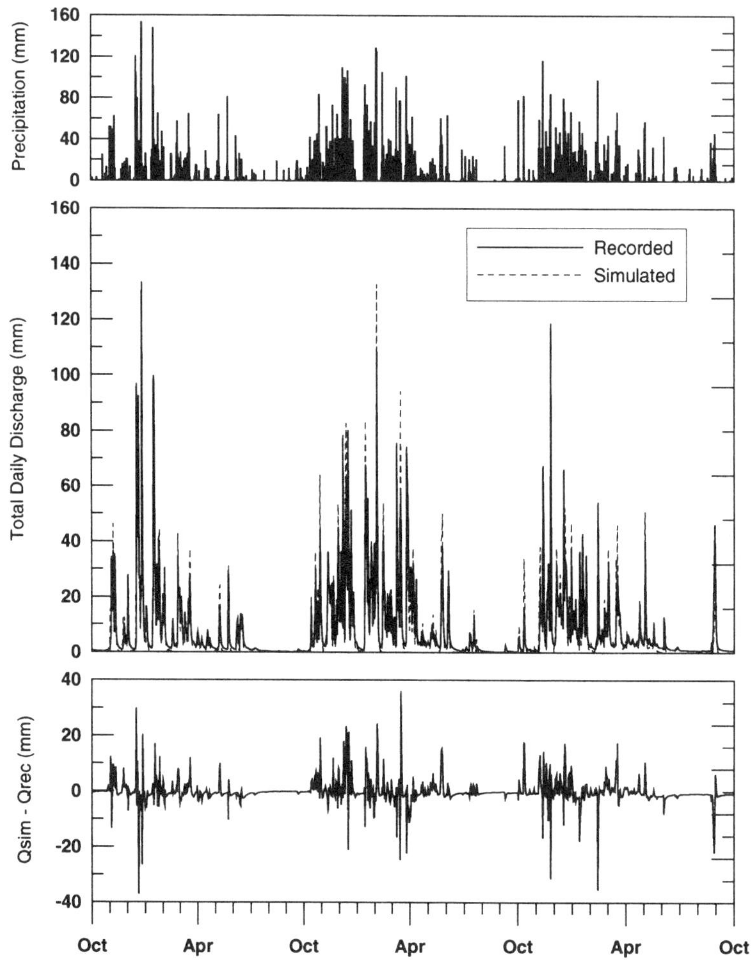

Figure 7. Simulated daily basin-average precipitation for Water Year's 1973 - 1975 (upper), simulated and recorded daily discharge at Weir B (middle), and simulated minus recorded discharge at Weir B.

developed from precipitation measured at Weir B and other stations in the watershed. There was insufficient information on vegetation and soil distributions to subdivide the watershed into multiple classes. As a result, a single vegetation class and a single soil class were used. Western Hemlock was the dominant tree species prior to harvest activities [Hartman and Scrivener, 1990] and was chosen to represent the overstory vegetation class. Soil thickness was assumed to be a constant 80-cm. The remaining soil parameters were adjusted during calibration. Soil moisture was taken equal to field capacity at the start of the simulation. The influence of initial conditions was damped out prior to the first runoff event in November.

Simulated and recorded daily streamflow for Water Year's 1973-1975 are presented in Figure 7. Total simulated runoff at Weir B is within 2% of recorded. The model is successful in capturing the timing of hydrograph rise, peak, and fall for most storms (daily $R^2 = 0.89$). Low flows (< 4 mm/d) are under-simulated, while moderate flows (15 - 55 mm/d) are over-simulated. There is no obvious bias in simulated peak flows (> 55 mm/d). Simulated and recorded hourly streamflow for the largest storm in Water Year 1974 and the largest storm in Water Year 1975 are presented in Figure 8 and Figure 9, respectively. The model reproduces the rain-dominated runoff hydrograph well for the February 2-3, 1974 storm (Figure 8). The simulated hydrograph for the large November 23-24 rain-on-snow event is below recorded. The model maintains about 2-mm of water equivalent in the snowpack during and after the storm (Figure 9). This may be part of the reason the simulated discharge is below recorded.

The model was then run using final calibration soil and vegetation parameters with the current road network. It was assumed that all roads were cut to bedrock and road drainage discharged at mid-slope ditch relief culverts was

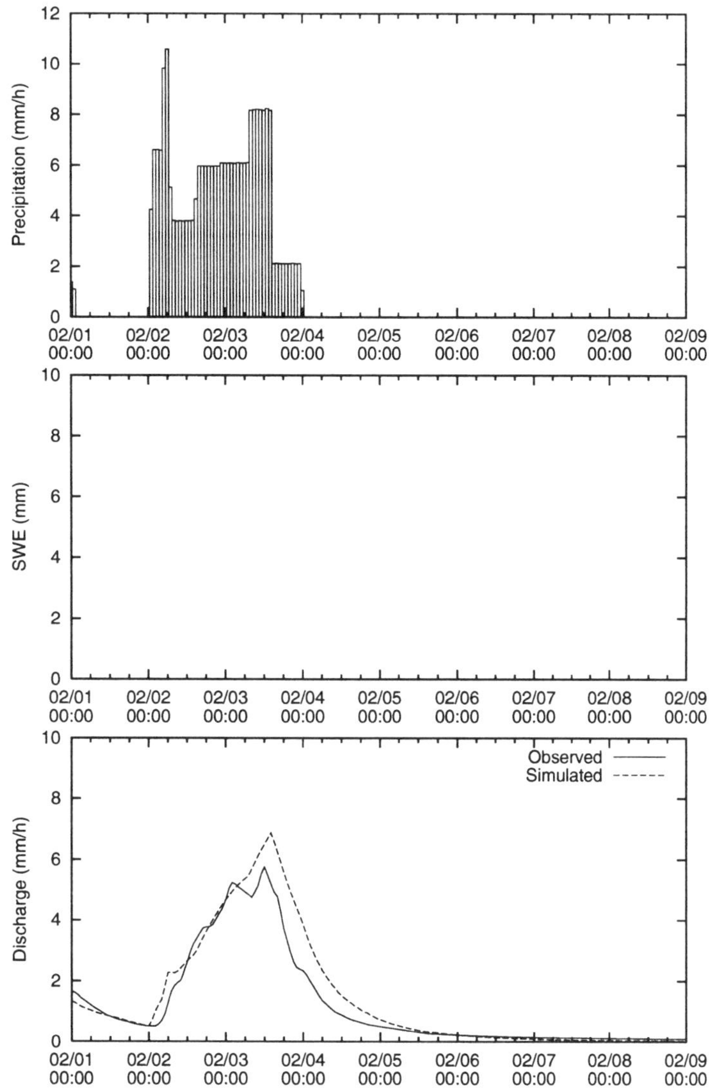

Figure 8. Simulated hourly basin-averaged precipitation for February 1-9, 1974 (upper), simulated basin-averaged snow water equivalent (middle), and simulated and recorded hourly discharge at Weir B (lower).

allowed to infiltrate downslope. The road drainage system diverts flows out of the basin in several locations. Simulated streamflow at Weir B with roads and without roads shows little difference during the three years. Although the integrated impact of the road network on streamflow at Weir B is minor, subsurface flow interception by the network alters significantly the distribution of soil moisture and runoff generation in many areas of the basin. The simulated spatial distribution of depth to the water table on December 16, 1972 with and without roads is shown in Plate 1. As would be expected, there is a correspondence between the channel network in Figure 6 and the zones of surface saturation (darkest blue, depth of zero) in Plate 1 with and without roads. The watertable is generally closer to the surface without the road network, which provides an efficient mechanism to help drain the watershed. The light colored bands of lower watertables below roads result from the interception of subsurface flow by the road drainage network.

The road network may effectively increase the contributing area of some channel segments and decrease the area draining to others. For example, stream segments 121 and 130 (Figure 6) are adjacent to each other and have natural drainage areas that are similar in size. The road that bisects the two channels captures some water that would normally flow to segment 130 and routs it to segment 121 instead. This is evident in Figure 10 where discharge in segment 121 has been increased greatly by the road

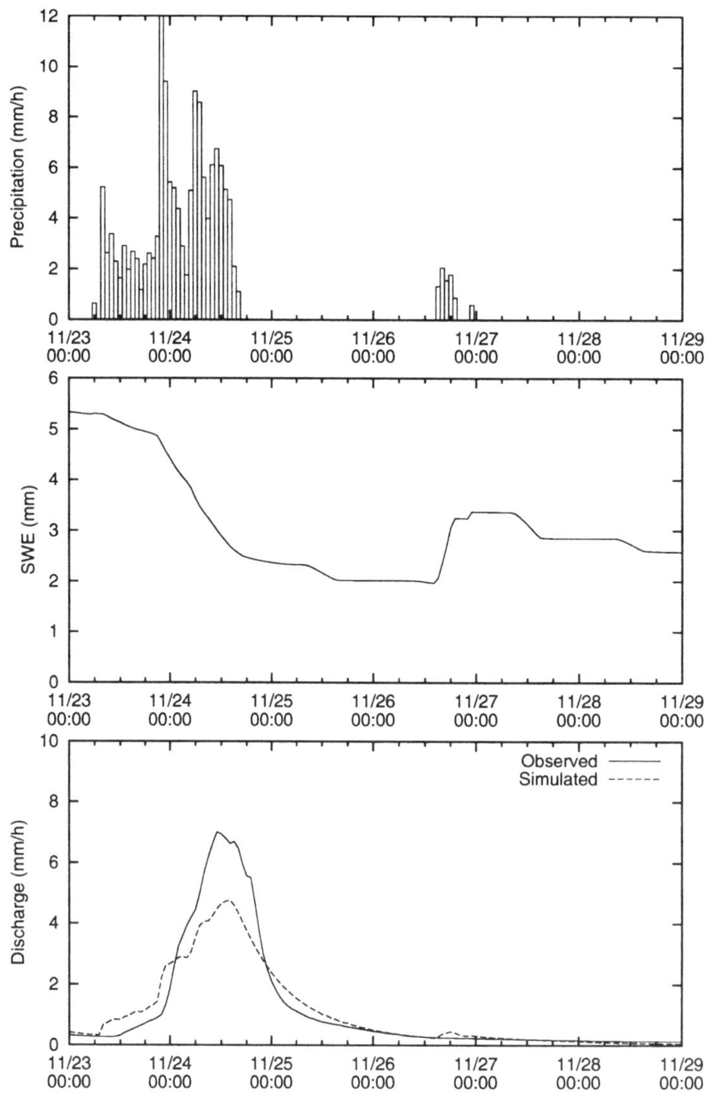

Figure 9. Simulated hourly basin-averaged precipitation for November 23-29, 1974 (upper), simulated basin-averaged snow water equivalent (middle), and simulated and recorded hourly discharge at Weir B (lower).

network, while discharge in segment 130 has been reduced. The impact of the road network on the effective contributing area can change in time depending on the storm characteristics, local geology, road design, and soil moisture conditions. A road does not have to increase the total drainage area of a channel to influence streamflow characteristics significantly. Consider the case of a road cut to bedrock, roughly perpendicular to a channel that bisects the channel drainage area. Early in a storm the areas contributing water directly to the road and channel expands as water from progressively more distant locations arrives. If water intercepted by the road is input to the channel, the current effective contributing area for the channel is increased by the road contributing area, causing a proportional increase in discharge. In this case the presence of a road would cause a sharper, higher peaked hydrograph, but would have little influence on the total volume of storm runoff. If the road design allows some water to pass under the road, there may be only minor impacts when the basin is relatively dry and little subsurface flow is intercepted by the road (contributing area to the road is small). The impact may increase as watertables rise and more subsurface flow is intercepted, increasing the effective road drainage area. The road system also has the ability to change the total (steady-state maximum) drainage area of a channel by redirecting flow into, or out of, the channel's natural drainage area.

Some of these mechanisms are illustrated in Figure 11, which presents simulated hourly discharge for channel segments 121 and 130 without roads, with roadcut depths

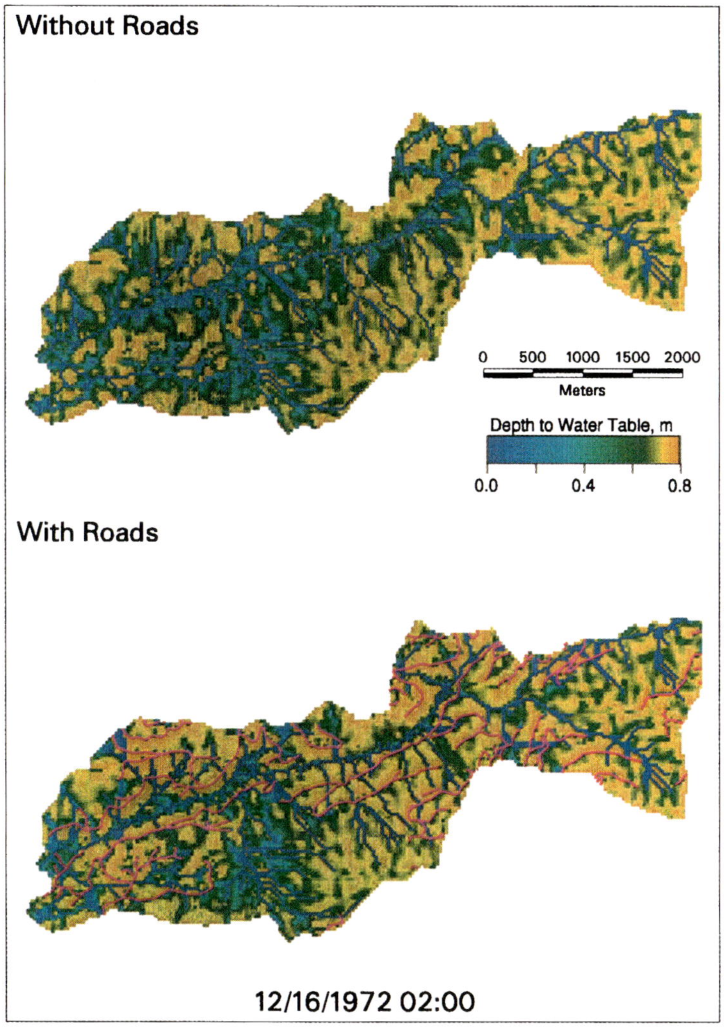

Plate 1. Simulated depth to the water table (m) near peak discharge on December 16, 1972 without roads (upper) and with roads (lower). The bands of lower watertables below roads result from the interception of subsurface flow by the road drainage network.

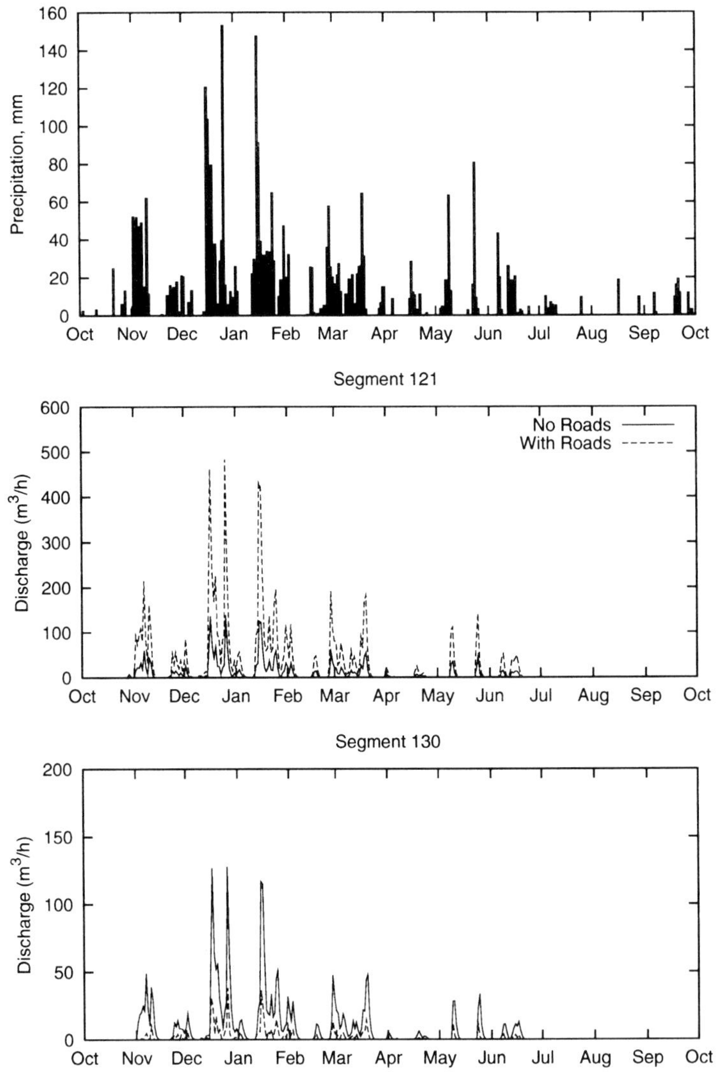

Figure 10. Daily basin-averaged precipitation for Water Year 1973 (upper) and simulated mean daily discharge (m^3/h) with and without roads for channel segments 121 (middle) and 130 (lower). Roads were assumed to have a roadcut depth of 0.8-m (the total soil thickness).

of 0.4 m (one-half the soil thickness) and 0.8 m (the total soil thickness). Discharge in segment 121 is increased greatly by a road network with 0.8 m cut depths. In this case the road system begins to intercept subsurface flow very early in the storm, increasing the effective contributing area (and discharge) relative to natural conditions. The road network also diverts water from adjacent areas as evidenced by the large increase in runoff volume. A road network with 0.4 m cut depths does not begin to impact channel flow until water tables rise above local road cut elevations, up to this point the road has no influence on channel flows. By the time significant flow interception occurs, the rainfall has reached its peak and decreases rapidly. As a result, the additional contributing area from the road causes a moderate increase in discharge relative to the non-road case.

Channel segment 130 provides an example where the road network diverts water away from the channel, reducing streamflow. In the first two hours of the storm (up to a discharge of ~ 70 m^3/h) discharge is the same in all three scenarios because the area contributing directly to the channel has not extended upslope to the road system. That is, rain falling above the road locations has not reached the channel in the non-road case, so the road network has no impact on streamflow. Later in the storm the contributing area in the non-road case continues to expand beyond the

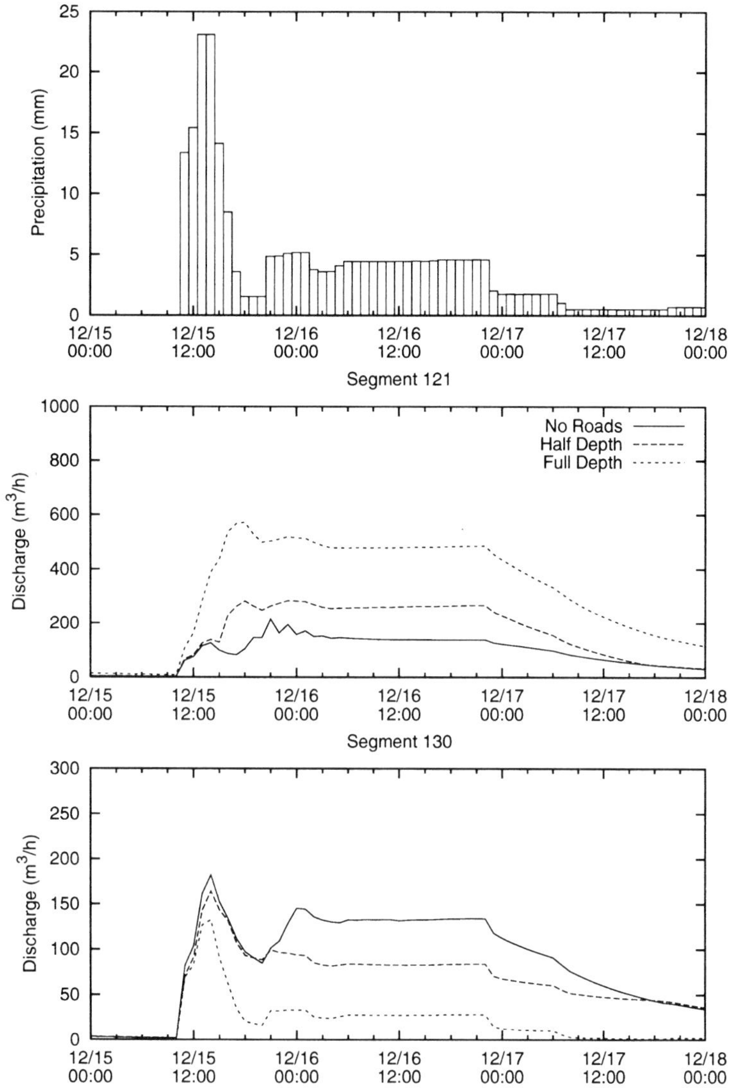

Figure 11. Hourly basin-averaged precipitation for December 15-17, 1972 (upper) and simulated hourly discharge (m^3/h) with and without roads for channel segments 121 (middle) and 130 (lower). Simulated discharge is presented without roads, with roadcut depths of 0.4 m (one-half the soil thickness), and roadcut depths of 0.8 m (the total soil thickness).

road locations and discharge continues to increase. With a 0.8 m cut depth the channel contributing area can not extend above the road locations, resulting in lower discharges than under natural conditions. The 0.4-m cut depth results in flows close to, but lower than the non-road case.

DISCUSSION AND CONCLUSIONS

DHSVM was restructured to represent the interception and redirection of surface and subsurface flows by roads and channels. A numerical simulation was conducted to evaluate the DHSVM representation of hillslope hydraulics associated with subsurface flow interception by a road-cut. Transient water table profiles both above and below the road cut simulated by DHSVM were in close agreement with an analytic solution to the kinematic wave equation. The model was then applied to the 10-km^2 Carnation Creek watershed located on the west coast of Vancouver Island, British Columbia. Model simulated discharge under natural conditions was in generally good agreement with observed over a three-year period. Model simulated peak flows show no obvious bias when compared to observed. However, moderate flows are over-simulated and low flows are under-simulated.

Rapid subsurface flow through soil macropores is an important runoff mechanism in many forested watersheds (Beven and Germann, 1982; McDonnell, 1990). DHSVM uses a matrix flow representation and does not consider explicitly more rapid macorpore flow. While the exponential decrease in saturated conductivity with depth mimics in some sense a decrease in macropores with depth, it is insufficient to capture many features of a two-domain matrix/macropore system. Modeling this type of system as matrix flow results in a tradeoff between peaks and baseflow. Higher soil conductivity increases the peaks, but reduces hydrograph recessions and baseflow. Portions of the Carnation Creek simulations reflect this type of compromise. The use of a single soil and vegetation type may also contribute to the problem. The watershed contains rock outcrops and exposed bedrock on some ridges. Modeling runoff from these areas as subsurface matrix flow rather than surface flow also requires the use of a high saturated hydraulic conductivity.

DHSVM was run using the existing road network to study the influence of network design on streamflow characteristics. It was demonstrated that even where the road network produces minor changes in streamflow at the basin outlet, subsurface flow interception by roads could alter significantly the spatial and temporal distribution of soil moisture and streamflow in tributary channels. Two channels with similar flow characteristics under natural conditions were examined in detail. The road network produced up to a five-fold increase in peak flows in one channel and up to a three-fold decrease in peak flows in the other. It should be noted however, that both segments drain relatively small headwater subbasins. Although the simulated discharge in segment 121 showed a dramatic increase with roads, the peak discharge was still relatively small (~ 500 m^3/h or 0.14 m^3/s) and the simulated flow depth remained below 3 cm in the 0.5-m wide channel.

The impact of the road network was shown to vary in time based on the road design, storm characteristics, topography, local geology, and soil moisture conditions. The impact occurs earlier in a storm when the road network diverts flow into the drainage rather than out of it. When flows are redirected into a channel, the road contributing area quickly increases the current total contributing area for the channel, causing a proportional increase in discharge. If the road design allows some water to pass under the road, there may be only minor impacts early in the storm with increasing impact as watertables rise and subsurface flow begins to be intercepted. When flows are redirected out of the natural drainage the impact is not felt until the channel contributing area extends upslope to the road system. If the storm ends prior to this point, removal of water by the road network may cause little if any reduction in peak flow relative to natural conditions.

REFERENCES

Beven, K., On subsurface stormflow: an analysis of response times, *Hydro. Sci. J., 4*, 505-521, 1982.

Beven, K. and P. Germann, Macropores and water flow in soils, *Water Resour. Res., 18(5)*, 1311-1325, 1982.

Bowling, L. C. and D. P. Lettenmaier, Evaluation of the effects of forest roads on streamflow in Hard and Ware Creeks, Washington, *Water Resources Series Technical Report No. 155*, University of Washington, pp. 189, 1997.

Bowling, L. C. and D. P. Lettenmaier, 1997. Evaluation of the effects of forest roads on flood flows in a mountainous maritime environment, *This Issue.*

Chow, V.T., D.R. Maidment, and L.W. Mays, *Applied Hydrology*, McGraw-Hill, Inc., New York, 572p., 1988.

Costa-Cabral, M.C. and S.J. Burges, Digital elevation model networks (DEMON): A model of flow over hillslopes for computation of contributing and dispersal areas, *Water Resour. Res., 30(6)*, 1681-1692, 1994.

Garbrecht, J, and G. W. Brunner, A Muskingum-Cunge channel flow routing method for drainage networks, *J. Hydraulic Engineering, ASCE, 117 (5)*, 629-642, 1991.

Harr, R. D., W. C. Harper, J. T. Krygier, Changes in storm hydrographs after road building and clear-cutting in the Oregon coast range, *Water Resour. Res., 11(3)*, 436-44, 1975.

Harr, R. D., and F. M. McCorison, Initial effects of clearcut logging on size and timing of peak flows in a small watershed in western Oregon. *Water Resour. Res., 15(1)*, 90-94, 1979.

Hartman, G. F., and J. C. Scrivener, Impacts of forestry practices on a coastal stream ecosystem, Carnation Creek, British Columbia, Department of Fisheries and Oceans, Ottawa, 1990.

Jones, J. A. and G. E. Grant, Peak flow responses to clear-cutting and roads in small and large basins, western Cascades, Oregon. *Water Resour. Res., 32(4)*, 959-974, 1996.

King, J. G. and L. C. Tennyson, Alteration in streamflow characteristics following road construction in North Central Idaho, *Water Resour. Res., 20(8)*, 1159-1163, 1984.

Lamarche, J. and D. P. Lettenmaier, Forest road effects on flood flows in the Deschutes River basin, Washington, *Water Resources Series Technical Report. No. 158*, University of Washington, 1998.

McDonnell, J. J., A rationale for old water discharge through macropores in a steep, humid catchment, *Water Resour. Res., 26(11)*, 2821-2832, 1990.

Megahan, W. F., Subsurface flow interception by a logging road in mountains of Central Idaho, National Symposium on Watersheds in Transition, Colorado State University, American Water Resources Association, 1972.

Megahan, W. F., Effects of clearcutting and wildfire on the hydrologic function of steep granitic slopes in Idaho. *Water Resour. Res., 19*, 811-819, 1983.

Rothacher, J., Streamflow from small watersheds on the western slope of the Cascade Range of Oregon. *Water Resour. Res., 1(1)*, 125-134, 1965.

Rothacher, J., Increases in water yield following clear-cut logging in the Pacific Northwest, *Water Resour. Res., 6(2)*, 653-658, 1970.

Storck, P., D.P. Lettenmaier, B.A. Connelly, and T.W. Cundy, Implications of forest practices on downstream flooding: Phase II Final Report, Washington Forest Protection Association, *TFW-SH20-96-001*, 100p., 1995.

Storck, P., L. Bowling, P. Weatherbee, and D. P. Lettenmaier, Application of a GIS-based distribution hydrology model for prediction of forest harvest effects on peak stream flow in the Pacific Northwest, *Hydrol. Process., 12 (6)*, 889-904, 1998.

Thomas, R. B. and W. F. Megahan, Peak flow responses to clear-cutting and roads in small and large basins, western Cascades, Oregon: A second opinion. *Water Resour. Res., 34(12)*, 3393-3403, 1998.

Wemple, B. C., Investigations of runoff production and sedimentation on forest roads, Ph.D. Thesis, Oregon State University, Corvallis, 1998.

Wemple, B. C., J. A. Jones, and G. E. Grant, Channel network extension by logging roads in two basins, Western Cascades, Oregon, *Water Res. Bull., 32(6)*, 1195-1207, 1996.

Wigmosta, M. S., and D. P. Lettenmaier, A distributed hydrologic model for mountainous catchments, in *Advances in Distributed Hydrology*, edited by R. Rosso, A. Peano, I. Becchi, and G.A. Bemporad, pp. 359-378, Water Resources Publications, Fort Collins, CO, 1994.

Wigmosta, M. S., and D. P. Lettenmaier, A comparison of simplified methods for routing topographically-driven subsurface flow, *Water Resour. Res., 35 (1)*, 255-264, 1999.

Wigmosta, M. S., and W. A. Perkins, A GIS-based modeling system for watershed analysis, Final Report to the National Council of the Paper Industry for Air and Stream Improvement, 1997.

Wigmosta, M. S., L. W. Vail and D. P. Lettenmaier, A distributed hydrology-vegetation model for complex terrain, *Water Resour. Res., 30(6)*, 1665-1679, 1994.

Wolff, C. G, and S. J. Burges, An analysis of the influence of river channel properties on flood frequency, *J. Hydrology, 153*, 317-337, 1994.

The Effects of Forest Roads and Harvest on Catchment Hydrology in a Mountainous Maritime Environment

Laura C. Bowling and Dennis P. Lettenmaier

Department of Civil Engineering, University of Washington, Seattle, Washington

Road networks in mountainous forest catchments have the potential to increase flood peaks via two mechanisms: interception and conversion of subsurface to surface runoff by road cuts, and generation of infiltration excess runoff from road surfaces. The quantity of runoff intercepted by forest roads was monitored in two small Western Washington catchments, Hard and Ware Creeks (drainage areas 2.3 and 2.8 square km, respectively), and subsequently was modeled using the Distributed Hydrology-Soil-Vegetation Model (DHSVM), a distributed hydrological model that simulates the land surface water and energy balance at the scale of a digital elevation model (DEM). Road densities were approximately 5.0 and 3.8 km/square km in Hard and Ware Creeks, respectively. Field observations indicated that the highest peak culvert discharges were associated with subsurface flow interception rather than road surface runoff. Model-simulated flows with and without roads were compared to observed flow from continuous recording gauges at the outlets of Hard and Ware Creeks, and to crest-recording gauges installed on 12 culverts for selected storms during the winter of 1995-96. The model generally simulated outlet peaks well, and culvert peaks approximately, for the selected storms. Based on retrospective simulations using eleven years of data, the mean annual floods in Hard and Ware Creeks were predicted to have increased by 11 percent, and the mean of 4 peaks over threshold were predicted to have increased by 8 and 9 percent, respectively, due to construction of logging roads. The simulated effects of road construction and forest harvest were approximately additive and roughly equal in magnitude.

INTRODUCTION

What amounts to a wide-spread land-use change experiment has taken place in the Pacific Northwest over the past half-century as most of the old growth timber has been harvested and replaced with stands of younger age and more uniform species composition. Large floods and resulting damage along the west slope of the Cascade Mountains in November 1990 and 1995, February 1996 and January 1997 have focused public attention on the possible role of forest harvest and road construction in flooding forested watersheds. There are two main mechanisms by which forest management might affect flood flows. The first is the effect of timber removal on surface energy and moisture fluxes. In particular, vegetation removal is known to result in increased accumulation of snow due to the absence of canopy interception and ablation and to enhanced melt during rain-on-snow periods due to increased turbulent transfer of latent and sensible heat at the snow surface [e.g. Berris and Harr, 1987; Harr, 1981]. The second mechanism is forest

Land Use and Watersheds: Human Influence on Hydrology and Geomorphology in Urban and Forest Areas
Water Science and Application Volume 2, pages 145-164

roads, which can have the effect of short circuiting the natural movement of (dominantly subsurface) flow to the channel network and thereby increasing flood peaks [Megahan, 1972]. Along with flood enhancement, slope instability is a potential side effect of these two mechanisms [e.g. Megahan, 1987; Montgomery 1994].

The effects of timber harvest on streamflow are varied at the catchment and basin scale. Many studies have demonstrated an increase in annual water yield following forest harvest, which has been attributed primarily to a reduction in evapotranspiration [Harr et al., 1982; Keppeler and Ziemer, 1990; Rothacher, 1965; 1970]. Statistically significant increases in peak flows have also been demonstrated [Harr; 1986; Jones and Grant, 1996; Rosencrantz et al., 1995; Storck et al., 1995; Wright et al., 1990]. In many cases the change in peak flow response depended on antecedent snow conditions [Harr, 1986; Rosencrantz et al., 1995]. These changes have often been attributed to increased meltwater supply during rain-on-snow events and/or increased runoff response due to a higher water table that results from decreased evapotranspiration.

During snow accumulation, interception of snow by the forest canopy results in smaller below-canopy snow accumulation as compared to clearings [Berris and Harr, 1987; Kattelmann, 1990; Megahan, 1983]. In addition, removal of vegetation can increase the turbulent transfer of energy and water vapor to the snow surface, thus increasing the rate of snowmelt during rainfall [Berris and Harr, 1987; Harr, 1981; USACE, 1956]. Greater accumulation and a faster rate of melt may result in increased peak discharges and decreased time to peak discharge during rain-on-snow events. The transient snow zone, usually assumed to lie between 350 and 1100 m in the Pacific Northwest, is defined as the range of elevations that accumulate and melt snow several times throughout the winter [Berris and Harr, 1987]. Because of the frequent accumulation and melt of snow, catchments in the transient snow zone are strongly affected by changes in the snowpack due to forest harvest.

Road corridors in forested catchments are permanently cleared, and although turbulent transfer effects are not as large on roads surrounded by trees, they may be a source of increased snow accumulation and melt. More importantly, roads constructed in forested catchments may affect both the timing and magnitude of basin response by redistributing runoff through the road drainage network. Roads incised into hillsides can capture and re-route runoff through two mechanisms: interception of subsurface flow by the road cutslope and direct runoff from the compacted road surface. Runoff intercepted by forest roads is routed through the ditch drainage system until reaching a natural stream or a ditch relief culvert. If diverted through a ditch relief culvert, runoff can either infiltrate downslope of the culvert outfall, or continue to follow a surface flow path, in many cases entering the natural drainage system as overland flow.

The compacted surfaces of forest roads have decreased infiltration capacity relative to the undisturbed forest floor [Foltz and Burroughs, 1990; Luce and Cundy, 1994; Megahan, 1972; Reid and Dunne, 1984]. The width of the compacted surface of forest roads in the Pacific Northwest is typically around 4.5 meters [Montgomery, 1994; Piehl et al., 1988; Reid and Dunne, 1984]. Infiltration rates through road surfaces varied between 0.0 to 21.5 mm/hr for two studies in the Pacific Northwest [Foltz and Burroughs, 1990; Reid and Dunne, 1984]. The average of all infiltration rates reported for these studies was 5 mm/hr. Observed infiltration rates depended on antecedent wetness, surfacing material and the degree of compaction [Foltz and Burroughs, 1990; Reid and Dunne, 1984]. Using the average infiltration rate, an intense one hour storm in the Pacific Northwest of 10 mm/hr will generate 22.5 m^3/hr of runoff per km of forest road. (During the February 1996 storm event, a precipitation intensity of 10 mm/hr was exceeded for approximately 2 hours in Hard Creek, WA.) For Hard Creek, this amounts to a runoff rate from roads of .11 mm/hr, compared to the total discharge of 3.5 mm/hr during the February storm. This suggests that the contribution of direct runoff from forest roads tends to be insignificant during peak flows.

The influence of vegetation on soil moisture distribution and quantity may enhance the effect of forest roads. Several studies have found an increase in annual water yield and the quantity of stored soil moisture during the growing season following timber harvest [Bosch and Hewlett, 1982; Keppeler and Ziemer, 1990; Ziemer, 1981]. By increasing soil moisture upslope of a road cut, clearcuts increase subsurface flow for interception by the road network. In contrast, removal of vegetation downslope of the road will replace soil moisture that has been diverted by the road network, thus diminishing the drying consequences of redistribution.

The ultimate effect of surface or subsurface runoff intercepted by forest roads on hydrograph timing and peaks will depend on the flow paths followed once the resulting runoff enters the road network. Forest road drainage networks in Washington, Oregon and California have been surveyed by Bilby et al. [1989], Montgomery [1994], Reid and Dunne [1984] and Wemple et al. [1996]. These studies indicate that between 34 and 68 percent of road runoff points (i.e. culverts, drainage bars or other points where runoff is released from the ditch network) discharge runoff directly to streams as surface flow. The subsequent decrease in the effective hillslope length may have a significant effect on hydrograph timing from individual catchments.

Land use activities in headwater catchments may have a cumulative effect, detectable at larger watershed scales. By changing the timing of flood hydrographs, changes in vegetation may desynchronize hydrographs from subcatchments and can, in some cases, reduce peak flows downstream [Harr, 1981]. Road construction and forest harvest near the catchment's centroid are more likely to desynchronize hydrographs by advancing the time to peak, while forest management activities upstream of the centroid are more likely to coincide with the accumulated hydrograph and increase peaks downstream. Most studies involving land use change have focused on runoff changes for small first and second order catchments [e.g. Bosch and Hewlett, 1982; Berris and Harr, 1987; Megahan, 1972; Rothacher, 1965; 1970]. These experiments are important for understanding the physical processes in either natural or disturbed landscapes. At catchment and basin scales (10 ha - 1000 km^2), it is difficult, if not impossible, to collect enough field data to understand the complex interactions of surface conditions and catchment runoff response induced by land use change. For this reason, an alternate approach to understanding the effects of forest roads and harvesting on catchment hydrologic response is pursued here, which is based on simulations of the natural system using a deterministic, spatially distributed hydrologic model. We evaluate the ability of such a model to predict the effects of forest roads on the magnitude and distribution of catchment runoff response based on one year of field investigation and 11 years of streamflow records for two small catchments in the western Cascades. If hydrologic models can be shown to capture the relevant hydrologic processes at this scale adequately, they can serve as tools to support land management decisions at the scale of management concern (e.g. 10^2-10^3 km^2 and up). The technique employed in this study can be adapted for application to river basins as shown by Lamarche and Lettenmaier [1998].

APPROACH

The Distributed Hydrology-Soil-Vegetation Model (DHSVM) [Wigmosta et al., 1994] was used to evaluate the hydrological effects of forest roads and harvest in conjunction with field studies of two headwater creeks in Western Washington. An explicit road network runoff algorithm [Wigmosta and Perkins, this issue] was tested by comparing simulated runoff for selected storms to observations of road runoff from ten road segments made during the winter of 1995 - 1996. Catchment discharge predictions were then compared for the period 1985 - 1996 with and without the road runoff algorithm.

Study Site Description

The study sites were Hard and Ware Creeks (2.3 km^2 and 2.8 km^2, respectively), two adjacent headwater catchments of the Deschutes River in Western Washington (Figure 1). The two catchments are separated by a steep southeasterly trending ridge and drain to the southwest. Steep, V-shaped valleys, with elevations between 457 - 1220 meters are found in both catchments [Sullivan et al., 1987]. Slopes are underlain by resistant and weathered andesite, basalt and breccia bedrock. Soils are stony and shallow, averaging 0.6 meters depth in Hard Creek and 1.0 meter depth in Ware Creek. Ware Creek soils are finer textured; derived from more highly weathered bedrock [Sullivan et al., 1987].

Average annual precipitation measured at the lowest elevation in the catchments, just upstream of the confluence of Ware Creek and the Deschutes River, is 2600 mm/year. The two catchments lie almost entirely within the transient snow zone. Thin snowpacks develop and melt several times throughout most winters. Snow accumulations at the catchment outlets amount to a few centimeters and usually melt quickly.

Hard and Ware Creeks lie entirely within the Weyerhaeuser Company's Vail Tree Farm where extensive logging and road construction have taken place, mostly since the 1970's (Table 1). Forest vegetation consists primarily of Douglas fir (*Pseudotsuga menziessii*), western hemlock (*Tsuga heterophylla*) and Pacific silver fir (*Abies amabilis*).

The majority of roads in Hard and Ware Creeks were constructed between 1975 and 1984 [Sullivan et al., 1987]. Construction of roads in steep terrain requires considerable excavation to produce a road surface of required width. Cutslopes range in height from 0 meters on the ridgetop roads to nearly 10 meters for some midslope roads. In many cases the road cuts intercept the entire soil column, exposing bedrock. Roadside ditches are lined with either very coarse rock fragments or vegetation. Infiltration rates are quite high in the crushed rock and intercepted subsurface flow often infiltrates into the ditch between storms.

As in most forested catchments in the Pacific Northwest, surface overland flow is rare on the hillslopes. However, concentrated flow in the top few inches of soil just below the root mass was frequently observed in clearcut areas. Runoff travels in this surface zone and accumulates in small topographic depressions. The resulting subsurface

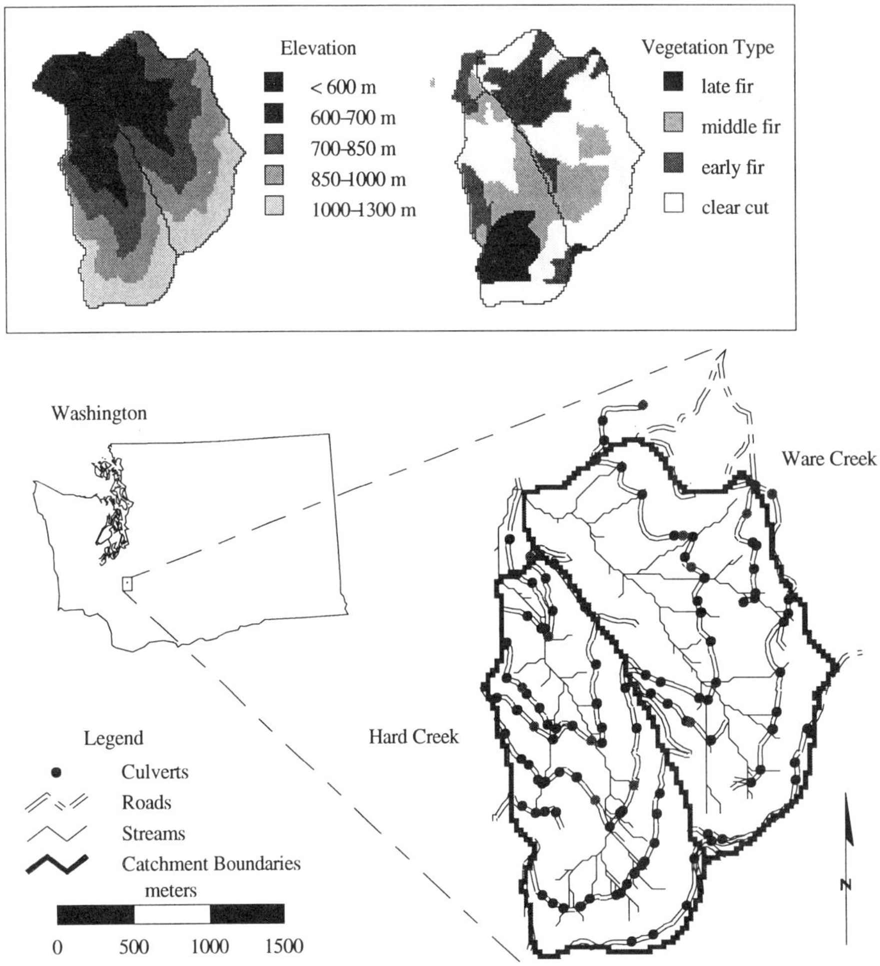

Figure 1: Location of Hard and Ware Creeks showing roads, stream network and culvert locations. Elevation and 1996 vegetation conditions are shown in inset.

streams are typically audible beneath the tangle of roots and surface vegetation, but are not visible until they emerge from the cutslope just below the vegetation layer. Continuous seepage of subsurface flow from the road cuts is visible in many parts of both catchments. Exposed bedrock outcrops in the road cuts often have a moist face with seepage visible between cracks. Based on field observations, roads in zones of locally deep soils intercept a larger volume of subsurface flow. Macropores appear to be the main mechanism for transport of this subsurface water. Several different points of exfiltration can be observed on exposed sandy banks.

Road Network Connectivity

As described above, runoff captured by the road network can travel to the basin outlet entirely as surface flow through one of two flow paths: a) by entering a stream crossing culvert or b) through a ditch relief culvert and eroded gully which extends to the natural drainage network. Therefore, the degree of connectivity of road segments to the stream network will influence the effect of forest roads on streamflow. To determine the degree of connectivity of the road and natural channel networks in Hard and Ware Creeks, a total of 111 ditch-relief and stream crossing culverts were field located using a hand-held portable Global Positioning System (GPS) during a three week period in August 1996. Each culvert was classified for its potential connectivity to the drainage network during an extensive field survey, using the categories of Wemple et al. [1996]: a) directly connected stream crossing, b) directly connected gully, or c) not connected. A culvert was considered to be directly connected through a gully if evidence of erosion and

TABLE 1. Hard and Ware Creek Land Use History

	Ware Creek	Hard Creek
Catchment area	2.8 km^2	2.3 km^2
Road construction	1975 - ??	1977 - 1980
Total road length	10.7 km	11.4 km
Road density	3.8 km/km^2	5.0 km/km^2
Harvesting	1979 - present	1984 - present
Harvested area (estimated)	66 %	35 %

TABLE 2. Culvert Classification Results

Culvert Class	Ware Creek		Hard Creek	
	Number	% of Total	Number	% of Total
Directly Connected (stream)	18	43	18	26
Directly connected (gully)	6	14	13	19
Not Connected	18	43	38	55
Total	42	100	69	100

overland flow existed from the culvert outlet until a permanent stream channel was encountered. Culverts which were difficult to classify were flagged for verification following rainstorms in the winter and spring of 1997 [Bowling and Lettenmaier, 1997]. Five percent of the culverts were reclassified following the verification sampling. The final results of the culvert survey are summarized in Table 2. These results were used to specify the locations of gullies between the roads and streams within the hydrology model, as described below.

Culvert Discharge Data Collection

Twelve road segments within the Hard and Ware Creek catchments were monitored during the winter of 1996. Road segments, defined as the length of road that drains to a specific culvert, were chosen to represent a variety of road positions, vegetation and soil conditions. Total runoff from the road segments was measured using crest-recording gauges constructed at the culvert entrances. Gauges consisted of a 4" (10.2 cm) diameter stilling well and inlet pipe [Bowling and Lettenmaier, 1997]. The corrugated metal culverts at each of these locations had a pool at the upstream end and a free fall at the culvert outlet, so there was no backwater influence on the water surface profile within the culvert. Culvert slopes were hydraulically steep, ranging between 0.02 and 0.14 m/m for the monitored road segments. These conditions create an inlet-controlled situation, allowing discharge to be estimated from the hydraulic head at the culvert inlet. Peak stages above the culvert lips were measured using floating cork in the stilling well and were recorded periodically from January - May 1996. The discharge rate was measured in triplicate for different stages by timing and capturing the discharge from the culvert outfall in a five-gallon bucket [Bowling and Lettenmaier, 1997].

For each of the monitored road segments, the pool at the culvert inlet was enlarged and deepened when the crest gauge was installed. Sediment was removed from the pool after large storms. For the range of stages measured in Hard and Ware Creeks, velocities in the roadside ditch were generally low and were further damped by the pool at the inlet to each culvert. Therefore, the velocity head of the pool can be neglected, and the headwater depth at the culvert inlet is equal to the stage above the lip of the culvert pipe.

A stage-discharge curve for each culvert type was derived using a combination of field measurements and experimental results from the Federal Highway Administration (FHA). Based on laboratory experiments the FHA has developed guidelines of discharge capacity for specific culvert diameters and entrance types [Normann et al., 1985]. Discharge values from the FHA chart were plotted in conjunction with observed values of stage and discharge for each of the four observed culvert categories. The culvert categories include: a) 18" diameter projecting, b) 18" diameter with headwall, c)15" diameter projecting and d) 15" diameter with headwall. A polynomial was fitted to each of these values to obtain a stage discharge curve for each culvert class [Bowling and Lettenmaier, 1997]. Discharge for recorded peak stages was obtained by reading from these stage-discharge curves. At no time did observations of peak stage exceed the height of the culvert.

THE DISTRIBUTED HYDROLOGY SOIL VEGETATION MODEL

DHSVM, originally developed by Wigmosta et al. [1994] and extended for use in maritime mountainous watersheds by Storck et al [1995], is a physically-based hydrologic model which explicitly solves the water and energy balances for each model grid cell. Recent enhancements to DHSVM are described in detail by Storck et al. [1997].

General Model Description

DHSVM combines a two-layer canopy representation for evapotranspiration, a two-layer energy-balance model for snow accumulation and melt, a one-dimensional unsaturated soil model, and a two-dimensional saturated subsurface flow model. An independent one-dimensional (vertical) water balance is calculated for each pixel.

Downward moisture flux in the unsaturated zone is calculated using Darcy's law. Soil moisture is removed from the unsaturated zone via evaporation and transpiration. The soil profile can consist of a variable number of rooting layers and a deep layer. The downward flux of moisture that is not removed through evapotranspiration serves to recharge the water table. Each DEM grid cell exchanges saturated subsurface flow with its four adjacent neighbors, by assuming that the local hydraulic gradient is equal to the local ground surface. Return flow and saturation overland flow are generated in locations where the grid cell water table intersects the ground surface.

Precipitation occurring below a threshold temperature is assumed to be snow. Snow interception by the overstory is calculated as a function of Leaf Area Index and is adjusted downward for windy or cold conditions [Schmidt and Troendle, 1992]. Intercepted snow can be removed from the canopy through snowmelt, sublimation, and mass release. Ground snow accumulation and melt are simulated using a two-layer energy-balance snow model. The model accounts for the energy advected by rain, throughfall or drip, as well as net radiation and sensible and latent heat.

Road and Channel Algorithm

A road and channel network algorithm was used to represent the formation of direct runoff from road surfaces and intercepted subsurface flow with its subsequent routing as open channel flow. Wigmosta and Perkins [this volume] describe the road and channel network algorithm in detail. A brief summary is presented here. Subsurface flow is discharged into road and stream networks based on the height of the local water table relative to the bottom of the channel or road. Once in the network, intercepted surface and subsurface flow are routed through roadside ditches and stream channels using a linear reservoir scheme. Each road segment is assumed to be crowned (i.e. half of the road surface runoff enters the ditch and half is immediately available for infiltration below the road). Open channel flow in the roadside ditches is released to the stream network at specified culvert locations, where the road runoff is added to the stream channel flow. If the specified culvert location does not correspond to a stream crossing, the runoff is applied to the soil surface as if it were precipitation, and is then available for reinfiltration. Surface water not in the channel or road networks is modeled as overland flow and can reinfiltrate into downslope model pixels.

Model Parameterization

Meteorological conditions were prescribed at a specified reference height well above the overstory. Precipitation and air temperature were collected by the Weyerhaeuser Corporation near the mouth of Ware Creek. Missing air temperature observations were filled in using data from Olympia Airport (about 50 km to the west of the basin). Solar radiation and vapor pressure were calculated using relationships described in Bowling and Lettenmaier [1997]. Accurate description of the surface wind field is one of the most difficult model inputs to obtain. Two-dimensional wind vectors for each model pixel were obtained using NUATMOS Version 6.0, a distributed wind model which produces a mass-consistent wind field based on interpolation of user-specified observations [Ross et al., 1988]. Boundary conditions for the wind model were obtained using wind speeds from the National Center for Environmental Prediction/National Center for Atmospheric Research (NCEP/NCAR) Reanalysis Project [Kalnay et al., 1996].

The Weyerhaeuser Company provided Arc/Info (GIS) coverages of forest stand boundaries for 1996 with accompanying attributes of LAI, tree height and canopy closure. The attributes were calculated based on literature values of LAI and reference height provided to Weyerheauser, which were linearly scaled by the 1996 tree heights. Stands with similar values were grouped into 13 classes and average values were calculated. Due to a lack of species-specific information, values for stomatal resistance, stomatal closure threshold, attenuation and albedo, were not changed for each vegetation class.

Soils data were extracted from the State Soil Geographic (STATSGO) Data Base. There are three STATSGO units within Hard and Ware Creeks, composed of several soil types. Areal extent, USDA soil texture, depth and percentage sand, silt and clay are available for each soil type and layer. The soil parameters required for DHSVM were derived from published empirical relationships using this data, as summarized in Table 3.

Road and culvert locations within Hard and Ware Creeks were surveyed on foot using a hand-held GPS unit. Stream network extent was derived from a 30 meter resolution

TABLE 3. Soil Parameters in the Top Soil Layer

Parameter	Soil 1	Soil 2	Soil 3
Porosity (%)	53	48	53
Field capacity	0.330	0.270	0.330
Wilting point	0.133	0.117	0.133
Density (kg/m^3)	1000.	840.	840.
Vertical sat. hydraulic cond. (m/s)	9.2x10^{-6}	1.3x10^{-6}	1.1x10^{-5}
Lateral sat. hydraulic cond. (m/day)	0.002	0.002	0.002
Max. infiltration rate (m/s)	1.68x10^{-6}	1.68x10^{-6}	1.68x10^{-6}

DEM using GIS software and a prescribed 2 ha threshold area. The derived stream network was compared with field observations of stream channel road crossings for accuracy. Channel width, depth and Manning's roughness coefficient for the stream channels and road-side ditches were estimated based on field observations. GIS coverages of the road and stream networks were processed to generate DHSVM input files which map the length, slope, aspect and channel characteristics of each stream segment to the appropriate model pixel.

MODEL CALIBRATION

DHSVM is a physically-based model and the parameters and inputs are meant, therefore, to be based on physically measurable quantities. For this reason, many parameters were not adjusted during the model calibration process. However, it is not feasible to make all of the exhaustive measurements that would be necessary to specify all parameters based on observations. Therefore, some calibration of the less well-defined parameters is required. The time period from July 1, 1993 through June 30, 1996 was selected for model calibration.

The calibration period encompassed the large rain-on-snow event of February 5-8, 1996, as well as the smaller April 28, 1996 event, in addition to the period from January 1, 1996 to June 30, 1996 for which field observations of culvert discharge were available. Initial conditions for the model on July 1, 1993 were estimated by running the model from October 1, 1992 to July 1, 1993. It was assumed that the basin was snow-free, with soil moisture at field capacity on October 1, 1992. Because the calibration period represents relatively recent harvest conditions, both the road and channel networks were implemented with the current (1996) road information.

Model calibration started with a general check of model-predicted snow accumulation and ablation. Estimates of the snow line elevation were made during field visits in 1996. In some cases, the modeled distribution of snow water equivalent (SWE) is more widespread than was observed (see Bowling and Lettenmaier [1997] for details.). The observed snow line was estimated as the location where continuous snow cover began, so model predictions of a thin, continuous snowcover most likely correspond, in fact ,to patchy cover. For example, on February 15th, patches of snow were observed down to 670 meters, although the snow line was recorded at 1067 m. This partially explains the higher modeled SWE equivalent relative to observed following the February 8, 1996 ROS event. The February 8 event was characterized by high winds and a high rate of turbulent mixing of the air. This results in fairly homogeneous, hydrostatic conditions with small or no temperature inversion. The boundary layer wind speeds used to drive the distributed wind model were quite high during this storm (between 16 and 25 m/s). However, the distributed wind model tends to attenuate the interpolated wind speeds at all but the highest elevations. Therefore, the exposed ridge top was predicted to have experienced high wind velocities and greater melt. The simulated wind attenuation and fixed temperature lapse rate together resulted in too little simulated melt in mid-elevation regions. Bowling and Lettenmaier [1997] have shown in a comparison of simulated snow extent that greater melt occurs in the mid-elevation region when the distributed wind model is not used. Despite the recognizable problems which occurred during this extreme ROS event, the simulated responsiveness to two previous accumulation and melt events which are consistent with observations is reassuring. In addition, the radiation-dominated melt-out in the spring is well-represented.

The model was subsequently calibrated to the two hour mean observed discharge at the outlets of Hard and Ware Creeks and the annual water balance by adjusting the lateral hydraulic conductivity, the exponent for decrease in lateral hydraulic conductivity with depth, the deep layer soil depth and the mean depth of the road cuts. Simulated and observed hydrographs for three storm events during the calibration period are shown in Figure 2. In general, the model tends to underpredict base flow slightly and overpredict [some] storm peaks and recessions following peaks. This suggests that not enough moisture is retained following storms for gradual drainage, which can be controlled through the total soil depth and lateral hydraulic conductivity. However, when model parameters were adjusted during model calibration to increase the amount of water retained, evapotranspiration increased and the

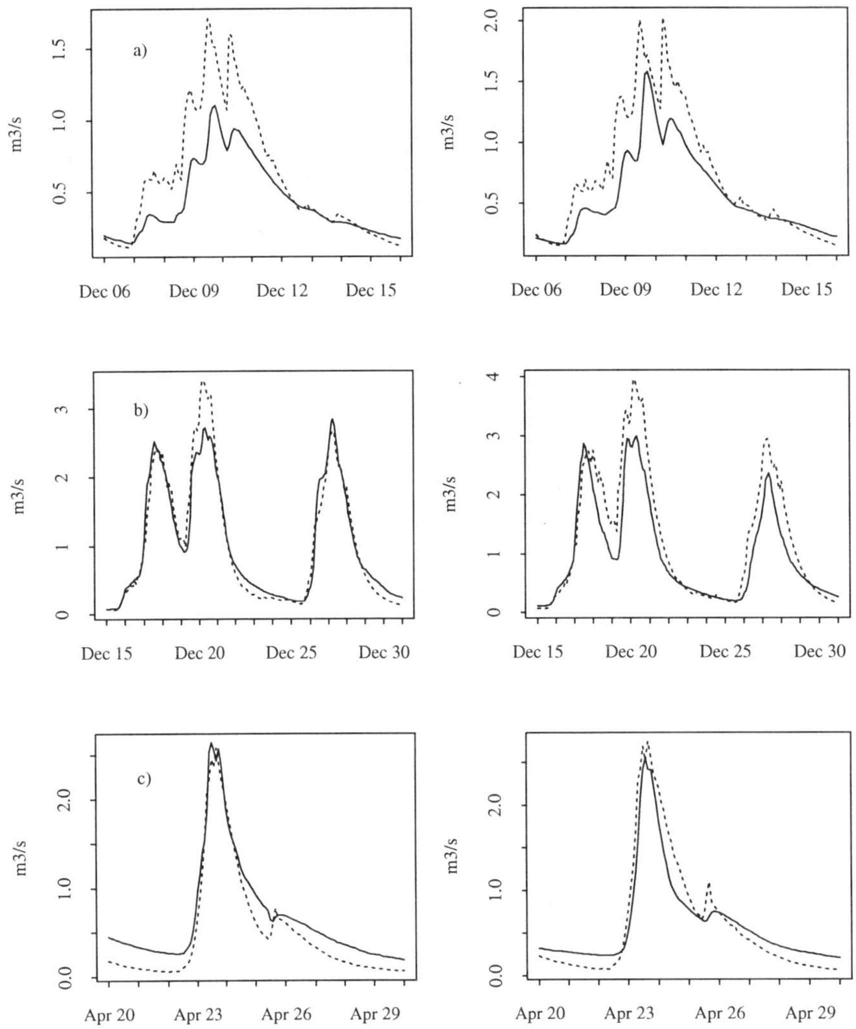

Figure 2: Observed vs. predicted Hard Creek (left) and Ware Creek (right) discharge during model calibration. Solid lines indicate observed discharge: a) December 6-16, 1993, b) December 15-31, 1994 and c) April 20-30, 1996.

annual water balance was less well-represented. The final hydrographs represent a compromise between accurate simulation of baseflow recessions and the total basin water balance. Improvement of these hydrographs would most likely require refined definition of the soil column and root zone dimensions to allow retention of soil moisture which is not immediately available for vegetation uptake. Total simulated Hard and Ware Creek discharge for the calibration period was 6.62 and 6.37 m, respectively (94 percent and 104 percent of observed). The difficulty in correctly simulating the water balance in both catchments most likely stems from precipitation differences due to microclimatology in these steep mountainous catchments and differences in soil properties. In particular, the aforementioned sensitivity to total soil depth is thought to be a major factor.

Following calibration, model performance was evaluated between October 1985 through June 1993 for Hard Creek and from October 1989 through June 1993 for Ware Creek (observed Ware Creek discharge prior to 1989 was lost). Three typical events from this period are shown in Figure 3. Total simulated discharge is within 11% of observed for Hard Creek and 1% of observed for Ware Creek between October 1989 through June 1993.

RESULTS AND DISCUSSION

Road Runoff Response

Based on observations of peak culvert stage between January and June 1996, the 12 monitored road segments

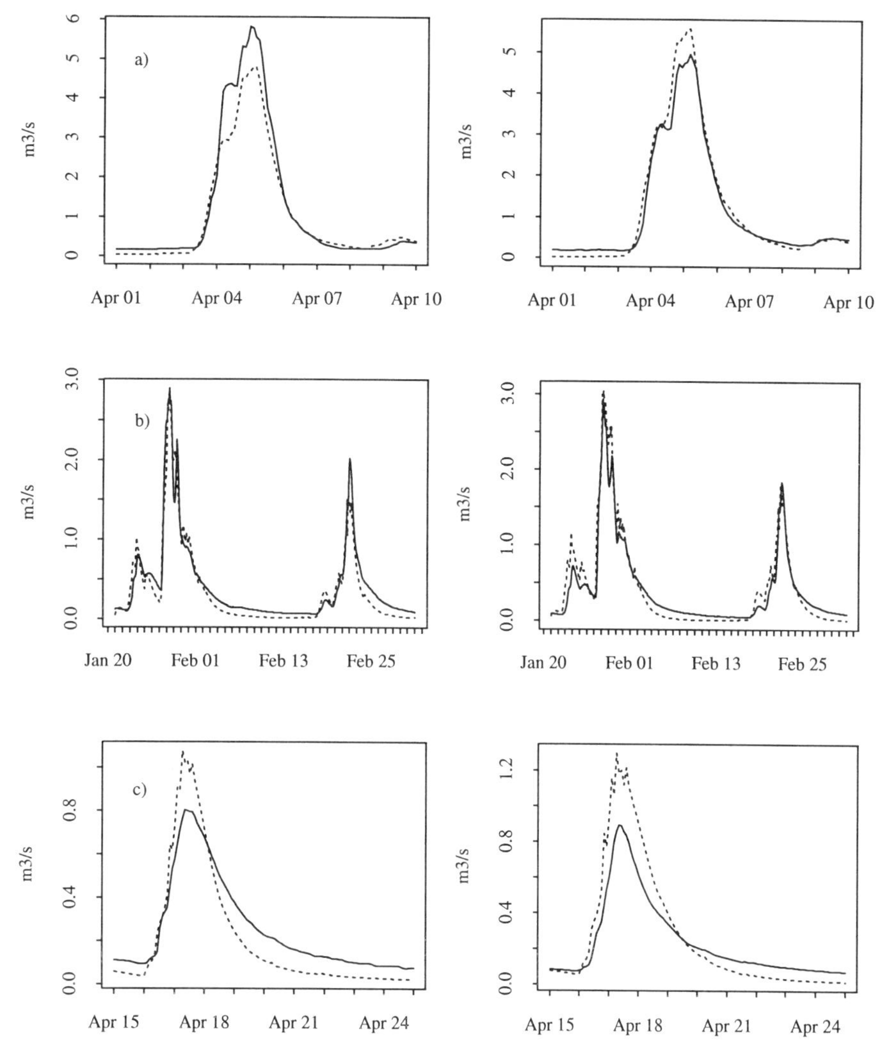

Figure 3: Observed vs. predicted Hard Creek (left) and Ware Creek (right) discharge for three events during model verification period: a) April 1-10, 1994, b) January 20-March 2, 1992, and c) April 15-25, 1992. Solid lines indicate observed discharge.

within Hard and Ware Creeks can be divided in three categories:

- Peak flow responding segments: Three road segments which only produce runoff during extreme rain events;
- Continuously responding segments: Three road segments which frequently continue to discharge over extended dry periods; and
- Frequently responding segments: Five segments which frequently respond to moderate rainfall events but do not continue to produce runoff during dry periods.

One culvert never registered a response. There are no clearly defined physical characteristics to distinguish these groupings, as discussed in Bowling and Lettenmaier [1999].

Inferences can be drawn regarding factors controlling road segment response based on the cross-section of physical characteristics represented by the various road segments. For the frequently responding culverts, the highest discharge values (normalized by contributing area) were associated with road segments draining slopes with immature vegetation and road segments which intercept subsurface flow, as observed in the field. The importance of subsurface flow interception relative to surface runoff is confirmed through observations of the relative flow volumes. The volume of rain falling on each road segment, for each storm monitored, was calculated for the hour of maximum precipitation rate. This volume of precipitation only accounts for 18% of the peak culvert runoff observed.

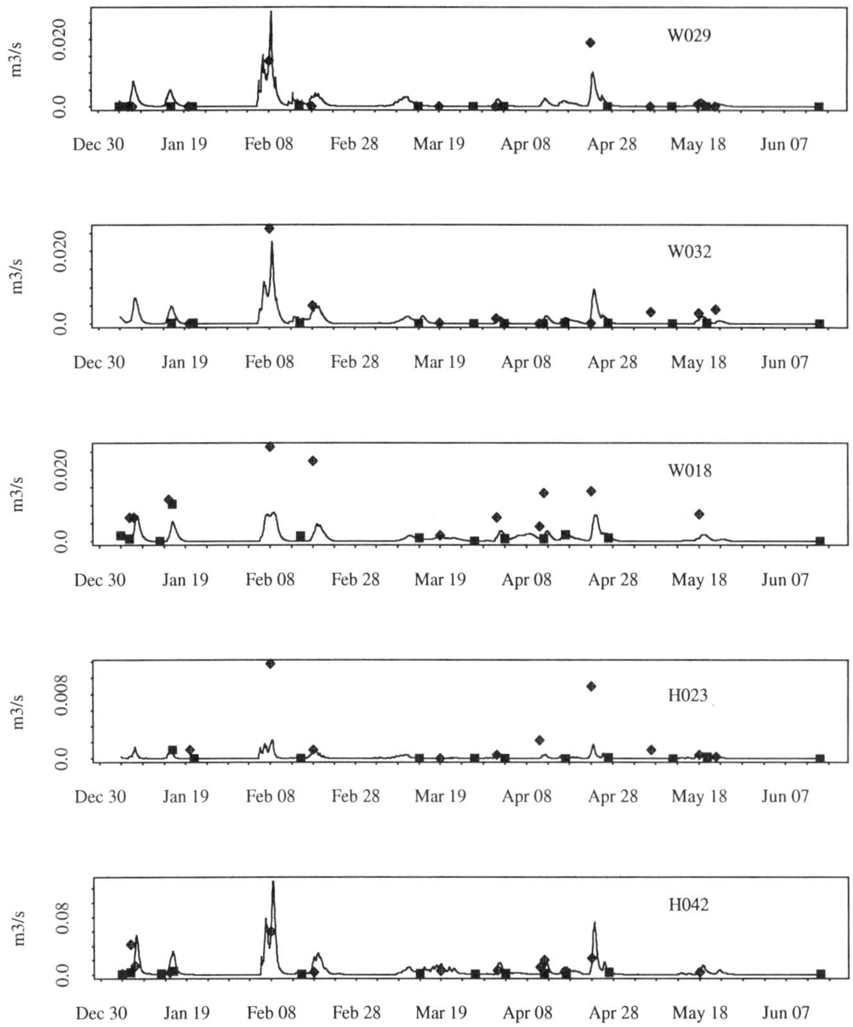

Figure 4: Monitored culvert discharge, January 4, 1996-June 15, 1996. Diamonds represent estimated peak discharges, squares are observed discharge and lines are the simulated culvert discharge.

If an average infiltration rate of 5 mm/hr is assumed, direct surface runoff can only account for 5% of the total observed culvert flow. The percentage varies greatly from culvert to culvert and storm to storm. In some cases, no culvert response is registered during precipitation events for road segments with contributing road surface areas. This indicates that either infiltration into the road and ditch or re-routing of surface flows from storm to storm tends to reduce the component of surface runoff intercepted by roads.

Simulated discharge from a representative cross-section of the monitored road segments is compared to observed point values in Figure 4. Diamonds represent peak stages measured between field visits (via crest stage recorders) and squares represent observed stages on field visit dates. The peak stages were assigned to the day of maximum rainfall, so some offset in timing may occur. The simulated discharge is seen to respond to individual rainfall events throughout the season, with little baseflow between events, which is consistent with field observations. Some of the culverts, such as H042, flow perennially throughout the year, which is also consistent with observations. Two of the culverts, W018 and H023, have much lower simulated responses than observed. H023 has a relatively small contributing area as delineated from the DEM. This culvert is located on a planar hillside with relatively little lateral relief. It is likely that the DEM is unable to capture the subtle distinctions in drainage topography along this hillside. In addition, the local soil depth relative to the road cut affects the amount of water storage in the pixel and therefore how quickly the water table can rise during a storm. It is likely that the underprediction of response from

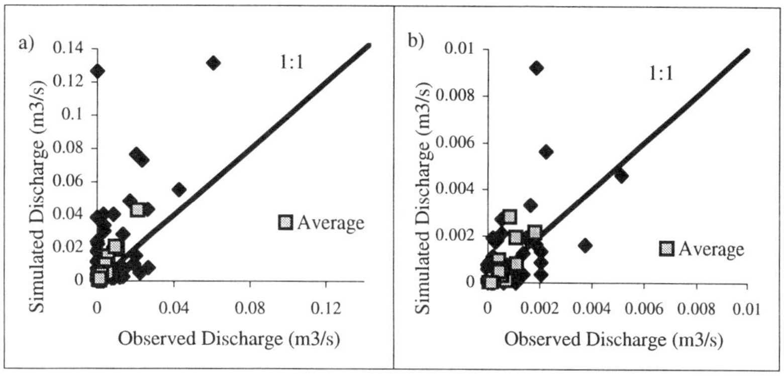

Figure 5: Simulated vs. observed culvert discharge: a) all peak flow events and b) all observed low flow events.

W018 could be corrected by adjusting either the soil depth or cut bank height for this location.

Simulations of road segments on the order of 100-200 m using pixels of 30 m width can at best be approximate. Therefore, it is important to evaluate road interception predictions in an aggregated way. Figure 5a compares modeled and observed discharge for peak events recorded between field visits. The gray boxes represent the average of all segments for each storm. This figure shows a tendency toward over-prediction of segment response as indicated by a higher than expected number of discharge peaks above the line of equality. The mean absolute error for all culverts and all events is on the order of 100%. Figure 5b shows a similar comparison for all low flow events in which discharge was measured directly . It is interesting to note that the overprediction in this case is smaller, with a mean absolute error of 37% for all events. This suggests that at least some of the apparent error in peak flow predictions may lie in the culvert stage discharge relationship. An alternative explanation would be that the model tends not to represent small scale peak flows as well as it does low flows.

In either case, the true model error appears to be less than suggested by Figure 5a. For the large storm response, the simulated hydrographs for culvert flow tend to be extremely flashy, with narrow peaks followed by very rapid recessions (see Figure 6). Although observed hydrographs that would allow comparison of total flow volumes are not available, data reported by LaMarche and Lettenmaier [1998] suggest that observed culvert hydrographs are somewhat less flashy than the simulated ones.

Road Network Effects on Peak Streamflow

The cumulative effect of the road network on peak flows was examined by analysis of both the peaks-over-threshold (POT) and annual maxima series (AMS) and through a frequency analysis. First the annual maxima peak flow series was extracted to analyze the effect of roads on the mean annual flood (average of the annual maxima). The base mean annual flood was determined using the model-simulated annual maxima series from October 1985 - June 1996 based on current (1996) vegetation and no imposed road network. Under current vegetation conditions, DHSVM indicates an 11 percent increase in both Hard and Ware Creeks' mean annual flood due to the road network. All but one of the annual maxima resulted from the same storm in the two simulations. The annual maximum for water year 1995 for Hard Creek changed from December 20, 1994 under the unroaded scenario, to November 30, 1994, when roads were simulated. Two of the eleven storms in Ware Creek and three in Hard Creek peaked between 2 and 6 hours faster with roads. In addition, two storms in Hard Creek peaked 18 and 20 hours earlier due to the road network.

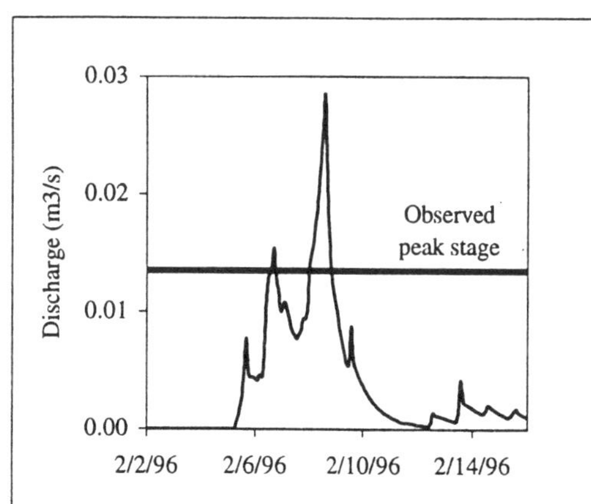

Figure 6: Simulated hydrograph of culvert H042, February2-16, 1996.

TABLE 4. Effect of Forest Roads on Four Largest Storms (1985 – 1996)

Storm Date	Peak magnitude w/o roads (m^3/s)	Peak magnitude w/ roads (m^3/s)	Percent change due to roads
	Ware Creek		
2/8/96	6.37	6.90	8.3
1/9/90	6.13	6.68	9.0
1/18/86	5.42	5.92	9.2
4/5/91	5.40	5.58	3.3
Average	5.83	6.27	7.5
	Hard Creek		
1/9/90	5.09	5.68	11.6
2/8/96	5.09	5.52	8.4
4/5/91	4.66	4.80	3.0
1/18/86	4.60	5.19	12.8
Average	4.86	5.30	9.1

TABLE 5. Effect of Forest Roads on Return Period

Return Period	Discharge w/o roads (m^3/s)	Discharge w/ roads (m^3/s)	Percent Change	Revised return interval w/ roads
		Hard Creek		
2	3.1	3.4	11.4	1.6 yrs
5	4.2	4.7	10.3	3.6 yrs
10	5.0	5.5	9.9	6.6 yrs
		Ware Creek		
2	3.6	4.0	11.8	1.6 yrs
5	5.0	5.5	9.4	3.7 yrs
10	6.0	6.5	8.4	7.0 yrs

To investigate road effects for higher flows, a threshold was next selected which would generate approximately one storm every two years for the 11-year record. Thresholds of 3.25 m^3/s for Hard Creek and 3.75 m^3/s for Ware Creek resulted in selection of four storm events based on the simulation without roads. The peak discharges for these four storms were compared to the peak discharges for the four largest storms generated for a simulation with roads (Table 4). For both catchments, the largest four storms were the same under both scenarios. This analysis indicated an 8 and 9 percent increase in the magnitude of the average peak discharge for the POT series for Ware and Hard Creeks, respectively.

A frequency analysis of the simulated annual flood series was performed by fitting an Extreme Value Type I distribution (EVI) to both the with- and without-road simulations. Parameters of the distribution were estimated using the method of moments. Goodness of fit was verified using the Kolomogorov-Smirnoff test. Although the length of the simulated discharge record limits analysis of extreme flows, the EVI distribution was used to investigate the effect of roads on events with return intervals of 2, 5 and 10 years. As summarized in Table 5, the forest road network as represented by DHSVM results in an increase of the ten year flood of 8 and 10 percent for Ware and Hard Creeks, respectively. Alternatively, this means that the flood with a ten-year return period in a harvested catchment without roads would have a return interval of 6.6 (7) years in Hard (Ware) Creek after road construction.

Table 5 indicates a slightly larger increase in discharge due to roads for the largest flows in Hard Creek as compared to Ware Creek, as would be expected from the higher road density in Hard Creek. For the two-year return interval flow and the mean annual flood, the two catchments show roughly the same change. Since these simulations were performed with 1996 (harvested) vegetation, Ware Creek contains a larger area with immature vegetation. It is possible that the elevated soil moisture in Ware Creek due to immature vegetation results in a greater response to roads in Ware Creek for lower return interval floods. For larger events, the soil moisture differences are less influential. This point is explored further in the next section.

Combined Forest Harvest and Road Construction Effects

DHSVM was also used to explore the link between forest harvest and road network effects on simulated peak discharge. Comparison of peak flows before and after timber harvest showed that the model predicted a larger percent increase for Ware Creek than for Hard Creek (Table 6). This is consistent with the harvest histories; the

TABLE 6. Effect of Roads and Harvest on Peak Discharge

	Increase in Mean Annual Flood				Increase in POT Series			
	Ware Creek		Hard Creek		Ware Creek		Hard Creek	
	m^3/s	%	m^3/s	%	m^3/s	%	m^3/s	%
Change due to roads w/ mature cover	0.39	11.7	0.38	12.6	0.43	8.4	0.44	9.7
Change due to roads w/ harvested (1996) cover	0.43	11.2	0.37	11.1	0.44	7.6	0.44	9.0
Change due to forest harvest w/o roads	0.50	14.8	0.29	9.7	0.62	12.0	0.34	7.6
Change due to forest harvest w/ roads	0.54	14.3	0.28	8.3	0.63	11.2	0.34	6.9
Change due to road construction and forest harvest	0.93	27.7	0.66	22.0	1.06	20.4	0.78	17.3

TABLE 7. Percent Increase in Peak Discharge for Various Return Intervals

	Ware Creek			Hard Creek		
Return Interval:	2 YR	5 YR	10 YR	2 YR	5 YR	10 YR
Change due to roads w/ mature cover	12.3	9.8	8.8	12.9	11.4	10.7
Change due to roads w/ harvested (1996) cover	11.8	9.4	8.4	11.4	10.3	9.9
Change due to forest harvest w/o roads	15.4	13.1	12.2	10.1	8.5	7.8
Change due to forest harvest w/ roads	14.9	12.7	11.8	8.6	7.5	7.0
Change due to road construction and forest harvest	29.0	23.7	21.7	22.6	20.0	18.5

extent of harvest in Ware Creek is much more than Hard Creek (66 vs. 35 percent, estimated). In addition, when comparing peak flows before and after road construction with a mature forest cover, the model predicted a larger percent increase for Hard Creek than for Ware Creek, consistent with the larger road density in Hard Creek (5 km/km^2 vs. 3.8 km/km^2). This proportional response means that the importance of roads relative to harvest depends on the area disturbed, although the scale of disturbance is markedly different for the causes (35-66% of basin area for harvest, 1.7-2.3% for roads). Although the simulated effects of harvesting and roads appear to be roughly equal, the magnitude of change for each treatment is quite different on a per-unit area basis.

Some previous paired catchment studies have suggested the possibility of a synergism between forest roads and timber harvest, meaning that the combined effect is greater than the additive effect of the individual components [Wright et al., 1990; Jones and Grant, 1996]. With respect to mean annual flow and the POT series, DHSVM simulations do not support this theory, although there is some suggestion that the existence of an interaction depends on the quantity of timber harvested. Specifically, the combined effect of forest harvest and roads is slightly larger than the sum of the individual effects in Ware Creek, while the combined effects are approximately equal to the additive effects in the less harvested Hard Creek catchment.

The effect of road construction and forest harvest on discharge return interval is summarized in Table 7. For the most part, the changes in mean annual flow and POT are consistent. The predicted increase in discharge due to roads in Hard Creek was larger than for Ware Creek and the predicted increase in discharge due to forest harvest in Ware Creek was larger than for Hard Creek. Again, the predicted increase in discharge due to roads and harvest combined was slightly larger than the sum of these individual treatments in Ware Creek. This apparent interaction was not present in the simulated Hard Creek response.

Road Effects on Catchment

The model simulations described in the previous sections show an approximately equal increase in peak flows due to both road construction and forest harvest. In this section we examine some of the causative mechanisms. The influence of the road network on simulated catchment soil moisture, as represented by the saturated zone thickness in Plate 1, is reflected in drier (green) areas beneath the roads.

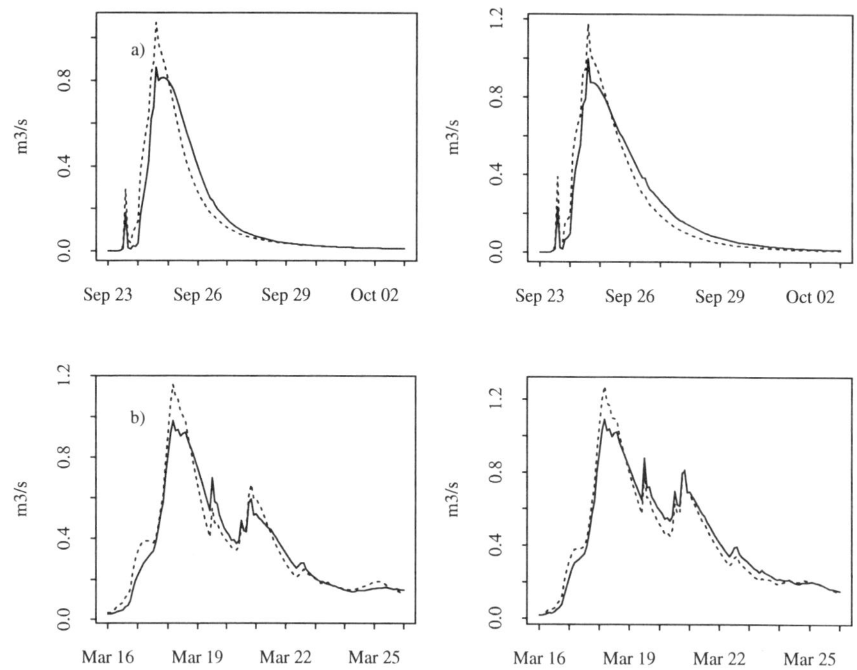

Figure 7: Simulated discharge with and without roads for Hard Creek (left) and Ware Creek (right). a) dry antecedent conditions, September 23, 1992 - October 3, 1992; b) wet antecedent conditions, March 16, 1994 - March 26, 1994. Solid lines indicate the simulation without the roads algorithm.

The first image shows conditions leading up to and following the extreme storm event of February 8, 1996. The points of culvert discharge also show up as localized areas of higher water table.

The simulated differences in soil moisture that develop during the storm decrease as the catchment dries out. The development of drier areas below the roads is likely to counteract the effect of the road network on streamflow, by decreasing predicted runoff from these hillslopes during storms. The diminishing differences in soil moisture between the roaded and non-roaded scenarios for drier conditions suggests that there could be a larger change in the timing and magnitude of peak flow due to roads for storms with dry antecedent conditions. Figure 7 shows simulated hydrographs for Hard and Ware Creek both with and without the road network for moderate storms with dry (September 23, 1992 - October 13, 1992) and wet (March 16, 1994 - March 26, 1994) antecedent conditions. Comparison of the storm hydrographs shows a slightly steeper ascent in the rising limb due to roads and an increase in peak discharge. However, the hydrograph rising limb does not seem to be substantially steeper due to roads under wet antecedent conditions than under dry conditions. In addition, the recession curve is steeper for the roaded scenario, reflecting the faster drainage of subsurface areas by conversion to surface flow. The simulated increase in peak discharge due to roads is approximately 17 percent for dry antecedent conditions and 14 percent for wet antecedent conditions. The smaller percent increase in peak for the March storm could be a reflection of the antecedent soil moisture, or the fact that the peak streamflow for the March storm is approximately 9 percent greater than the September 1992 peak.

The predicted road network – related changes in stored soil moisture have a small effect on the simulated response of the catchment to forest harvest (Plate 2). The left hand column of Plate 2 shows the difference in total soil moisture between harvest and no harvest scenarios for the case where both simulations had roads. In the right hand column both simulations were performed without the road algorithm. The simulated patterns of soil moisture are nearly identical in both cases, suggesting that road network effects do not interact with the harvest effects due to the difference in timing between the two. Maximum differences in soil moisture due to forest harvest occur at the beginning of the storm and largely result from decreased evapotranspiration and interception losses; differences due to roads occur during the storm and diminish as the catchment drains. The offset in timing is visible as the catchment dries out. Following forest harvest alone (Plate 2, right-hand column), areas adjacent to stream channels are slightly wetter due to the gradual

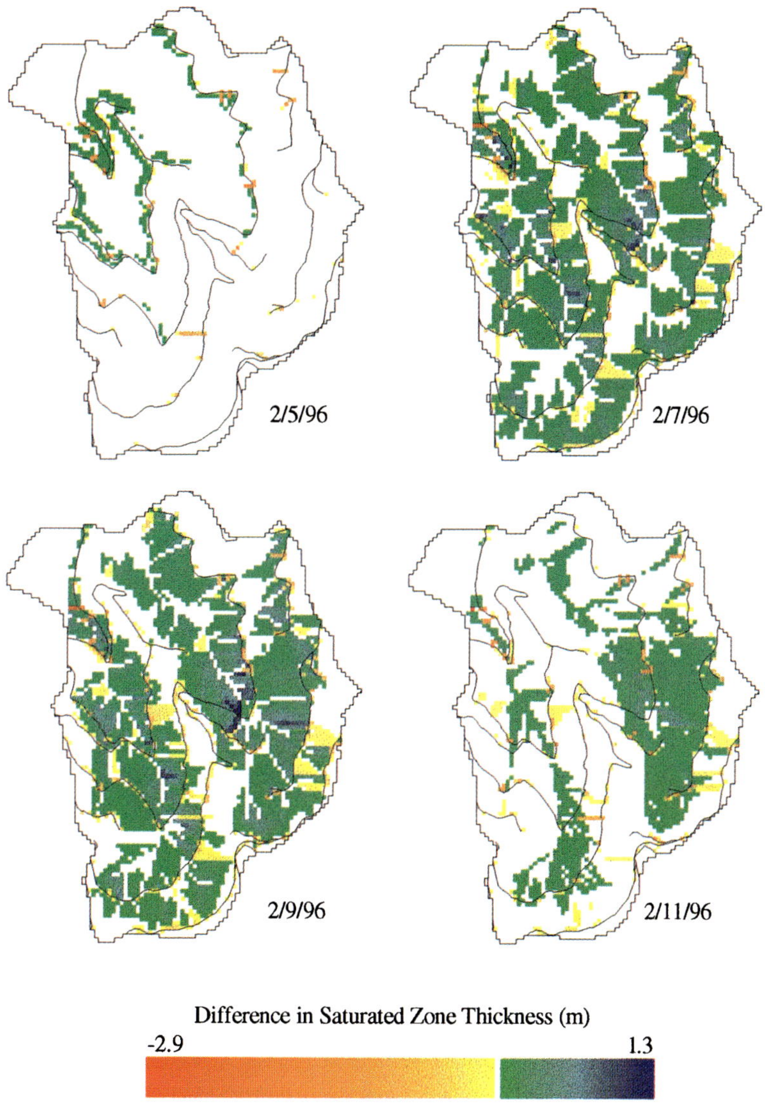

Plate 1: Difference in the saturated zone thickness (no road – roads) during the February 1996 storm. Positive differences (green) indicate areas that are drier following road construction.

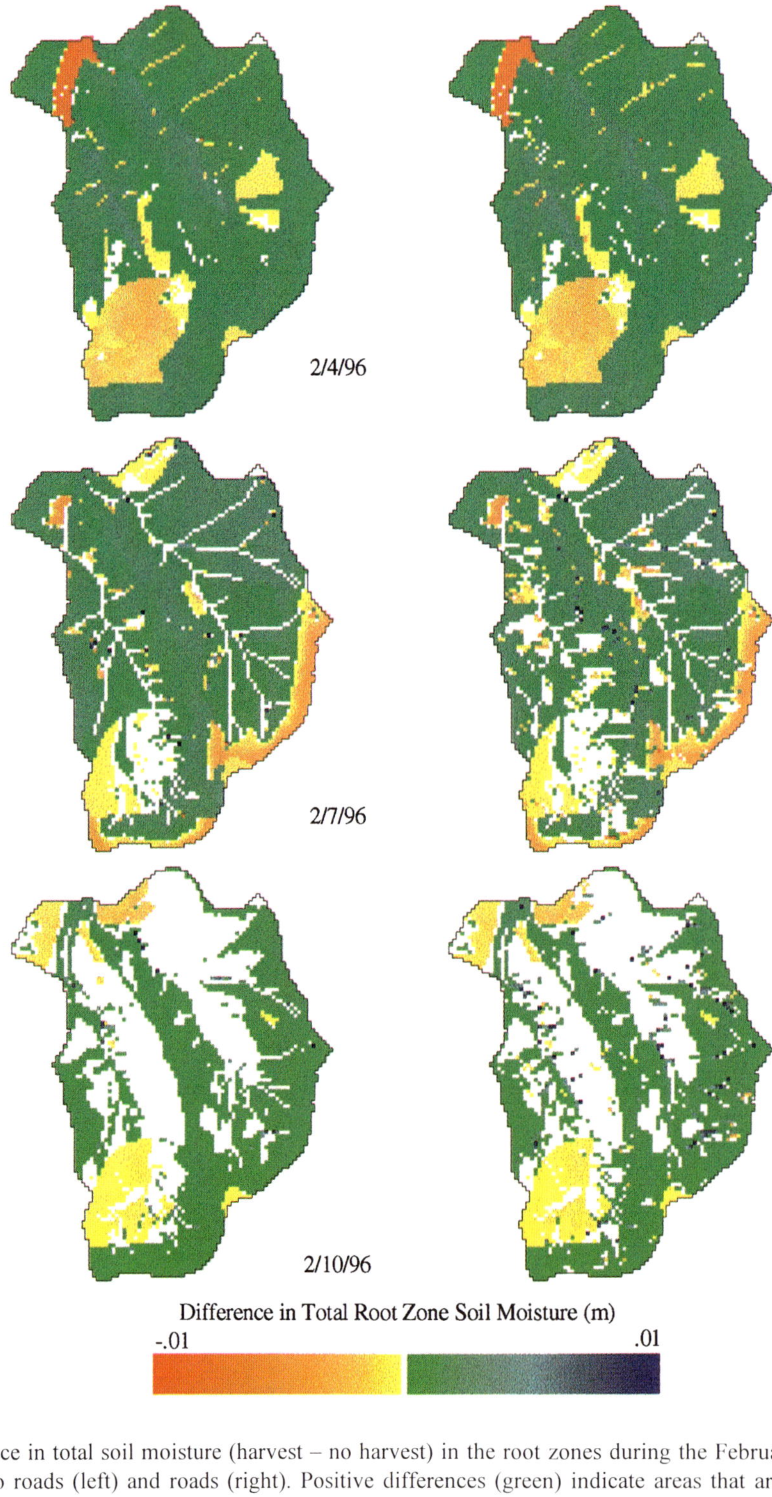

Plate 2: Difference in total soil moisture (harvest – no harvest) in the root zones during the February 1996 storm for two scenarios, no roads (left) and roads (right). Positive differences (green) indicate areas that are wetter following forest harvest.

draining of the pulse of subsurface recharge due to snowmelt. The reverse is true after road construction (Plate 1), by rerouting runoff through the drainage network the catchment is robbed of subsurface flow and the near-stream areas are drier following the storm. Therefore, under the scenario with both roads and harvest (Plate 2, left-hand side), the residual moisture along the stream channel on February 7th and 10th is not visible.

In addition to the initial soil moisture deficit, simulated increases in peak flow due to forest harvest are in large part due to modeled differences in snow accumulation and melt (Plate 3). The sequence of images in Plate 3 demonstrate that initially greater accumulation was modeled in much of the catchment following forest harvest, but as the storm progressed greater melt was predicted in the cleared areas. The black lines on these figures represent the boundaries of different forest stands.

Investigation of state variables at several points throughout a major rain-on-snow storm suggests that soil moisture deficits and enhanced snow accumulation and melt combine to increase peak flows following forest harvest. The forest road network, as represented by DHSVM, acts to redistribute soil moisture. This results in drier areas downslope of roads, but the timing is such that it does not cause a synergistic effect on peak flows with forest harvest.

CONCLUSIONS

The effect of forest roads on streamflow in a Pacific Northwest maritime mountainous catchment was evaluated using field observations and model predictions. Runoff intercepted by 12 road segments was monitored in Hard and Ware Creeks, two headwater catchments in the Western Cascades. It was observed that the magnitude of culvert response was controlled by subsurface flow interception rather than road surface runoff. The observed quantity of intercepted subsurface flow was higher beneath harvested hillslopes. A comprehensive survey of culvert drainage paths in Hard and Ware Creeks revealed that 45 and 57 percent of culverts in Hard and Ware Creeks, respectively, are directly connected to the drainage network.

The cumulative effect of the forest road network on Hard and Ware Creek streamflow was explored using the Distributed Hydrology-Soils-Vegetation Model (DHSVM) [Wigmosta et al. 1994, Storck et al. 1995, 1997], as modified to represent interception of subsurface flow and surface runoff by the road network [Wigmosta and Perkins, this volume]. The model was calibrated for the period July 1, 1993 through June 30, 1996. Comparison with catchment discharge and point observations of peak culvert discharge indicated that the model approximately represented the hydrological processes of Hard and Ware Creeks.

Multiple 11-year simulations (1985-1996) of Hard and Ware Creeks were performed using DHSVM for harvested, roaded and natural scenarios. Through the retrospective simulations it was possible to infer the following about peak flow increases:

- The increase in peak stream flow due to logging roads and forest harvest independently are approximately equal in magnitude. However, the magnitude of change for each treatment is quite different per unit area (8-15% increase in the mean annual flood per 35-66% area disturbed for forest harvest, 11-12% for roads per 2% area disturbed.
- Both the increases in peak flows due to forest harvest and due to the road network decrease with increasing return interval.
- The combined effects of forest harvest and road networks appear to be additive rather than synergistic.

Utilization of hydrologic models to test road design scenarios is an area of research that is still largely unexplored. However, a few management implications are readily apparent from our work. The model simulations suggest that culvert connectivity is a key contributor to peak flow increases. If all culverts are assumed to be disconnected from the channel network, the simulated average peak increase is only 6.9% in Ware Creek and 6.6% in Hard Creek (4.3% and 4.6% less, respectively, than if connected). Within Hard and Ware Creeks, 75% of culverts within 100 m of the stream were found to be connected (53% directly, 22% through gullies). All connected culverts were within 170 m of the inferred natural stream system. Lamarche and Lettenmaier [1999] and Wemple et al. [1996] examine some other causative factors for culvert connectivity. Further research will be required to evaluate how other factors that are susceptible to design, like road hillslope position and culvert spacing, affect peak flows in roaded forest catchments.

Acknowledgements. Funding for this research was provided by a Valle Scholarship to the first author, and by the Washington State Department of Natural Resources. In addition, the National Council for Air and Stream Improvement (NCASI) provided financial support for the field study. Hard and Ware Creek lie within the Weyerhaeuser Company Vail Tree Farm; Weyerhaeuser Co. allowed access to the sites and provided the long term hydrometeorological and land cover data used in this study.

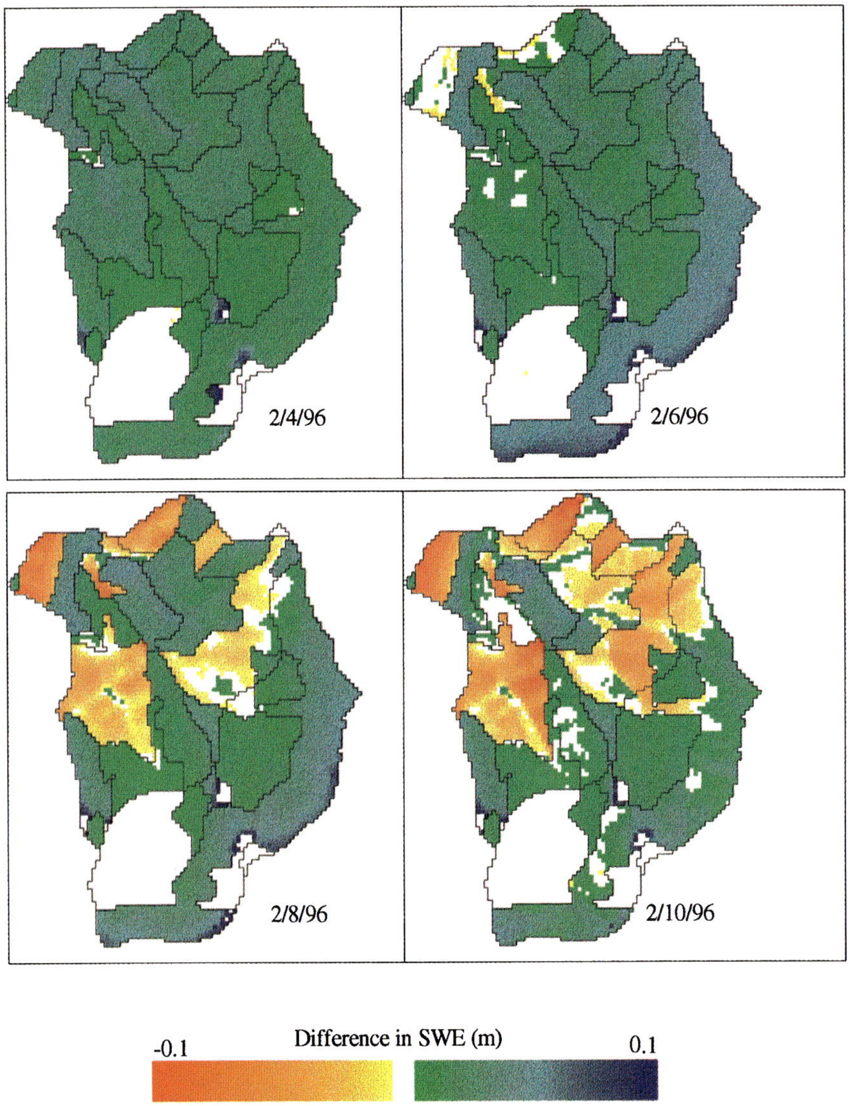

Plate 3: Difference in distribution of SWE (harvest – no harvest) during the February 1996 storm due to forest harvest. Positive differences (green) indicate areas with more snow following forest harvest.

REFERENCES

Berris, S.N. and R.D. Harr, Comparative snow accumulation and melt during rainfall in forested and clear-cut plots in the western Cascades of Oregon, *Water Resures Research*, 23 (1), 135-142, 1987.

Bilby, R.E., K. Sullivan and S.H. Duncan, The generation and fate of road-surface sediment in forested watersheds in southwestern Washington. *Forest Science*, 35 (2), 453-468, 1989.

Bosch, J.M. and J.D. Hewlett, A review of catchment experiments to determine the effects of vegetation changes on water yield and evepotranspiration, *Journal of Hydrology*, 55, 3- 23, 1982.

Bowling, L.C. and D.P. Lettenmaier, Evaluation of the effects of forest roads on streamflow in Hard and Ware Creeks, Washington, Water Resources Series Technical Report No. 155, University of Washington, 1997.

Folz, R.B. and E.R. Burroughs, Sediment production from forest roads with wheel ruts, *Watershed Planning and Analysis in Action*, Symposium Proceedings of IR Conference, ASCE, 266-275, 1990.

Harr, R.D., Some characteristics and consequences of snowmelt during rainfall in western Oregon, *Journal of Hydrology*, 53, 277-304, 1981.

Harr, R.D., Effects of clear-cut logging on rain-on-snow runoff in western Oregon: A new look at old studies, *Water Resources Research*, 22(7), 1095-1100, 1986.

Harr, R.D., A. Levno and R. Mersereau, Streamflow changes after logging 130-year-old Douglas fir in two small watersheds, *Water Resources Research*, 18, 637 – 644, 1982.

Jones, J.A. and G.E. Grant, Peak flow responses to clearcutting and roads in small and large basins, western Cascades, Oregon, *Water Resources Research*, 32 (4), 959-974, 1996.

Kalnay, E., M. Kanamitso, R. Kistler, W. Collins, D. Deaven, L. Gandin, M. Iredell, S. Saha, G. White, J. Woolen, Y. Zhu, A. Leetma, R. Reynolds, M. Chelliah, W. Ebisuzaki, W. Higgins, J. Janoviak, K.C. Mo, R. Jenne, and D. Joseph, The NCEP/NCAR 40-Year Reanalysis Project, Bulletin of the American Meteorological Society, 77:437-471, 1996.

Kattelmann, R., Effects of forest cover on a snowpack in the Sierra Nevada, in *Watershed Planning and Analysis in Action*, ASCE, New York: New York, 276-284, 1990.

Keppeler, E.T. and R.R. Ziemer, Logging effects on streamflow: water yield and summer low flows at Caspar Creek in nortwestern California, *Water Resources Research*, 26 (7), 1669-79, 1990.

LaMarche, J. and D. Lettenmaier, Forest road effects on flood flows in the Deschutes River basin, Washington, Water Resources Series Technical Report No. 158, University of Washington, 1998.

Luce, C.H. and T.W. Cundy, Parameter identification for a runoff model for forest roads, *Water Resources Research*, 30, 1057-1069, 1994.

Megahan, W.F., Subsurface flow interception by a logging road in mountains of central Idaho, paper presented at Syposium on watersheds in transition, American Water Resources Association, Ft. Collins, Colorado, June 1972.

Megahan, W.F., Hydrologic effects of clearcutting and wildfire on steep granitic slopes on Idaho, *Water Resources Research*, 19 (3), 811-819, 1983.

Megahan, W.F., Effects of forest roads on watershed function in mountainous areas, Environmental Geotechnics and Problematic Soils and Rocks, Balasubramaniam et al. (eds.), Balkema, Rotterdam, 1987.

Montgomery, D.R., Road surface drainage, channel initiation and slope stability, *Water Resources Research*, 30, 1925-1932, 1994.

Normann, J.M., R.J. Houghtalen and W.J. Johnston, Hydraulic Design of Highway Culverts, Federal Highway Administration Technical Report No. FHWA-IP-85-15, 1985.

Piehl, B.T., R.L. Beschta and M.R. Pyles, Ditch relief culverts and low-volume forest roads in the Oregon Coast Range, *Northwest Science*, 62 (3), 91-98, 1988.

Reid, L.M. and T. Dunne, Sediment production from forest road surfaces, *Water Resources Research*, 20 (11), 1753-1761, 1984.

Rosencrantz, S.D., P. Storck and D.P. Lettenmaier, Statistical analysis of logging effects on flooding in the Snoqualmie River basin, Washington, poster presented at the fall meeting of the American Geophysical Union, San Francisco, CA, 1995.

Ross, D.G., I.N. Smith, P.C. Manins, D.G. Fox, Diagnostic wind field modeling for complex terrain: model development and testing, *Journal of Applied Meteorology*, 27, 785-796, 1988.

Rothacher, J.. Streamflow from small watersheds on the western slope of the Cascade Range of Oregon, *Water Resources Research*, 1, 125 – 134, 1965.

Rothacher, J., Increases in water yield following clear-cut logging in the Pacific Northwest, *Water Resources Research*, 6(2), 653-658, 1970.

Schmidt, R.A. and C.A. Troendle, Sublimation of intercepted snow as a global source of water vapor, Proceedings, 60th Western Snow Conference, 1992.

Storck, P., D.P. Lettenmaier, B.A. Connelly and T.W. Cundy, Implications of forest practices on downstream flooding, Phase II final report, 1995.

Storck, P., T. Kern and S. Bolton, Measurement of differences in snow accumulation , melt and micrometeorology between clear-cut and mature forest stands, Proceedings of the Western Snow Conference, Banff, Alberta, Canada, 1997.

Sullivan, K., S.H. Duncan, P.A. Bisson, J.T. Heffner, J.W. Ward, R.E. Bilby and J.L.Nielsen, A Summary Report of the Deschutes River Basin, Sediment, Flow, Temperature and Fish Habitat, Weyerhaeuser Company, Technical Report, Paper No.044-5002/87/1, 1987.

U.S., Army Corps of Engineers, Snow Hydrology; Summary report of the snow investigations, Portland, OR, NNorth Pacific Division, Corps of Engineers, 1956.

Wemple, B.C., J.A. Jones and G.E. Grant, Channel network

extension by logging roads in two basins, Western Cascades, Oregon, *Water Resources Bulletin*, 32 (6), 1195 – 1207, 1996.

Wigmosta, M.S. and W.P. Perkins, Simulating the impacts of road drainage in a distributed hydrologic model. *This issue*.

Wigmosta, M.S., L.W. Vail and D.P. Lettenmaier, A Distributed Hydrology-Vegetation Model for Complex Terrain. *Water Resources Research*, Vol. 30, No. 6, 1665-1679, 1994.

Wright, K.A., K.H. Sendek, R.M. Rice and R.B. Thomas, Logging effects on streamflow: storm runoff at Caspar Creek in northwestern California, *Water Resources Research*, 26 (7), 1657-1667, 1990.

Ziemer, R.R., Storm flow response to road building and partial cutting in small streams in northern California, *Water Resources Research*, 17, 907-917, 1981.

Spatial and Temporal Patterns in Erosion from Forest Roads

Charles H. Luce and Thomas A. Black

Rocky Mountain Research Station, U.S. Forest Service, Boise, Idaho

Erosion from forest roads is an important contribution to the sediment budget of many forested basins, particularly over short time scales. Sediment production from 74 road segments was measured over three years to examine how road slope, segment length, cutslope height, and soil texture affect sediment production and how these relationships change with time. In the first year, differences in sediment production between plots could be explained by differences in sediment transport capacity of the plots. With time, differences between plots of different slope, length, cutslope height, and soil were reduced as all plots produced less and less sediment. Recovery was rapid with around 70% recovery between by the second year and 90% recovery by the third year.

1. INTRODUCTION

1.1 The Role of Road Surface Erosion in the Sediment Budget

Networks of forest roads traverse many mountain and forested regions in the western United States. Roads impose substantial local changes to soil properties and hydrologic behavior and commonly alter sediment budgets of the basins they pass through. The net effect of the changes is to increase the supply of sediment to streams through surface erosion on the bare road surfaces and mass erosion and gullying below the road.

Evaluation of the sediment budget is an important step in analyzing potential problems caused by the increased sediment supply from roads. Reid and Dunne [1984] suggest that the sediment budget be used to answer two questions with respect to forest roads: (1) "How much is contributed by surface erosion on roads?" and (2) "How important is it relative to other sources?" The answers to these two questions do not well serve the purposes of managing and reducing sediment inputs from roads unless we also ask, (3) "Where is the contribution coming from?" and (4) "When is the sediment contributed?"

The question of the relative magnitude of sediment sources (2, above) has been raised several times [Gilbert, 1917; Dietrich and Dunne, 1978; Reid, 1981; Reid and Dunne, 1996] and is one of the primary purposes for calculating a sediment budget. In some regions, roads may promote gully formation or headward channel migration and decrease slope stability through concentration of water [e.g. Fredriksen, 1970; Sidle et al., 1985; Furniss, 1991; Montgomery, 1994]. In other regions, surface erosion from the bare soil surfaces comprising roads is the dominant mode of sediment contribution from roads. The question of relative magnitude (2) has several levels of importance. At one level it can assist in prioritizing erosion abatement practices; at another, it can assist in prioritizing information needs for constructing sediment budgets. Similarly, resources for acquiring information to improve sediment budgets would best be spent reducing uncertainty in estimates for the greatest contributors. Given these practical goals for sediment budgets, questions (1) and (2) should be answered in the context of clearly defined temporal and spatial scales.

We need to know the spatial (3, "from where") and temporal (4, "at what time") patterns of sediment

Land Use and Watersheds: Human Influence on Hydrology and Geomorphology in Urban and Forest Areas
Water Science and Application Volume 2, pages 165-178

Published in 2001 by the American Geophysical Union

production from forest roads relative to other sources for small areas and short time scales as well as for long time frames and whole watersheds. These questions suggest prioritization based on relative contributions in time and relative contributions from different locations. Reasons for prioritization of erosion abatement and information needs at small spatial scales (location) are considered "common sense"; they are practically oriented and based in part on budgets and the capability of human engineering.

Short time scales (on the order of one to a few years) may be important in constructing sediment budgets intended to identify and prioritize opportunities for erosion abatement. Long-term sediment budgets often have the aim of quantifying relative magnitudes of the various processes contributing to landscape evolution [Dietrich and Dunne, 1978]. Large-magnitude low-frequency sediment delivery events may well dominate the long-term sediment delivery to stream systems, and therefore landscape form and stream sediment volume and structure, in places like the Oregon Coast Range [Benda and Dunne, 1997; Benda et al., 1998]. These events are largely driven by large winter rainstorms with recurrence intervals on the order of one to a few decades. "Recovery" of streams in temperate regions following a major flood/erosion event may occur within a few years to two decades [Wolman and Gerson, 1978; Megahan et al., 1980]. If major mass wasting erosion events are sufficiently spaced in time that "recovery" occurs on shorter time scales, processes that introduce sediment on a more frequent basis, like road surface erosion, become important concerns in determining the most common stream condition in time. Changes to stream condition from these higher-frequency low-magnitude effects may include changes in the surface bed composition (e.g. cobble embeddedness, surface fines) and turbidity. Biologists refer to the large-magnitude low-frequency events as "pulsed" [Yount and Niemi, 1990] as opposed to the chronic (high-frequency) lesser magnitude "press" effects resulting from road erosion. While many aquatic species are adapted to large-scale infrequent disturbances, press disturbance above "natural" levels may pose special problems for species persistence.

1.2 Estimating the Contribution of Road Surface Erosion to the Sediment Budget.

Distributed modeling of sediment production and delivery from forest roads [e.g. Cline et al., 1984; USDA Forest Service Northern Region, 1991; Washington Forest Practices Board, 1995; Dubé et al., 1998] estimates sediment production from short segments of road based on characteristics of each segment. Estimates from individual segments are commonly based on traffic level and road use [Reid and Dunne, 1984; Bilby et al., 1989], soil texture, surfacing [Swift, 1984; Burroughs and King, 1989; Foltz and Elliot, 1997], time since construction [Megahan, 1974], and surface protection on cut and fill slopes [Burroughs and King, 1989; Megahan et al., 1992].

Technological advances in remote sensing, global positioning technology, and geographic information systems increase the ease with which we can collect and assimilate data about road networks. As information about the spatial distribution of road slope, cutslope height, cutslope condition, soil texture, maintenance practices, culvert location, become more readily available, it is important to understand their effects so that those effects may be described in distributed road erosion models. Because road segments with recently constructed or cleared ditches can produce so much more sediment than older roads [Megahan, 1974; Luce and Black, 1999], it is also important to understand how the effects of road characteristics such as slope, texture, and cutslope height may change over time.

Observations on the effects of road slope on sediment production [e.g. Vincent, 1985; Burroughs and King, 1989; MacDonald et al., 1997; Black and Luce, 1999; Luce and Black, 1999] are increasing in quantity and are beginning to note that the effect of slope on sediment production changes with time [Black and Luce, 1999]. Similarly, there are few publications on the effects of cutslope height, road segment length, soil texture, and maintenance practices and how those relationships change with time. These are important attributes of forest roads and observations are needed to describe these effects in distributed empirical models and verify predictions of physically based models such as ROSED [Simons et al., 1980], KINEROS [Woolhiser et al., 1990], or WEPP [Flanagan and Nearing, 1995; Tysdal et al., 1999].

This study was organized to collect observations on the relationships between sediment production and segment length, road slope, cutslope height, soil texture, and time since disturbance. Results from the first year of measurements are presented in Luce and Black [1999]. We also desired to explore changes of erosion from roads with time and test the applicability of Megahan's [1974] model.

In this paper we 1) Describe the relationship of sediment production to a few road characteristics (slope, cutslope height, length of drainage unit, and soil texture) that can be mapped on a road network, 2) Describe how these relationships change with time following disturbance for a short time period, and 3) Use insights gained from these examinations to suggest a conceptual model of the sediment "micro-budget" of the road-tread-ditch combination.

2. THEORY

Some expectations about the relative production of sediment from road segment to road segment and from time period to time period can be derived from our existing understanding of sediment transport and soil erosion processes. Mass conservation dictates

$$E = \nabla \bullet Q_s \quad (1)$$

where E is the change in storage of soil in an area (Erosion) and Q_s is the sediment transport rate. This equation means that for a small volume above a small area on the ground (infinitesimally small in both cases) the amount of sediment leaving the volume is the same as the amount flowing into the volume plus any erosion that occurs over the small area. For a small watershed (such as the cutslope, tread, and ditch of a road), equation 1 can be evaluated as

$$E = Q_{s(in)} - Q_{s(out)} \quad (2)$$

Where $Q_{s(in)}$ is 0, and $Q_{s(out)}$ depends on the transport capacity and supply to the exit point. If there is too much supply, material will be deposited before the exit point, and the output will match the transport capacity. If the supply does not meet the transport capacity at that point (the more common case) plot erosion is dictated by the amount that arrives at that point. Such a calculation requires estimating detachment along the slope. Detachment at some point along the slope is not necessarily related to transport capacity. Foster and Meyer [1972; 1975] and Lei et al. [1998] describe methods to account for the difference between transport capacity and actual sediment flux. In general, however, on long plots with easily detached non-cohesive materials (such as a ditch immediately following a grading operation), there is a close relation between transport capacity at the end of hillslope and the actual sediment discharge [Kirkby, 1980].

Consequently, immediately following disturbance, we would expect sediment production to be tied to transport capacity. Over time, processes of armoring, pavement formation, and vegetation growth reduce availability of fine sediments. We would expect to see relative sediment production tied to relative indexes of availability, such as vegetation growth or armoring [e.g. Megahan, 1974].

2.1 Length and Slope

At a point sediment transport capacity can be defined by one of two models:

$$Q_s = k(\tau - \tau_c)^{n_\tau} \quad (3)$$

$$Q_s = k(\Omega - \Omega_c)^{n_\Omega} \quad (4)$$

where k is some index of mobility of the sediment, τ is shear stress, τ_c is the critical shear stress for incipient motion, n_τ is an exponent between 1.5 and 1.9 [Foster and Meyer, 1975; Kirkby, 1980], Ω is the stream power and Ω_c is the critical stream power for incipient motion, and n_Ω is an exponent between 1.1 and 1.5 [Govers, 1992; Bagnold, 1977]. Shear stress, τ, is given by

$$\tau = \rho_w gds \quad (5)$$

where g is gravity, d is the depth of flow (alternatively hydraulic radius), and s is the water surface slope, usually accepted to be the same as the bed slope. Stream power, Ω, is given by

$$\Omega = \rho_w gqs \quad (6)$$

where q is the flow per unit width. Bringing in a simple relationship for the hydrology of a particular event, considering a nearly impermeable road at steady state flow,

$$q \propto x \quad (7)$$

where x is distance downslope, and

$$d \propto \sqrt{q} \quad (8)$$

[Dunne and Dietrich, 1980] yields two approximations for sediment transport at the end of the ditch. By the argument stated earlier regarding the close relationship between transport capacity and sediment flux, road segment sediment production from a segment of length, L:

$$E \propto k\left(s\sqrt{L} - \tau_c\right)^{n_\tau} \quad (9)$$

$$E \propto k\left(sL - \Omega_c\right)^{n_\Omega} \quad (10)$$

In the first model, based on shear stress transport, erosion is roughly proportional to the slope times the square root of length, and the in the other it is proportional to slope times length. Both equations suggest a statistical interaction effect for length and slope. Luce and Black [1999] found that sediment production from recently disturbed plots (E) was proportional to length (L) times slope (S) squared,

$$E \propto LS^2 \quad (11)$$

which exhibits the interactive behavior suggested by theory and agrees well with observations [e.g. Wischmeier and Smith, 1978; McCool et al., 1987; Burroughs and King, 1985; Vincent, 1985; Renard et al., 1994].

Over time, as the initially available material is eroded and surface armoring and plant growth limit the availability of sediment, there is an expectation that transport capacity would be less important in determining sediment yield leaving the plot. Even when vegetation regrowth is slow, we would expect local particle size distributions to shift towards larger particles in locations with steeper slopes and larger contributing areas. After some period of erosion, then, contributing length and slope would not only affect local shear stress and stream power, but also the critical value required to initiate sediment movement.

2.2 *Soil Texture*

Depths of flow and turbulence are not so great from a 100-m long road segment that all soil particles travel as suspended load. Larger particles move more slowly than smaller particles in saltating transport. Burroughs et al. [1992] attempted to describe forest soil erodibility as primarily a function of soil texture and found that erodibility (for 0.6-m^2 plots under a rainfall simulator in a laboratory) was low for both clay dominated and sand dominated soils and higher for soils with a high silt fraction. They also noted a dependence on clay minerology. Other authors [e.g. Carling et al., 1997] have seen similar behavior with respect to soil texture, and the general trend seems to be reflected in erodibility estimates published in soil surveys. Because the ditch is commonly set in the native soil, we expect that road segment sediment production would initially be greater on silty soils than on sandy soils. As discussed in the previous section selective transport in the ditch may reduce the importance of native soil texture with time.

2.3 *Cutslope Height*

The effects of cutslope height on sediment production must be considered in light of the relative roles of transport capacity and sediment supply. Conceptually, flow and transport in the ditch control the sediment yield of a road segment. Most of the water probably comes from the road tread, where infiltration capacities are low compared to most rainfall events. The cutslope contributes mostly loose material to the ditch through a variety of processes, including soil creep, sheet wash, rilling, raveling, and slumping, and we would expect more material to come from higher cutslopes. If the ditch already has a large loose sediment supply, separate from the cutslope contribution (for example from a recent ditch blading operation), we would expect that the cutslope height would make little difference. Once armoring begins in the ditch, material added from the cutslope should become an important contribution to the loose material supply. So initially, following disturbance to the ditch and cutslope, we would expect little difference in sediment production from plots with different cutslope heights. In later years, however we would expect to see some effect from cutslope height.

2.4 *Time Following Construction or Disturbance*

Megahan [1974] describes a model of erosion over time following disturbance, t [days], based on the idea of an initially available amount of soil, S_0 [tons/mi^2], a rate constant, k [day^{-1}], and a long term constant erosion rate, ε_n.

$$\varepsilon(t) = \varepsilon_n + kS_0e^{-kt} \qquad (12)$$

from this, we can see that the initial erosion rate

$$\varepsilon_0 = \varepsilon_n + kS_0 \qquad (13)$$

and that the total soil eroded in increase over the long term rate is S_0. The S_0 parameter should relate to soil characteristics like the volume fraction taken by non-transportable rock fragments. S_0 should also related to the transport capacity, as more strongly flowing water has access to a greater portion of the particle size distribution of the soil. The k parameter describes how quickly erosion rates move from high initial rates toward the background rate. It may, in part, be a function of transport capacity. With a high k, the initial rate will be high, and armoring will occur rapidly. Lower k values predict a longer decay period with overall lower erosion rates.

In some situations, the effect of varying transport capacity from plot to plot will appear more strongly in the k parameter and in others more strongly in the S_0 parameter. If the volume of transportable material is entirely transportable by all flows regardless of the transport capacity, the variation will manifest in the k parameter, with high k values and rapid declines associated with higher transport capacities and low k values and gentle declines associated with low transport capacity. An example of this might be if uniform fine sand were the material to be transported. Alternatively, if the material to be transported were heterogeneous in character, the volume to be transported, S_0, would also be a function of the transport capacity. Variations in S_0 do not alter the relative

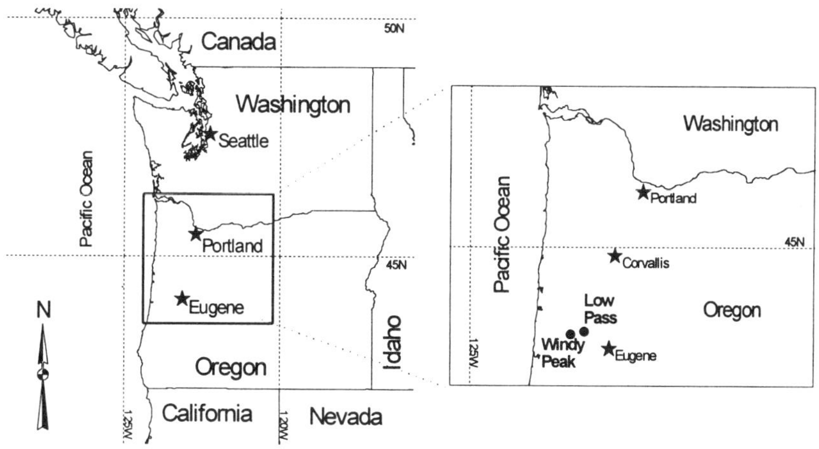

Figure 1. Location map with inset of northwestern Oregon showing the location of the two study areas, Low Pass and Windy Peak.

timing of the recovery. It is important to know how these two parameters might change with changing transport capacity (segment length or slope) on forest roads.

3. FIELD MEASUREMENTS

3.1 Description of Study Sites

We measured annual sediment production from 74 road segments in the Oregon Coast range (Figure 1) over three years to test these expectations. The central Oregon Coast Range receives between 1800 and 3000 mm of rainfall annually, with drier portions being further inland and wetter portions near the crest [Miller et al., 1973]. Winters are mild and wet; summers are warm and dry. Plots are located between 250 m and 600 m in elevation, below elevations where snow commonly accumulates. Soils are derived from sedimentary and metasedimentary rocks through most of the Coast Range with some igneous dikes in the inland foothills. The Tyee arkosic sandstone formation is the dominant bedrock throughout this part of the Coast Range. Douglas-fir and Western Hemlock forests cover much of the Coast Range.

Two field areas were used to examine sediment production on two soil textures. Many of the plots were located near Low Pass, Oregon. These sites were on the finer textured soils of the inner Coast Range. Soil series at Low Pass were Jory and Bellpine silty clay loams. The Jory soil is a clayey, mixed, active, mesic Palehumult; and the Bellpine soil is a clayey, mixed, mesic Xeric Haplohumult. The other plots were located near Windy Peak, 15 km west of Low Pass and had coarser soils. The soils at Windy Peak were the Bohannon gravelly loam, a fine-loamy, mixed, mesic Andic Haplumbrept, and the Digger gravelly loam, a loamy-skeletal, mixed, mesic Dystric Eutrochrept.

All road segments were selected in clearcut areas to prevent differences in precipitation due to variable interception by forest cover. All segments had high quality basalt aggregate surfacing and received infrequent administrative and recreational traffic from light vehicles.

3.2 Measurement Procedures

Large tank style sediment traps (~1.5 m^3) similar to those described by Ice [1996] collected sediment produced from insloped road segments with rubber-flap waterbars bounding the top and the bottom on the road surface (Figure 2). The plots were installed in the summer of 1995. Measurements of erosion were taken on a water year (beginning October 1) basis, so they cover Water Years 1996, 1997, and 1998.

At the end of each water year, we weighed the sediment traps containing sediment (M_{ts}) and again filled with just clean water (M_{tw}). For both weighings, the tank was topped off with water so that the volume was always the same. These weights reflect the contents of the tank:

$$M_{ts} = M_t + V_s\rho_s + V_w\rho_w \quad (14)$$

$$M_{tw} = M_t + \rho_w(V_s + V_w) \quad (15)$$

where M_t is the mass of the tank, V_s is volume of soil solids, ρ_s is the density of soil solids (we assumed a value of 2.65 Mg/m^3), ρ_w is the density of water (1.0 Mg/m3),

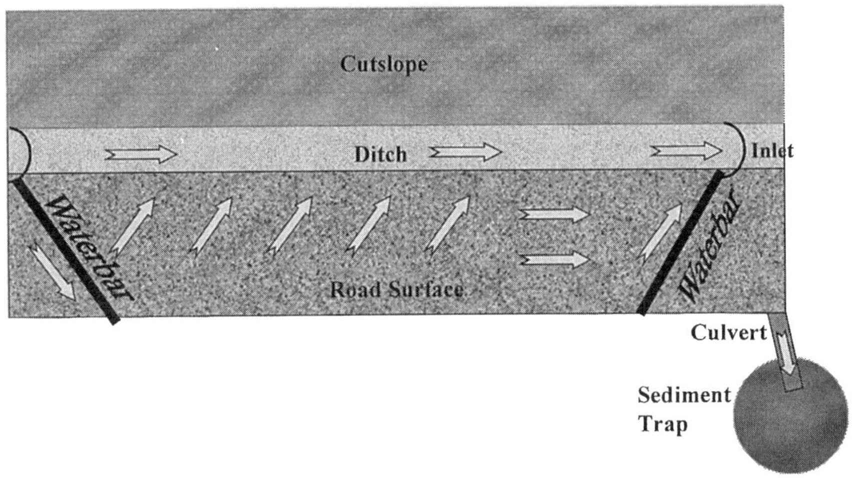

Figure 2. Typical plot layout. Water flows through 8-inch plastic pipe under road surface from inlet to sediment trap. Arrows indicate approximate patterns of water flow.

and V_w is the volume of water in the tank. Note that $V_s + V_w$ is the volume of the tank. The difference in weight between the tank containing sediment and the tank full of clean water is the submerged weight of sediment.

$$M_{ts}-M_{tw} = V_s(\rho_s-\rho_w) \qquad (16)$$

and the mass of sediment (M_s) is just $\rho_s V_s$, yielding

$$M_s = (M_{ts}-M_{tw})\rho_s/(\rho_s-\rho_w) \qquad (17)$$

A few measurements and modeling indicated that for the range of flows we observed and soil particle size distributions of the sites measured that the traps probably caught at least 90% of the sediment entering [Luce and Black, 1999]. Three tanks were overtopped by sediment in the first year; we estimated volumes of two of them from local deposits on the concrete pads holding the tanks.

We also measured climate variables at two weather stations near the plots. Data collected included precipitation, temperature, humidity, incoming solar radiation, wind speed, and wind direction. Precipitation was measured in a 0.01-inch-per-tip tipping bucket gage. Half-hourly binned data were used to estimate the annual Erosivity Index (EI) [Wischmeier and Smith, 1978] values for each of the three years.

4. EXPERIMENTAL DESIGN AND DATA ANALYSIS

The plots were managed in 5 groups according to type and timing of disturbance to observe how erosion changed with time following disturbance.

Group I:	No Treatment. (N=5, All at Low Pass)
Group II:	Road surface graded at beginning of WY 1996. No treatment thereafter. (N=18, 9 at Low Pass, 9 at Windy Peak).
Group III:	Road surface graded and ditch and cutslope scraped clear of vegetation at the beginning of WY 1996, no treatment thereafter. (N=31, 22 at Low Pass, 9 at Windy Peak)
Group IV:	Road surface graded at beginning of WY 1996. Road surface graded and ditch and cutslope scraped clear of vegetation at the beginning of WY 1997, no treatment thereafter. (N=15, 10 at Low Pass, 5 at Windy Peak)
Group V:	Road surface graded at beginning of WY 1996. Road surface graded and cutslope scraped clear of vegetation at the beginning of WY 1997, and again at beginning of WY 1998. (N=5, All at Low Pass)

This organization of plots was designed to isolate the effects of clearing the ditch and cutslope from the effects of changes in weather between years. The effect of weather alone is reflected in the comparison between years for a subsample of five plots from Group III for WY 1996 and the five plots in Group V for WY 1997 and 1998. Although the plots changed between WY 1996 and WY 1997, all five plots in Group V had the same length-slope combinations as the subsample of five plots from Group

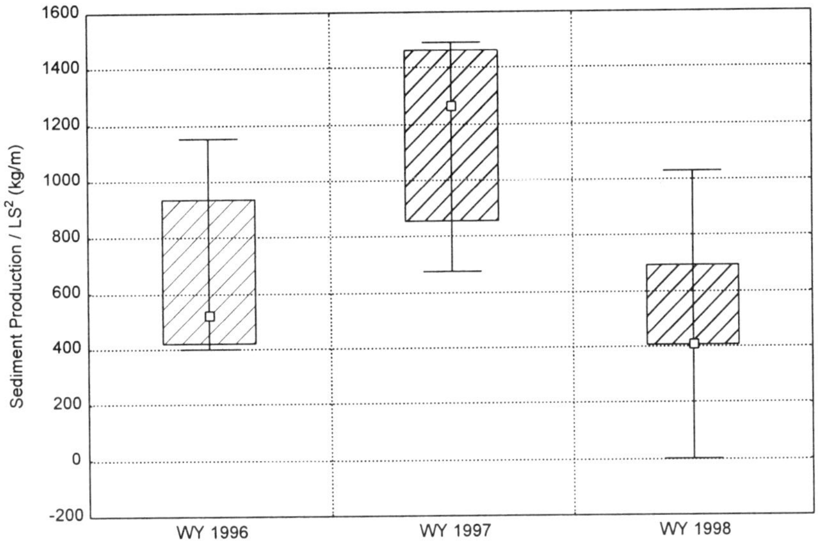

Figure 3. Differences in erosion induced by differences in weather from year to year. Sediment production was normalized by length x slope squared to account for differences in slope and length among the five plots. Whiskers indicate the minimum and maximum; the top and bottom of the box are at the 75th and 25th percentile, and the square is at the median.

III. This set of plots was used for examination of weather effects so that changes in sediment availability caused by site recovery would not affect the year to year changes in sediment production. All rainfall plots were on the finer soils at Low Pass. There is some variation in slope and length among the five plots, so erosion was normalized for length and slope variation among the plots (LS^2) using the results of Luce and Black [1999] for recently disturbed plots. Remaining comparisons between groups over time were done graphically. No normalization for length or slope was done for the other groups because per unit length and slope adjustments valid in the initial year may not be valid in later years. As a consequence, only comparisons between temporal trends are logically sound.

The plots in Group III had a special purpose beyond looking at the effects of ditch clearing. These plots were organized into two experiments:

Experiment 1. To test the effects of changing length and slope on sediment production.

Experiment 2. To test the effects of changing soil texture and cutslope height on sediment production.

Experiment 1 was arranged with 18 plots, all at Low Pass. For these plots, length was varied between 40 and 120 m. Six plots were close to 40 m in length; six others were close to 60 m; and the other six were about 120 m long. Slope was varied between 3 and 12 %, again broken down with equal numbers of plots in each of three classes (3-6%, 6-11%, and 11-13%). Plots were constructed and selected so that there were two plots within each length and slope class combination to remove colinearity effects between length and slope. All of these plots had cutslope heights in the range of 2-4 m. These data were examined by multiple regression in a manner similar to that in Luce and Black [1999] for comparison to equations 9 and 10. We also estimated the k parameter from the Megahan [1974] model by dividing the WY 1997 and WY 1998 sediment production by the WY 1996 sediment production and back calculating the k parameter from

$$k_{97} = -\ln(E_{97}/E_{96}) \qquad (18)$$

$$k_{98} = -\tfrac{1}{2}\ln(E_{98}/E_{96}) \qquad (19)$$

where k is the time decay constant in equation 11 and E is the sediment production. Subscripts indicate respective water years.

Experiment 2 was arranged with 15 plots to test the effects of cutslope height and soil texture. Soil was treated as a categorical variable with two values, and cutslope height was varied between 0 and 6 m. Plots were placed so that the plot-average cutslope heights were evenly distributed through three classes (0-2 m, 2-4 m, and 4-6 m) on each soil texture. Six plots at Low Pass (fine texture) and nine plots at Windy Peak (coarse texture) were analyzed in the cutslope height and soil texture experiment.

TABLE 1. Several indexes of the relative contribution of weather to the erosion from each water year. EI is calculated from ½ hour binned precipitation data by method of Wischmeier and Smith [1978]. Erosion from five plots disturbed at beginning of each year is normalized by dividing by length times slope squared.

Water Year	Annual Precipitation (mm)	Erosivity Index (EI) Annual ($MJ\ mm\ ha^{-1}\ hr^{-1}$)	Erosivity Index (EI) 1-Storm Maximum ($MJ\ mm\ ha^{-1}\ hr^{-1}$)	Normalized Erosion Median (kg/m)	Normalized Erosion Average (kg/m)
1996	1955	1574	274	524	685
1997	2086	2060	425	1265	1149
1998	1496	1033	116	410	507

Because all plots in the study had basalt aggregate surfacing, we expected to see increasing differences between soils with increasing cutslope height. Analysis of the cutslope height data was done using analysis of covariance (MANCOVA) with soil as a categorical variable and cutslope height as a covariate for each of the three years. This tested both the significance of the differences between soils with cutslope height effects removed and the regression of sediment production against cutslope height.

5. CHANGES IN EROSION WITH TIME FOLLOWING DISTURBANCE

5.1 Effects of Rainfall Differences

Water year 1997 showed the greatest normalized erosion from the rainfall plots; water year 1998 showed the least, and WY 1996 was intermediate (Figure 3, Table 1). Indexes of erosion based on precipitation data, including maximum single storm EI, total EI for season, and total precipitation for the water year all show the same ranking among the three years (Table 1). Water years 1996 and 1997 are known for the particularly large flood events in the Willamette River and streams in the Oregon Coast Range. WY 1998 had no particularly notable events and had close to the annual average precipitation for this area.

5.2 Effects of Site Recovery

The average erosion volumes for each of the treatment Groups (I-V) at Low Pass do not follow this same trend (Figure 4). The general behavior seen is a dramatic increase in erosion immediately following disturbance followed by a swift reduction in erosion the following year. Group III saw a 70% reduction between WY 1996 and WY 1997 and an 87% reduction between WY 1996 and WY 1998. Erosion from Group IV was 78% less in WY 1998 than WY 1997. Differences between lines approximately follow expected behavior and echo the results of Luce and Black [1999] for the first year showing that grading only the road tread gives no statistically significant increase in erosion, whereas clearing the ditch and cutslope yields a seven-fold increase in sediment production. The Group V plots which were disturbed at the beginning of WY 1997 and 1998 are the same plots shown in Figure 3, so echo the decline in erosion caused by weather changes. The road segments with no treatment or with only the road surface treated show a steady decline through time as do the more fully treated plots of Group III. We interpret this to mean that interannual differences caused by disturbance and recovery may be greater than interannual differences caused by differences in weather. However, it must be noted that we have not seen a full range of variability in climatic driving during these three years. Recovery for each disturbance set is rapid. We see tremendous decreases in sediment yields within one to two years. This is similar to Megahan's [1974] results for small watersheds and for plots on fillslopes.

6. SEDIMENT PRODUCTION OVER TIME FROM THE LENGTH-SLOPE EXPERIMENT

6.1 How Variations in Transport Capacity are Represented by the Exponential Recovery Model

The plots used to estimate the effects of segment length and slope on erosion show the same general decrease over time shown by the average of all Group III plots at Low Pass and the general form of Megahan's [1974] model of exponential recovery (Figure 5). The trajectory of individual plots, however, may differ from the trend. Those plots not following the recovery trend follow a pattern similar to the precipitation effects in Figure 3 suggesting that armoring may not have occurred on those plots. Visual observation of minor gullying and bare soil in the ditch of these plots confirms the suspicion, although there are no other plot characteristics that separate them from the

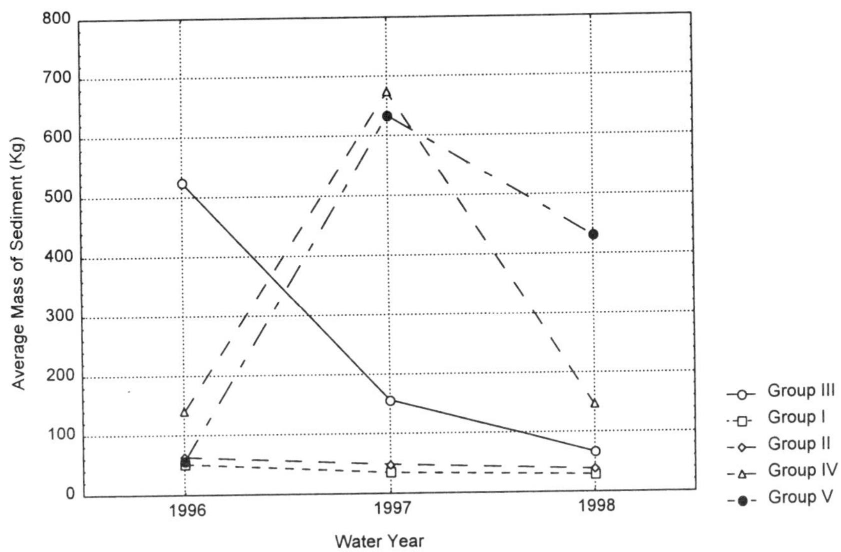

Figure 4. Average erosion from each of the five groups over the three year study:
Group I: No Treatment.
Group II: Road surface graded at beginning of WY 1996, no treatment thereafter.
Group III: Road graded & ditch and cutslope cleared before WY 1996, no treatment thereafter.
Group IV: Road graded before WY 1996, Road graded, ditch and cutslope cleared before WY 1997, no treatment thereafter.
Group V: Road graded before WY 1996, Road graded, ditch and cutslope cleared before WYs 1997 and 1998.

remainder of the group. Sediment production is not high on these plots in spite of the abnormal activity (Figure 5).

The remaining 14 segments (data from two segments was lost in the first year of the study) were used to examine the relative role of k and S_0 in the model of Megahan [1974]. If the dominant effect in changing transport capacity from plot to plot is in changing the k parameter (high transport capacity yields high k) we would expect to see the curves in Figure 5 crossing. Because they are somewhat parallel in their decline and the initial erosion rate scales with transport capacity (LS^2) [Luce and Black, 1999], we can conclude that the main effect of changing transport capacity is in changing the amount of available material for transport between ditch disturbances, S_0. We can also examine this problem by estimating k values implicit in the change between years from equations 18 and 19. We found no correlation between the first year sediment production and the estimated k for WY 1997 or WY 1998 (r = 0.07 for both), confirming the visual analysis above. The relative dominance of the S_0 parameter highlights the role of particle size heterogeneity in the sediment supply and the importance of armoring as a process in recovery over time.

6.2 How the Relationship Between Length, Slope, and Sediment Production Changes with Recovery

In the first year, the proportionality between sediment production and LS^2 [Luce and Black, 1999], suggested a close relationship between sediment transport capacity and sediment yield. The expectation was that as sediment availability decreased (Group III, Figure 4), the proportionality based on sediment transport would become a poorer approximation.

In year two, we found through a cross validation procedure that the proportionality of equation 11 was still the best predictive relationship for plots with less than 10% vegetation (Figure 6). While the cross validation procedure reduces the influence of the point in the upper right, its influence is still noticeable, and inspection of the scatter around the regression excepting this point reveals that the relationship is not strong. The constant of proportionality was much lower in WY 1997 than WY 1996, in spite of a very small increase in vegetation cover and a slight increase in rainfall erosivity, suggesting that armoring may be an important factor in the general trend from WY 1996 to WY 1997. Because detachment capacity equations are

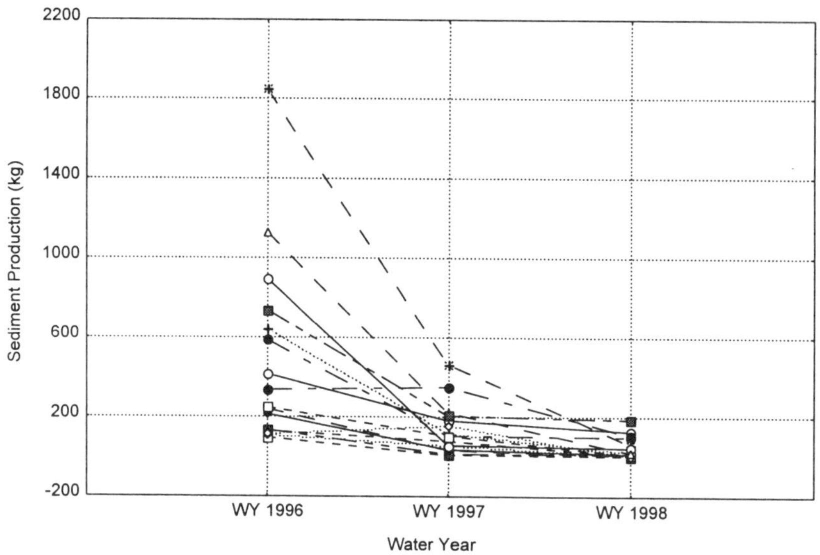

Figure 5. Erosion over the three-year study from each of the plots in Experiment 1 (length and slope) that operated all three years (N=14). All plots were treated at the beginning of WY 1996 by clearing the ditch and cutslope and grading the road surface. In WY 1996 differences are related to length times slope squared.

similar in nature to transport capacity [e.g. Nearing, 1991; Govers, 1992] and net erosion is a convolution of the transport and detachment capacity down the plot [Kirkby, 1980; Lei et al., 1998] it may not be possible to tell whether sediment production is transport limited or detachment limited from the fact that factors controlling transport capacity and net erosion are correlated. Erosion from plots with > 10% vegetation in WY 1997 and erosion from plots in WY 1998 (all plots > 80% vegetation) show no correlation between LS^2 and sediment production. There were too few plots with > 10% vegetation in WY 1997 (and little variation in cover among them) to separate the relative effects of vegetation and LS^2.

These results do not seem to be helpful in establishing guidelines for waterbar spacing. On recently disturbed segments, sediment production is proportional to length; so the per-unit-length sediment production is the same on a 40-m segment as it is on a 120-m segment. On "recovered" segments there is no scaling with length, and per-unit-length sediment yield may actually decrease as length increases. From this data, there is no clear threshold of length and slope beyond which erosion will become unacceptable, so the decision must be based on other information. Megahan and Ketcheson [1996] show that runoff leaving the road prism at concentrated points, such as cross-drains, is more likely to deliver sediment to streams than diffuse drainage. Montgomery [1994] argues that concentration of drainage by roads is the primary reason for gully development and landsliding below roads. Given these insights, we can see that total per-segment sediment and water outputs of the road may be as important as the per-unit-length outputs. A greater segment length leads to greater runoff, greater sediment yield at the outlet, a higher probability of delivering eroded material to streams, and a greater likelihood of inducing gullies or landslides below the road. Montgomery [1994] recommends limiting road segment length based on slope of the hillside to which the segment drains. Megahan and Ketcheson [1996] suggest spacing to limit delivery distance across hillslopes above streams.

7. SEDIMENT PRODUCTION OVER TIME FROM THE CUTSLOPE HEIGHT AND SOIL EXPERIMENT.

The first year of sediment production data showed no correlation between cutslope height and sediment yield. Based on the idea of the plots being transport limited, this was a reasonable expectation. Given that by WY 1997, erosion was no longer transport limited we expected some effect of greater availability of sediment from higher cutslopes. No relationship was found between sediment production and cutslope height for either soil in any of the three years of the study. Logic and other observations [e.g. Megahan et al., 1983] suggest that erosion from cutslopes is a large contributor to road sediment production, particularly in the long term. There may be several reasons

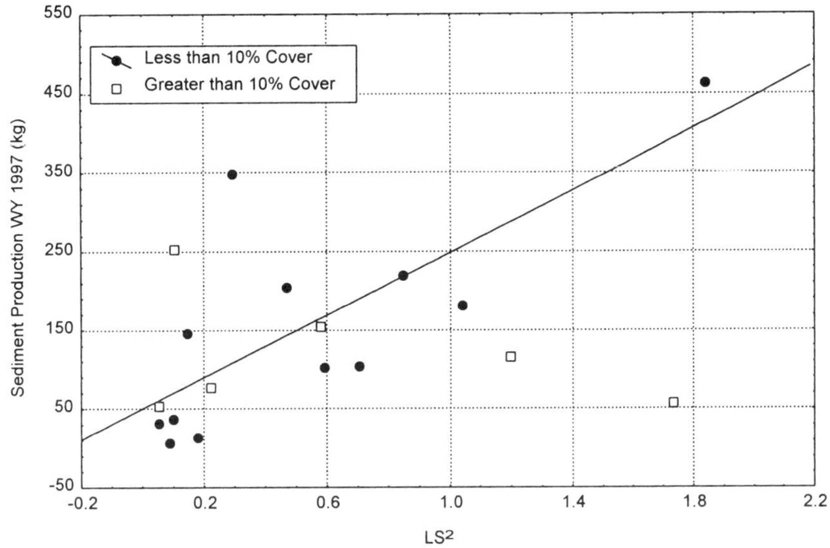

Figure 6. Sediment production versus Length X Slope Squared for WY 1997. Plots are separated into those that have greater than 10% vegetation cover and those that have less.

that variations in cutslope height from nearly zero to more than 4 m produced no systematic variation in sediment yield for road segments. The simplest explanation may be that sediment production from cutslopes is not a function of cutslope length. [Megahan et al., 1999] found no correlation between cutslope length and cutslope sediment production on decomposed granite cutslopes and reasoned that most sediment production came from the A and B horizons, not the less erodible C horizon. The effect is magnified by the fact that the A and B soil horizons tend to be thinner where roads intersect ridges, and this is also where the cutslopes are highest. It may also be that for these soils, the signal strength is smaller than other sources of noise in our data. One avenue that we did not explore is the possibility that much of the reduction observed between years is due to stream capture by the ruts in the road surface. This would take runoff that is bound for the relatively erodible ditch (into which the cutslope contributes material) and run it down shallow ruts in the aggregate road surface, which on our roads was nearly non-erodible. If much of the runoff produced in storms comes from the nearly impermeable road surface, this effect might be sufficient to reduce or mask cutslope effects.

Differences in erosion between the two soil textures were large in WY 1996 and declined with time (Figure 7). Initially there is higher erosion from the silty clay loam and there is a larger decline in erosion over time on the finer soil. Our results follow the general expectation that armoring of the soil surface and growth of vegetation will make the two soil surfaces more similar with time.

8. CONCLUSIONS

This paper addressed questions and about "where" and "when" road surface erosion occurs in a basin. The basis was observations from 74 road segments in the Oregon Coast Range monitored for three years.

With respect to question (where), sediment production is greater where slope is greater in proportion to the square of the slope of the road segment; longer segments produce more sediment individually, but no more per unit length; and segments on more erodible soils produce more sediment. Segments with higher cutslopes did not produce more sediment.

With respect to question (when), erosion is greatest immediately after disturbance, and there is a decline in erosion following initial disturbance that is exponential in shape, consistent with Megahan [1974]. Recovery is rapid; within 1 to 2 years most plots experienced at least a 50% percent reduction in erosion. On recently disturbed roads, there is more erosion in years with more precipitation and with higher single storm or total EI values.

More interesting results address both and together – the where of the when. The dependence of erosion on length and slope is only important for the first two years following disturbance. Initially, sediment production can

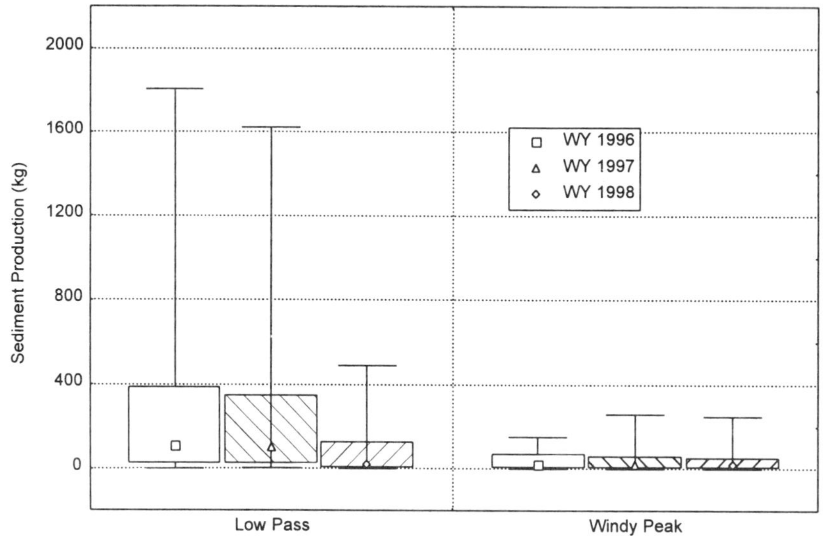

Figure 7. Sediment production over three years from plots in Experiment 2, Cutslope Height and Soil Effects. Low pass has silty clay loam soils, and Windy Peak has gravelly loam soils. Whiskers indicate the minimum and maximum; the top and bottom of the box are at the 75th and 25th percentile, and the symbol (square, triangle, or diamond) is at the median.

be tied to indexes of transport or detachment capacity of flowing water, but the dependence seems to vanish with increasing vegetation cover and armoring. Similarly, the importance of the soil properties in determining erosion is reduced with increased armoring and vegetation. The initial availability parameter, S_0, of Megahan's [1974] exponential decay model varies more strongly with transport capacity than the rate parameter, k, indicating that selective transport of finer sediments from heterogeneous soils may be an important process in reduced erosion with time.

There were a few "odd" plots in this study that did not follow the general rules outlined above. We can develop and fit process-based and empirical models to fit the answers to questions and outlined above, but those models will not predict the "exceptions" to the rule. The rule sets above and earlier findings on cutslopes suggest that roads that do not recover become the greatest contributors of sediment in the long run. We need to learn what road characteristics increase the risk of non-recovery.

Erosion from forest roads is a large component of the sediment budget in some forested basins. Because of its frequency in time, it is probably an important component with respect to aquatic habitat quality in many basins. Road erosion varies on short time scales and over short distances, creating a challenge for agencies and companies managing a large road mileage. Rules that increase our ability to estimate the variation in road erosion at small time and space scales will improve prioritization of erosion abatement, information gathering, and watershed planning. The rules examined in this study relate directly to simple information that can, in many cases, be obtained through existing maps and record keeping. In addition, the methods of this study can be used with relative ease in other locations for understanding local soils and climate.

Acknowledgments. This research was supported by the Oregon State Office of the U.S. Department of the Interior Bureau of Land Management and by the National Council of the Paper Industry for Air and Stream Improvement (NCASI). We would like to thank Barry Williams, Clif Fanning, and Walt Megahan for logistical support and many helpful suggestions. We also extend thanks to the many other people who have given insightful suggestions for our work. And finally thanks to David Watts, Dennis Fletcher, Brent Gennaro, Michael Dewey, and Daniel Cohnstaedt, for their field assistance. We thank Jim Clayton, Walt Megahan, and an anonymous reviewer for their thoughtful comments on earlier drafts of this work.

REFERENCES

Bagnold, R. A., Bed Load Transport by Natural Rivers, *Water Resour. Res.*, *13*(2), 303-312, 1977.

Benda, L., and T. Dunne, Stochastic forcing of sediment routing and storage in channel networks, *Water Resour. Res.*, *33*(12), 2865-2880, 1997.

Benda, L. E., D. J. Miller, T. Dunne, G. H. Reeves, and J. K. Agee, Dynamic Landscape Systems, in *River Ecology and*

Management, edited by R. J. Naiman and R. E. Bilby, pp. 261-288, Springer Verlag, New York, New York, 1998.

Bilby, R. E., K. Sullivan, and S. H. Duncan, The generation and fate of road-surface sediment in forested watersheds in southwestern Washington, *Forest Science, 35*(2), 453-468, 1989.

Black, T. A. and C. H. Luce, Changes in erosion from gravel surfaced forest roads through time, in *Proceedings of the International Mountain Logging and 10th Pacific Northwest Skyline Symposium*, edited by J. Sessions and W. Chung, pp. 204-218, International Union of Forestry Research Organizations and Oregon State University, Corvallis, Oregon, 1999.

Burroughs, E. R., Jr. and J. G. King, Surface erosion control on roads in granitic soils, in *Watershed Management in the Eighties*, edited by E. B. Jones and T. J. Ward, pp. 183-190, American Society of Civil Engineers, New York, 1985.

Burroughs, E. R., Jr. and J. G. King, Reduction of soil erosion on forest roads, *General Technical Report INT-264*. U.S. Department of Agriculture, Forest Service, Intermountain Research Station, Ogden, Utah, 21 pp., 1989.

Burroughs, E. R., Jr., C. H. Luce, and F. Phillips, Estimating interrill erodibility of forest soils, *Transactions ASAE, 35*(5), 1489-1495, 1992.

Carling, P. A., M. S. Glaister, and T. P. Flintham, The Erodibility of Upland Soils and the Design of Preafforestation Drainage Networks in the United Kingdom, *Hydrological Processes, 11*(15), 1963-1980, 1997.

Cline, R., G. Cole, W. Megahan, R. Patten, and J. Potyondy, *Guide for predicting sediment yield from forested watersheds,* U.S. Forest Service Northern Region and Intermountain Region, Missoula, Montana and Ogden, Utah, 1984.

Dietrich, W. E. and T. Dunne, Sediment budget for a small catchment in mountainous terrain, *Z. Geomorph. Suppl., 29*, 191-206, 1978.

Dunne, T. and W. E. Dietrich, Experimental investigation of Horton overland flow on tropical hillslopes, 2, Hydraulic characteristics and hillslope hydrographs. *Z. Geomorphol. Suppl., 33*, 60-80, 1980.

Flanagan, D. C. and M. A. Nearing, USDA-Water Erosion Prediction Project: Hillslope profile and watershed model documentation, *NSERL Report 10*, National Soil Erosion Laboratory, West Lafayette, Indiana, 1995.

Foltz, R. B. and W. J. Elliot, The impact of lowered tire pressures on road erosion, *Transportation Research Record, 1589*, 19-25, 1997.

Foster, G. R. and L. D. Meyer, A closed form soil erosion equation for upland areas, in *Sedimentation, Symposium to Honor Professor H. A. Einstein*, edited by H. W. Shen, pp. 12.1-12.19, Colorado State University, Fort Collins, Colorado, 1972.

Foster, G. R. and L. D. Meyer, Mathematical simulation of upland erosion by fundamental erosion mechanics, in *Present and Perspective Technology for Predicting Sediment Yields and Sources – Proceedings of Sediment-Yield Workshop*, USDA Agricultural Research Service Report ARS-S-40, pp. 190-206, United States Department of Agriculture Sedimentation Laboratory, Oxford, Mississippi, 1975.

Fredriksen, R. L., Erosion and sedimentation following road construction and timber harvest of unstable soils in three small western Oregon watersheds, *USDA Forest Service Research Paper PNW-104*, Pacific Northwest Forest and Range Experiment Station, Portland, Oregon, 15 pp., 1970.

Furniss, M. J., T. D. Roelofs, and C. S. Yee, Road construction and maintenance, in *Influences of Forest and Rangeland Management on Salmonid Fishes and Their Habitatats*, edited by W. R. Meehan, pp. 297-323, American Fisheries society, Bethesda, Maryland, 1991.

Gilbert, G. K., Hydraulic Mining Débris in the Sierra Nevada, *U.S. Geological Survey Professional Paper 105*, 154 pp., 1917.

Govers, G., Evaluation of transporting capacity formulae for overland flow, in *Overland Flow Hydraulics and Erosion Mechanics*, edited by A. J. Parsons and A. D. Abrahams, pp. 243-273, Chapman & Hall, New York, New York, 1992.

Ice, G. G., A study of the effectiveness of sediment traps for the collection of sediment from small forest plot studies, *NCASI Technical Bulletin 483*, NCASI, New York, 27 pp., 1986.

Kirkby, M. J., Modelling water erosion processes, in *Soil Erosion*, edited by M. J. Kirkby and R. P. C. Morgan, pp. 183-216, John Wiley and Sons, Chichester, 1980.

Lei, T., M. A. Nearing, K. Haghigi, and V. F. Bralts, Rill erosion and morphological evolution: A simulation model, *Water Resour. Res., 34*(11), 3157-3168, 1998.

Luce,C. H. and T. A. Black, Sediment production from forest roads in western Oregon. *Water Resour. Res., 35*(8), 2561-2570, 1999.

MacDonald, L. H., D. M. Anderson, and W. E. Dietrich, Paradise threatened: Land use and erosion on St. John, US Virgin Islands, *Environmental Management, 21*(6), 851-863, 1997.

McCool, D. K., L. C. Brown, G. R. Foster, C. K. Mutchler, and L. D. Meyer, Revised slope steepness factor for the Universal Soil Loss Equation, *Transactions ASAE, 30*(5), 1387-1396, 1987.

McCool D. K., G. R. Foster, C. K. Mutchler, L. D. Meyer, Revised slope length factor for the Universal Soil Loss Equation, *Transactions ASAE, 32*(5), 1571-1576, 1989.

Megahan, W. F., M. D. Wilson, and S. B. Monsen, Sediment Production from Granitic Cutslopes on Forest Roads in Idaho, USA. Submitted to *Earth Surface Processes and Landforms*, 1999.

Megahan, W. F., Erosion over time: A model, *USDA For. Serv. Res. Pap. INT-156*, 14 pp., USDA Forest Service, Intermountain Research Station, Ogden, UT, 1974.

Megahan, W. F., and G. L. Ketcheson, Predicting downslope travel of granitic sediments for forest roads in Idaho, *Water Resources Bulletin, 32*(2), 371-382, 1996.

Megahan, W. F., S. B. Monsen, M. D. Wilson, N. Lozano, D. F. Haber, and G. D. Booth, Erosion control practices applied to granitic roadfills for forest roads in Idaho: Cost effectiveness evaluation, *Journal of Land Degradation and Rehabilitation, 3*, 55-65, 1992.

Megahan, W. F., W. S. Platts, and B. Kulesza, Riverbed improves over time: South Fork Salmon, in *Proceedings, symposium on watershed management, Boise, Idaho*, Vol. 1, pp. 380-395, American Society of Civil Engineers, New York, 1980.

Megahan, W. F., K. A. Seyedbagheri, and P. C. Dodson, Long-term erosion on granitic roadcuts based on exposed tree roots, *Earth Surface Processes and Landforms, 8*, 19-28, 1983.

Miller, J. F., R. H. Frederick, and R. J. Tracey, Precipitation Frequency Atlas of the United States, *NOAA Atlas 2*, U.S. Department of Commerce, National Weather Service, 1973.

Montgomery,D. M., Road surface drainage, channel initiation, and slope instability, *Water Resour. Res., 30*(6), 1925-1932, 1994.

Nearing, M. A., A probabilistic model of soil detachment by shallow turbulent flow, *Transactions ASAE, 34*, 81-85, 1991.

Reid, L. M., Sediment production from gravel-surfaced forest roads, Clearwater basin, Washington. *Publ. FRI-UW-8108*, 247 pp., University of Washington Fisheries Research Institute, Seattle, Washington, 1981.

Reid, L. M., and T. Dunne, Sediment Production from Forest Road Surfaces, *Water Resour. Res., 20*(11), 1753-1761, 1984.

Reid, L. M., and T. Dunne, *Rapid evaluation of sediment budgets*, Catena Verlag, Reiskirchen, Germany, 164 pp, 1996.

Renard, K. G., J. M. Laflen, G.R. Foster, and D. K. McCool, The Revised Universal Soil Loss Equation, in *Soil Erosion Research Methods*, edited by R. Lal, pp. 105-124, Soil and Water Conservation Society, Ankeny, Iowa, 1994.

Simons, D. B., R. M. Li, and T. J. Ward, Application of road sediment models to natural and simulated rainfall runoff sites, *Colorado State University Report CER79-80DBS-RML-TJW66*, Colorado State University, Fort Collins, Colorado, 40 pp., 1980.

Swift, L. W., Jr., Gravel and grass surfacing reduces soil loss from mountain roads, *Forest Science, 30*(3), 657-670, 1984.

Tysdal, L. M., W. J. Elliot, C. H. Luce, and T. A. Black, Modeling Erosion from Insloping Low-Volume Roads with WEPP Watershed Model, *Transportation Research Record, 1652*(2), 250-256, 1999.

U.S. Departmnt of Agriculture (USDA) Forest Service Northern Region, WATSED, 125 pp., USDA Forest Service, Region 1, Missoula, Montana, 1991.

Vincent, K. R., Runoff and erosion from a logging road in response to snowmelt and rainfall, M.S. Thesis, 60 pp., University of California, Berkeley, California, 1985.

Washington Forest Practices Board, Standard methodology for conducting watershed analysis, Version 3.0, November 1995, pp. B1-B52, Washington State Department of Natural Resources, Olympia, WA, 1995.

Wischmeier, W. H. and D. D. Smith, Predicting rainfall erosion losses, *Agriculture Handbook No. 537*, 58 pp., United States Department of Agriculture, Washington, D.C., 1978.

Wolman, M. G. and R. Gerson, Relative scales of time and effectiveness of climate in watershed geomorphology, *Earth Surface Processes and Landforms, 3*, 189-208, 1978.

Woolhiser, D. K., R. E. Smith, and D. C. Goodrich, KINEROS, A Kinematic Runoff and Erosion Model, Documentation and User Manual, ARS-77, 130 pp., USDA Agricultural Research Service, Tucson, Arizona, 1990.

Yount, J. D. and G. J. Niemi, Recovery of lotic communities and ecosystems from disturbance—a narrative review of case studies, *Environmental Management, 14*, 547-570, 1990.

Evaluation of the Temporal and Spatial Impacts of Timber Harvesting on Landslide Occurrence

Roy C. Sidle

Department of Geography, National University of Singapore, Singapore

Weimin Wu

OnLink Technologies, Inc., Redwood City, California

Timber harvesting activities in steep terrain can affect the stability of hillslopes by reducing the cohesion associated with tree roots. Additionally, local hydrogeomorphic and topographic attributes control the susceptibility of harvested sites to landsliding. These factors act together with the probabilistic influence of a rainstorm that generates a critical pore water pressure in the soil mantle to trigger slope failure. Other triggering mechanisms include rapid snowmelt and seismic activity. Several methods are available to assess the effects of timber harvesting on shallow landslide potential. At the most general and qualitative levels are terrain mapping and evaluation procedures that utilize topographic and geologic information to provide broad categories of landslide hazard related to potential harvesting activities. In regions where good site data and landslide records are available the effect of land use (including harvesting) can be evaluated by weighted multi-factor overlays. Both of these methods are subjective either in the development or application phases and much weight is placed on expert knowledge. Recently, distributed modeling has been applied to mountainous terrain to analyze the potential for shallow rapid landslides. The distributed, physically based model (dSLAM) which is based on an infinite slope model, a kinematic wave groundwater model, and a continuous change vegetation root strength model is discussed and applied to several real and simulated forest harvest scenarios. The model has the flexibility of utilizing either actual storm records or synthesizing a random Monte Carlo series of storms. Applications of dSLAM in two managed forested basins of coastal Oregon produced accurate estimates of landslide numbers and volumes for a major storm in 1975. Simulations showed that most landslides were clustered in the period of 3-15 yr after clearcutting; these findings are consistent with field observations. Although data demands are quite high, distributed, physically based modeling of shallow landslide hazards appears to be a promising tool that can be applied to intensively managed forest sites.

INTRODUCTION

Steep hillslopes with shallow soil mantles are typically unstable especially in regions of high precipitation and recent tectonic activity. In many of these regions

Land Use and Watersheds: Human Influence on Hydrology and Geomorphology in Urban and Forest Areas
Water Science and Application Volume 2, pages 179-193

worldwide, shallow landslides are the dominant erosion processes [Swanston, 1974; Dietrich and Dunne, 1978; Kienholz et al., 1984; Aniya, 1985; Sidle et al., 1985; Maharaj, 1993; Bangxing et al., 1994; DeRose, 1996]. For such steep terrain, the delicate balance between resistance of the shallow soil mantles to failure and the gravitational forces tending to move the soil downslope can easily be upset by external triggering factors such as rainfall [O'Loughlin et al., 1982; Sidle and Swanston, 1982; Tsukamoto et al., 1982], snowmelt [Megahan, 1983; Anderson et al., 1985; Wieczorek et al., 1989], and earthquakes [Keefer, 1984; Fukuda and Ochiai, 1993; Murphy, 1995; Harp and Jibson, 1996], as well as the effects of vegetation management [Gray and Megahan, 1981; Sidle et al., 1985; Marden and Rowan, 1993]. Precipitation induced landslides are more common in forested terrain; however, seismically active regions such as Papua New Guinea experience widespread landsliding during major earthquakes, especially when accompanied by rainfall [Pain and Bowler, 1973; Garwood et al., 1979]. An increase in shallow landsliding after clearcutting has been demonstrated in mountainous terrain worldwide [Bishop and Stevens, 1964; Fujiwara, 1970; O'Loughlin, 1972; Swanson and Dyrness, 1975; O'Loughlin and Pearce, 1976; Megahan et al., 1978; Gresswell et al., 1979; Schwab, 1983]. Earthquake-induced landslides in south-central Chile were found to be associated with areas where forest land had been converted to farmland [Wright and Mella, 1963]. Although this paper focuses on shallow landslides triggered by rainfall, it is apparent that forest harvest can affect the frequency of landslides triggered by other mechanisms.

Debris slides and avalanches are common shallow mass failure types on steep forested hillslopes. Often these failures mobilize into debris flows as they enter headwater streams, but debris flows can also occur on open hillslopes. In many sites these translational hillslope failures initiate as a slower moving debris slide. The failure mass may then transform into a more rapidly moving debris avalanche (0.3 m/min to >3 m/s). With increasing liquefaction and channelization, the failure may then become a very rapid (>3 m/s) debris flow [Temple and Rapp, 1972; Sidle et al., 1985]. Such linked sequences of mass movement are primary sources of sedimentation in forest streams. Sediment and large woody debris movement associated with landslides and debris flows can have a large influence on channel substrate and form [Benda and Cundy, 1990; Harvey, 1991; Rice and Church, 1996; Gomi et al., 1998] and related aquatic habitat [Keller and Swanson, 1979; Swanson et al., 1982a; Pearce and Watson, 1983]. Accelerated inputs of sediment and large woody debris from timber harvesting activities can adversely affect aquatic habitat [Meehan, 1974; Swanson et al., 1982b; Ralph et al., 1994].

During major rainstorms a pore water pressure builds up in the soil mantle usually just above the lithic contact or other hydrologic impeding layer [Sidle and Swanston, 1982; Tsukamoto et al., 1982; Onda et al., 1992]. A common hydrologic sequence for slope failure initiation involves wet antecedent conditions followed by a prolonged period of rainfall with a burst of high intensity [e.g., Sidle and Swanston, 1982; Finlay et al., 1997]. The wet antecedent conditions provide the necessary soil moisture to promote an accretion of a perched water table in unstable sites; the prolonged rainfall gradually builds up pore water pressures; and the high intensity burst causes pore water pressure to increase rapidly to a threshold the initiates slope failure [e.g., Sidle and Swanston, 1982; Sidle and Tsuboyama, 1992; Torres et al., 1998]. Alternatively, shorter duration rainfall events with very high intensities can trigger landslides. In either case, the landslide initiates during the actual storm event, often just after the peak rainfall intensity.

HYDROGEOMORPHIC AND SOIL FACTORS

For thin, noncohesive soils with planar failure surfaces and fixed groundwater conditions factor of safety (FS) for slope failure, a ratio of shear strength (S) to shear stress (τ), is highly dependent on slope gradient. In such cases, FS can be evaluated by a simplified version of the infinite slope model [O'Loughlin, 1974; Swanston, 1974] as follows

$$FS = \frac{(z\gamma_m \cos^2 \alpha - u)\tan\phi}{z\gamma_m \cos\alpha \ \sin\alpha} \qquad (1)$$

where γ_m is the unit weight of soil at field moisture content (kN/m^3), z is the vertical soil thickness (m), α is the slope gradient (degrees), and ϕ is the effective internal angle of friction of the soil (degrees). In certain localized settings where such generalizations can be made, slope gradient has been successfully used as the sole indicator of landslide potential [e.g., Lohnes and Handy, 1968; Carson and Petley, 1970; Swanston, 1974; Ballard and Willington, 1975]. However, because of the complicating influences of variability in regolith depth, cohesion, climate (and therefore pore water pressure), slope shape, and vegetation (including land use effects) found in many areas, other parameters must be considered in the geotechnical analysis of landslides.

Variations in soil mantle thickness on natural forest slopes play an important role in determining landslide potential. Soil thickness in hillslope depressions can change appreciably over a century as the result of infilling and episodic evacuation [Dietrich and Dunne, 1978; Lehre, 1982; Sidle, 1987]. Observations of repeated failure of shallow soils over competent bedrock [Mark et al., 1964; Schweinfurth, 1967] indicate that for some combinations of gradient and soil properties, there exists a threshold of soil thickness beyond which failure must occur. Thus, in steep terrain with rapidly developing or changing soil mantles, soil thickness and regolith properties must be considered in addition to slope gradient in an analysis of potential landslide hazards [Sidle et al., 1985].

Concave hillslope elements known as geomorphic hollows or zero-order basins are frequent sites of shallow, rapid failures [Dietrich and Dunne, 1978; Okunishi and Iida, 1981; Tsukamoto et al., 1982]. These unchanneled depressions serve to concentrate shallow groundwater either via piping or more diffuse preferential or matrix flow [Tsukamoto et al., 1982; Sidle, 1984a; Tsukamoto and Ohta, 1988; Montgomery et al., 1997]. Once these hollows fail as shallow debris slides, debris avalanches or debris flows, they gradually fill in by localized sloughing around the headwall, soil creep, weathering of exposed bedrock or till, surface erosion, biogenic processes, and recruitment of organic debris [Dietrich and Dunne, 1978; Okunishi and Iida, 1981; Tsukamoto et al., 1982]. The periodicity of evacuation of these hollows depends on the long-term rates of infilling and the timing of episodic storms [Shimokawa, 1984; Sidle et al., 1985; Sidle, 1988; Dunne, 1991]. Even though accretion of soil in the hollow bottoms occurs after failure [Dengler and Montgomery, 1989], it is difficult to generalize about the spatial distribution of soil depth in these depressions due to the influences of the time since last evacuation and localized failures and erosion [Sidle, 1984a; Dunne, 1991; Onda et al., 1992; Tsuboyama et al., 1994a]. Soil depth in these geomorphic hollows is therefore a function of a complex array of inter-linked geomorphic, hydrologic, and biologic processes.

Sensitivity analyses for forest soils based on the infinite slope equation, indicate that FS is highly influenced by effective cohesion [Gray and Megahan, 1981; Sidle, 1984b]. Small values of effective cohesion in the order of 1 to 2 kPa have been shown to differentiate between unstable and marginally stable hillslopes on steep forested sites [Wu and Sidle, 1995]. Thus, FS values calculated by (1) might greatly underestimate the stability of sites with 'essentially' cohesionless soils in cases where a minor cohesive component exists in such soils.

Pathways of subsurface water movement in unstable hillslope soils are important determinants of landsliding. The rapid increases in pore water pressure noted in hillslope hollows suggest that preferential flow pathways, including soil pipes, macropores, and bedrock fractures, play a role in transporting subsurface water in these sites [Pierson, 1980; Sidle, 1984a; Tsukamoto and Ohta, 1988; Montgomery et al., 1997]. The rapid flux of subsurface water can generate zones of positive pore water pressure if water does not continue to flow downslope at similar rates [Pierson, 1983; McDonnell, 1990]. Thus, discontinuities in soil pipes or convergence zones of macropore systems may be sites of pore water pressure accretion. Macropores and soil pipes have been observed around headscarps of landslides suggesting a linkage with landslide initiation [Blong and Dunkerley, 1976; Tsukamoto et al., 1982; Sidle et al., 1999]. Recent research has suggested that macropore systems in forest soils expand as a hierarchical network and become self-organized when antecedent conditions in the vicinity of macropores become wet [Tsuboyama et al., 1994b; Sidle et al., 1995; 1998; 2000]. This phenomenon could explain the large and rapid increases in pore water pressure observed in unstable sites [O'Loughlin and Pearce, 1976; Rogers and Selby, 1980; Montgomery et al., 1997]. Alternative explanations for the rapid piezometric responses noted during large storms include the so-called "capillary-fringe" response in nearly saturated soils [Gillham, 1984], pressure wave propagation [Torres et al., 1998], and displacement of "old water" from soils by newly infiltrating rainwater [Pearce et al., 1986].

VEGETATIVE FACTORS

Vegetation can affect the stability of slopes in two ways: (1) modification of the soil moisture regime through evapotranspiration; and (2) imparting a root cohesion component to the soil mantle. The first factor is not particularly important for shallow landslides that occur during an extended rainy season. For such conditions, soils are near saturation and the reduced evapotranspiration would have a minimal effect on soil water content [Sidle et al., 1985]. However, evapotranspiration, or the lack thereof, could alter the "window of susceptibility" for shallow landsliding if a large storm occurred near the beginning or end of the rainy season. Also, when large landslide-producing storms occur during drier conditions, evapotranspiration can be a factor related to shallow landslide initiation.

A more important stabilizing factor associated with vegetation is the effect of rooting strength. A number of field investigations in steep forested terrain worldwide have noted an increased frequency of shallow landslides during the period of about 3 to 15 yr after timber harvesting [Bishop and Stevens, 1964; Endo and Tsuruta, 1969; O'Loughlin and Pearce, 1976; Megahan et. al., 1978; Wu and Sidle, 1995]. These studies have associated this accelerated mass erosion with the period of minimum rooting strength at the site following initial root decay and prior to substantial regeneration. Independent studies of the effects of timber harvest on rooting strength based on mechanical straining of roots [Burroughs and Thomas, 1977; Ziemer and Swanston, 1977; Wu et al., 1979; Abe and Ziemer, 1991] have confirmed these empirical observations.

Root systems that vertically anchor through the soil mantle into a more stable substrate such as bedrock or till provide the most effective mechanism for stabilizing shallow soils prone to landslides [Wu et al., 1979, Gray and Megahan, 1981]. Dense networks of medium to small roots may also act as a membrane of lateral strength that stabilizes the soil [Sidle et al., 1985]. Sidle [1991] summarized root cohesion data for a number of forest types and incorporated these into models of root strength changes following various timber harvesting practices. Root strength decay was simulated by a negative exponential relationship

$$D = e^{-kt^n} \qquad (2)$$

where D is the dimensionless root cohesion parameter for decay, t is the time since harvesting in years, and k and n are empirical constants. Root strength regrowth after harvest has been quantified by Sidle [1991] as

$$R = (a + be^{-kt})^{-1} + c \qquad (3)$$

where R is the dimensionless root cohesion parameter for regrowth, and a, b, c, and k are empirical constants which define the sigmoid regrowth curve. The sum of the root regrowth and decay curves represents the relative net root cohesion for the site. This relative net root cohesion (also dimensionless) multiplied by the maximum root cohesion for the specified vegetation type (ΔC_{max}), yields the actual root cohesion at any given time after timber harvest [$\Delta C(t)$]. These relationships assume that most of the anchor roots penetrate the potential failure plane. In cases where soils are deeper, root strength must be reduced to account for the lesser number of effective anchor roots.

A landslide model applied to coastal Alaska showed that shorter successive management rotations in a Sitka spruce–western hemlock forest impart a cumulative increase in the probability of slope failure [Sidle, 1992]. This increased landslide potential is caused by the inability of regenerating forests to reach their maximum potential root strength prior to the next harvest entry. For managed forests with very short rotations, such as Radiata pine in New Zealand, such cumulative effects could be of even greater concern. Sidle [1992] also showed that maintaining a viable understory root strength during harvesting cycles can greatly reduce the probability of slope failure. With the destruction of half of the understory rooting strength during timber harvest (from 2 kPa to 1 kPa) the maximum probability of failure was increased approximately two-fold; however, both cases were unstable compared to an uncut scenario.

The influence of the weight of vegetation on slope stability is generally trivial [Sidle, 1992]. The increase in vegetation surcharge (W, kPa) after timber harvest can be described by the same general sigmoid function as root strength regrowth (3). Tree surcharge values for coastal forest types in the Pacific Northwest range from 1 to 5 kPa [Bishop and Stevens, 1964; O'Loughlin, 1974; Wu et al., 1979].

OVERVIEW OF TERRAIN EVALUATION PROCEDURES

Most terrain assessment or evaluation methods are not focused on predicting specific landslide sites or triggering conditions, but rather on the general susceptibility of contiguous parcels of terrain to landslide hazard as well as the management implications of the hazard. Two examples of terrain evaluation related to forest management in mountainous regions will be briefly discussed: (1) the terrain stability classification used by the British Columbia Ministry of Forests; and (2) several examples of stability hazard assessments (applied at different levels of detail) that use weighted factor overlays.

Terrain Stability Classification, British Columbia

Terrain stability maps provide a relative ranking of the landslide hazard potential of an area that is large enough to delineate on an aerial photograph or map. As developed in British Columbia (BC) the mapping criteria are based on the origin of surface materials, the surface expression of the landform, and the geological processes that are modifying either the surficial materials or landforms [Howes and Kenk, 1988]. The surface materials can further be modified by dominant textural categories or organic components. Both geological processes and surface materials can also be qualified with regard to their current state of activity. Data required to develop these maps are typically obtained from aerial photographs, topographic maps, geologic maps, vegetation and ecosystems maps, and general field reconnaissance. Because of regional variations in climate, geology, soils, vegetation, and hydrology, specific criteria that define terrain stability classes across BC have not been developed [Ministry of Forests, 1995]. Such criteria are typically developed by terrain mappers and applied to specific areas within the Province. Thus, the quality of the resultant terrain hazard maps is highly dependent on the knowledge and experience of the mapper.

Although there are five levels of terrain survey recognized by the BC Ministry of Forests, the two most common applications are 'reconnaissance' and 'detailed' mapping of terrain stability. Reconnaissance mapping typically covers large areas without regard for specific forest management operations. These maps can be effective tools to identify where more detailed mapping is required. Reconnaissance mapping is typically performed at scales of 1:20,000 to 1:50,000 depending on terrain characteristics, vegetative cover, and available air photo coverage. Very little on-site verification is employed at the reconnaissance mapping level; from 0% to a maximum of 25% of the delineated polygons are field-checked at the low intensity level (low intensity infers from 0 to 1 ground checks per 1000 ha; essentially no ground checking) [Ministry of Forests, 1995]. Terrain polygons should be mapped in a single slope category and in one of three reconnaissance terrain stability classes. The original version of this terrain stability classification developed in BC included five categories of landslide potential [Howes and Kenk, 1988]. The first two classes contained no stability problems and the third class contained only minor stability problems. Hence, for reconnaissance mapping these three classes have been subsequently grouped into a category "S" for stable terrain. The next class (formerly IV and now "P") represents areas with a moderate probability of landsliding following road construction and logging. The most unstable areas (formerly class V and now "U") are likely to experience landslides following road construction or logging [Ministry of Forests, 1995]. Terrain categories P and U both require a field inspection by a qualified engineer or geoscientist prior to any activities that may adversely affect the stability of the areas. However, prior to any intensive field inspections, a detailed mapping of terrain stability may be initiated on the more unstable categories, especially if they are directly impacted by proposed management practices. Because reconnaissance maps are generally quite conservative in terms of delineating unstable polygons, they usually show a higher percentage of unstable land area than would be mapped at a more detailed scale.

Detailed terrain stability maps identify areas that require on-site assessments of landslide potential prior to approval of specific forest management activities. Mapping is done on 1:15,000 to 1:20,000 air photos and the final terrain stability interpretations are typically presented on topographic base maps of similar scales. Slope gradient is still presented as a range, but field measurements or slope data from 1:5000 scale topographic maps increase the accuracy of these data compared to methodology used for estimating slope categories in reconnaissance mapping. General soil drainage classes as described in soil inventories are also included. Approximately 25-50% of the mapped polygons are field-checked at an intensity of 5 to ≥ 10 ground checks per 1000 ha. Ground checks verify landform and surface material properties that were initially inferred from aerial photographs and maps, as well as any recent landslide activity.

It should be emphasized that the actual criteria that are used to classify a parcel of terrain into a given stability class is largely based on professional judgement and available information. Several empirical investigations have been undertaken in BC to link specific terrain attributes to post-logging landslide occurrence [e.g., Howes and Kenk, 1988; Rollerson, 1992; Rollerson et al., 1997]. Such multi-variate analyses have identified slope gradient class; occurrence of natural landslides; topographic position, shape, and morphology; soil drainage class; and various bedrock attributes as potential indicators of post-harvesting landslide susceptibility on the west coast of Vancouver Island [Rollerson et al., 1997]. Such analyses together with expert knowledge, climatological data, and landslide inventories, provide the basis for the development of weighted factor overlay models that are discussed briefly in the next section. While such a degree of standardization is desirable to remove aspects of human bias, in some cases certain empirical indicators of slope instability can only be used within limited geographic regions.

At either the reconnaissance or detailed levels of the BC terrain-mapping system, the implied accuracy of addressing forest management operations is very broad and basically equates timber harvest issues with those related to road construction. While this general procedure is useful for conservative slope stability analysis, it is apparent this analysis cannot be used to delineate specific localities of high landslide risk nor can it evaluate even crude magnitudes of road- or harvest-related landslide potential. Additionally, this procedure gives no indication of the frequency, magnitude, or potential impact of the landslide hazard and as does not address the fine-scale spatial and temporal aspects of timber harvesting operations on landslide initiation. The terrain stability maps can be utilized as a first step in identifying susceptible terrain that needs further analysis and field investigation prior to management actions. In contrast, A Watershed Analysis Methodology used in Washington provides greater detail slope form, soil thickness, sediment delivery, and landslide history and thus provides insights into these spatially and temporally distributed management issues, especially if coupled with long-term records of landslides.

Stability Hazard Assessment

Numerous examples of stability hazard assessment or landslide hazard zoning that employ multiple factor overlays can be found in the literature. Early research that predated sophisticated Geographic Information Systems (GIS) utilized basically the same concepts as incorporated in the more advanced processing systems of today. For example, a conceptually advanced slope hazard assessment was developed for the San Francisco Bay region in the 1970's using detailed information derived from maps of slope gradient, previous landslide deposits, and surficial and bedrock geology [Nielsen et al., 1979]. A key to the success of this stability mapping was the intensity of field data that had been compiled over a number of years in this highly populated region. Later research established rainfall intensity and duration thresholds for initiation of debris flows on susceptible slopes in various parts of the Bay area [Cannon and Ellen, 1985]. Keefer et al. [1987] utilized these relationships between rainfall and landslide initiation together with geological indicators of slope instability and real-time rainfall data to develop a warning system for landslides during major storms in the region. While such an advanced warning system is dependent on spatially distributed, accurate, and timely dissemination of rainfall data, it is possible that similar applications could be successful in densely populated regions where local governments make long-term commitments to support regional networks of telemetered rain gages.

In stark contrast to the advanced data acquisition system employed in the San Francisco region is the case of stability hazard assessment in the Himalayas of Nepal and northern India [e.g., Kienholz et al., 1984; Gupta and Joshi, 1990]. In this region landslides are triggered by both earthquakes and rainfall. However, because of the paucity of spatially distributed data in this area (especially in mountain regions) these causative factors cannot be directly included in a factor overlay analysis of landslide hazard potential. A very general approach to landslide hazard assessment was conducted in the Ramganga catchment in the Lower Himalayas [Gupta and Joshi, 1990]. The approach involved mapping recent and old landslides from air photos and overlaying this information on geologic maps, remotely sensed maps of land use, and maps of major faults and thrust zones. Four geoenvironmental criteria were used to evaluate landslide hazard: (1) five general lithology groups; (2) five categories of land use; (3) eight distance ranges from major tectonic features; and (4) eight slope aspect classes. The distribution of landslides in various slope aspects is a surrogate for the influence of rock structure and physiography on landslide occurrence. Landslides tend to be more frequent in the direction of the dip of the bedrock. Since the hazard assessment only incorporated recent and older failures (i.e., not potential failures), slope gradient was not incorporated in the analysis. This important parameter would obviously improve the GIS hazard zonation especially if inferences on future land use changes and predictive capabilities are desired. Percentages of landslide occurrence in all the geoenvironmental subcategories were computed and compared to the average landslide incidence; subcategory values >33% higher than the overall average were weighted as high risk (2), values <33% lower than the average were weighted as low risk (0), and values in the range of ± 33% of the mean were weighted as moderate risk (1) [Gupta and Joshi, 1990]. Each of the four geoenvironmental criteria (lithology, land use, distance from tectonic features, and slope aspect) were equally weighted in this preliminary analysis [Gupta and Joshi, 1990]. More detailed hazard analysis should consider the potential of weighting each of the criteria based on local knowledge and relationships of parameters to landslide initiation. The overall methodology is illustrated in Figure 1 along with suggestions for improvement. Although forest harvest is not explicitly addressed in this hazard mapping, it can be seen that areas where vegetation is either sparse or totally removed are most susceptible to landslides. The landslide hazard zonation proposed for the Kathmandu-Kakani area of Nepal [Kienholz et al., 1984] is more detailed and includes information on regolith thickness, slope length and gradient, topographic expression, hydrogeology, and certain land use practices in addition to the geologic and land use/vegetation indicators employed in the Ramganga catchment study. Nonetheless, specific triggering factors such as rainfall amounts and intensities and seismicity could not be included in this more detailed assessment due to lack of distributed data.

To significantly improve such factor-overlay based landslide hazard assessments, two significant issues need to be overcome: (1) methods that can be applied in broader geographic areas or in areas that experience multiple failure types (e.g., slump-earthflows, debris avalanches) need to be developed; and (2) there is a need to clearly focus on the underlying processes relating to slope failure. The first problem can partly be addressed by developing separate mapping rules for distinctly different failure types. For example, deep-seated slump-earthflow hazards would evoke different geologic, soils, and hydrologic indicators compared to shallow rapid failures. The second issue is difficult to implement in data sparse regions but as remotely sensed data becomes more available and accurate it may be possible to

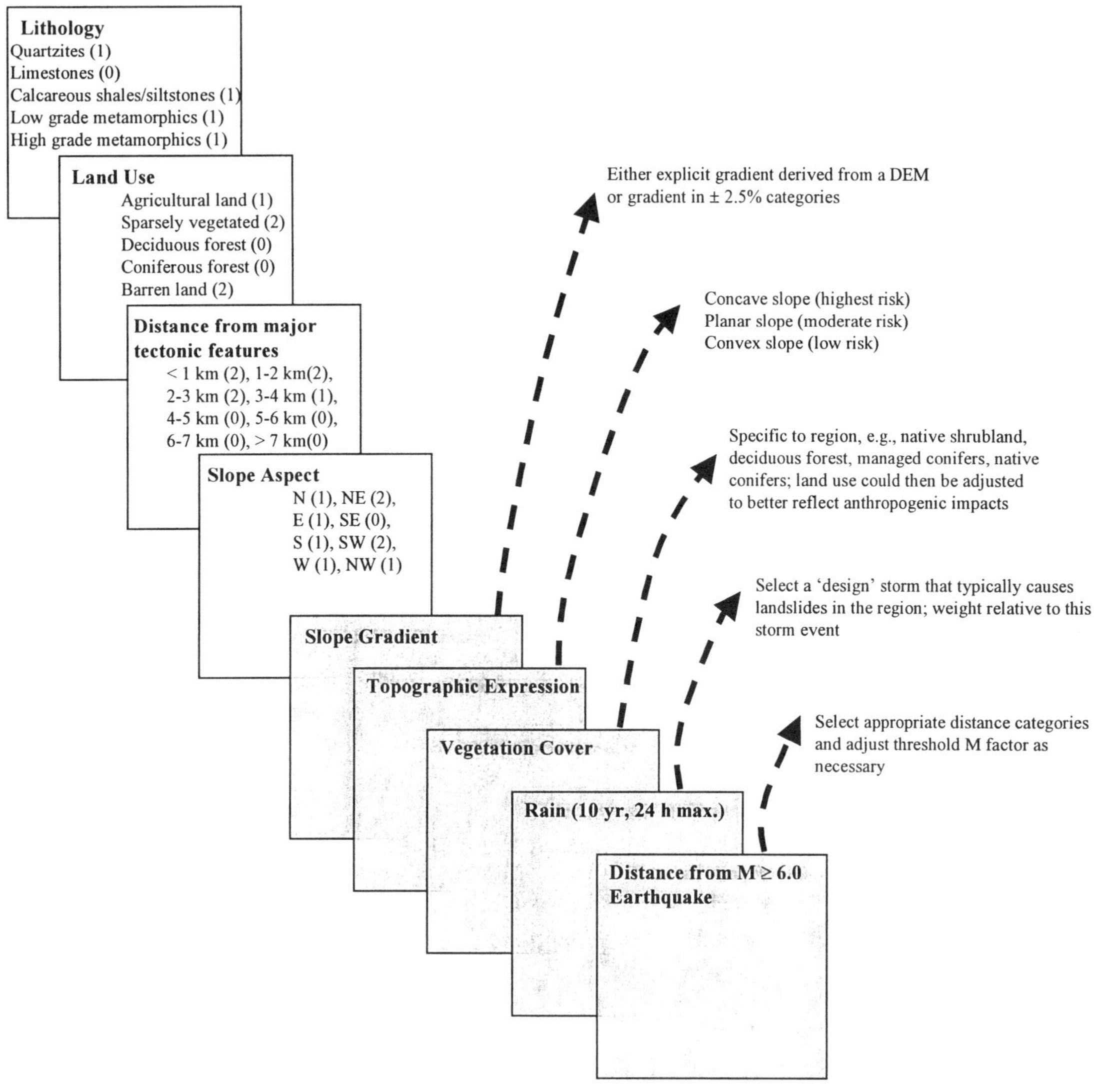

Figure 1. Schematic of the GIS-based landslide hazard zonation proposed by Gupta and Joshi [1990] for the Ramganga catchment in the Lower Himalayas. More detailed hazard analysis should include factors outlined in the lower five stippled 'boxes'.

incorporate such important factors as rainfall distribution, detailed vegetation cover, digital elevation models (DEMs) and geomorphic features into hazard assessments. It is important to focus on specific factors that are intrinsic to landslide initiation in the region of concern – e.g., high rainfall or snow accumulation areas, topographic regions of shallow groundwater accretion, steep slopes, hummocky topography, and areas with weak or decaying root strength. Such data that are focused on improving process understanding will also benefit the application of landslide hazard assessments to both multiple failure types and larger regional applications. Additionally, slope stability inferences related to forest operations and other land uses will be enhanced.

MODELING EFFORTS

Model Background

Most research on shallow landslides has focused on assessing on-site conditions and processes [e.g., Wu et al., 1979; Schroeder and Alto, 1983; Sidle, 1992]. To better predict landslide potential in larger drainage basins and to design appropriate land management strategies, it is necessary to evaluate the temporal and spatial attributes of watersheds as they relate to slope stability [Ward et al., 1982; Shasko and Keller, 1991]. When distributed, physically-based modeling is applied to landslide analysis, not only are the distributed properties of the parameters of concern, but also the model output represents a spatial problem, because we need to know the locations of landslides. Although GIS technology is highly regarded as a tool for landslide analysis in terms of spatial data extraction and display [Shasko and Keller, 1991; Leroi et al., 1992; Dikau et al., 1996], very little research has been done on incorporating distributed, physically-based slope stability modeling with GIS. A recent physically based model for shallow landslide analysis developed by Montgomery and Dietrich [1994] couples digital terrain data with near surface through flow [i.e., TOPOG, O'Loughlin, 1986] and slope stability models. The model (later called SHALSTAB) has not been tested in drainage basins where actual landslides were triggered during specific storm events. However, a recent application of SHALSTAB in the Pacific Northwest showed that it accurately predicted locations at highest risk for shallow landslides (Montgomery et al., 1998).

Another physical model was developed to analyze both the effects of seismic acceleration and rainfall-induced groundwater levels on slope failure probability [van Westen and Terlien, 1996]. This GIS-based analysis employs a simple two-dimensional groundwater model that calculates relative saturation of the soil mantle on a daily basis. Variances of soil cohesion, internal angle of friction, and soil thickness are also considered. The model does not assess the effects of rooting strength of vegetation or any related changes in land use. The model was applied to a mountainous area in central Columbia, surrounding the city of Manizales located on the western flank of the Cordillera Central, near the Nevado del Ruiz volcano. The final hazard map was checked with general occurrences of landslides, but the model was not tested for specific rainfall and seismic events that generated such failures.

To illustrate the potential for a distributed, physically based model to predict shallow landslides, the rather data intensive dSLAM model is discussed in this paper [Wu, 1993; Wu and Sidle, 1995, 1997; Sidle and Wu, 1999]. This model allows us to examine the spatial and temporal effects of timber harvesting on slope stability. The basis for dSLAM is a conceptual, non-distributed shallow landslide model [Sidle, 1992] that incorporates: (1) infinite slope analysis; (2) continuous temporal changes in root cohesion and vegetation surcharge; and (3) stochastic influence of rainfall on pore water pressure. This model quantifies the effects of various timber harvesting strategies on temporal failure probability of a particular site. Only average values of input parameters are used for various spatially distributed components in dSLAM.

Other models have been developed to evaluate the effects of vegetation management on landslide occurrence [Benda and Zhang, 1990, Ziemer et al., 1991]; these along with Sidle's [1992] model are all non-distributed and time-oriented. The spatial aspect of the slope stability problem caused by timber harvesting, such as locations of failures and changes in the spatial distribution of safety factor due to harvesting strategies in a forest rotation, has not been investigated in great detail.

Model Structure and Functional Components

Based on the model by Sidle [1992], a distributed, physically based slope stability model (dSLAM) was developed to analyze shallow rapid landslides at the watershed scale [Wu, 1993; Wu and Sidle, 1995; 1997]. This model can assess the effects of timber harvesting on slope stability in both the temporal and spatial dimensions. The root strength model developed by Sidle [1991] that simulates combined root strength decay (2) and regrowth (3) following timber harvest is used in dSLAM. A similar vegetation surcharge model is also used [Sidle, 1992]. Various scenarios of clearcutting and partial cutting can be simulated including varying rotation length, different levels of partial cutting, root regrowth and decay characteristics, and cumulative effects of multiple timber harvests. Thus, dSLAM can be used to assess the both the long-term effects of current logging practices as well as proposed logging scenarios.

To apply the concept of distributed modeling to landslide analysis, the drainage basin must be partitioned into relatively homogenous elements and then topographic parameters for each element should be calculated. Moore et al's [1988] TAPES-C stream tube model is adapted in this topographic analysis since it is consistent with the hydrologic and geomorphic processes in the subsurface (Figure 2). The stream tube approach simplifies the modeling of groundwater levels and assumes that water enters an element orthogonal to the upslope contour and exits orthogonal to the downslope contour line (Figure 2). A one-dimensional form of the kinematic wave equation is used to route subsurface water through the flow tubes [Takasao and Shiiba, 1988]. For our application in forest soils of coastal Oregon, we assume that Hortonian overland flow does not occur, thus, surface runoff is generated only when a stream tube element is fully saturated. In that case, surface runoff is routed using Manning's equation. Calculated groundwater levels are then incorporated into the factor of safety equation to determine the dynamic stability of individual elements.

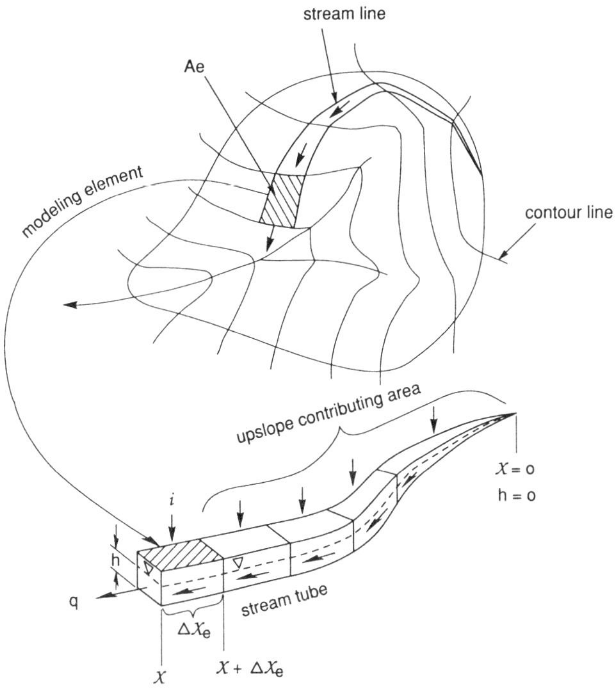

Figure 2. Diagram of Wu and Sidle's [1995] modification and adaptation of the TAPES-C model; an example stream tube is shown.

The dSLAM model accepts only event-based rainfall inputs. These inputs can either be actual rainstorm records or rainfall events generated from historical data or from theoretical distributions using Monte Carlo techniques. The temporal distribution of rainfall intensity can be used to select the time step in the model; in the examples presented a 1-h time step was used. The vector digital elevation model (DEM) for TAPES-C is derived from the GIS (ARC/INFO) contour coverage. After the contour line-based network is generated by TAPES-C it is imported into ARC/INFO to generate network polygon coverage as the basic structure for modeling and data interpretation. With the vector-based partitioning of the basin into relatively homogenous elements, the infinite slope equation (used in dSLAM) is employed to compute FS for each element in the basin using topographic information derived from TAPES-C, groundwater levels simulated by dSLAM, and soil and vegetation data extracted from ARC/INFO coverages so that unstable elements (FS<1.0) can be located (Figure 3). Spatially distributed data on the natural variability in soil properties, particularly soil depth and cohesion, may be difficult to obtain for many sites of interest. This dilemma poses a limitation for the application of distributed models, although certain algorithms that relate these properties to topographic attributes and more readily available data (e.g., soil texture) may prove useful. The model estimates landslide volume based on the average soil depth of each element incorporated in a failure path.

To more accurately assess the potential for a shallow, rapid failure equation (3) needs to be expanded to consider the effects of soil cohesion, root strength, differences in unit weights of soil above and below a water table, and surcharge due to weight of vegetation. Assuming slope parallel flow in vegetated hillsides, a more complete version of the infinite slope equation can be expressed as

$$FS = \frac{C + \Delta C + [((Z-h)\gamma_m + h\gamma_{sat} - h\gamma_w)\cos^2\beta + W\cos\beta]\tan\phi}{[(Z-h)\gamma_m + h\gamma_{sat}]\sin\beta\cos\beta + W\sin\beta} \qquad (4)$$

where C is the effective soil cohesion (kPa), ΔC is the cohesion attributed to root strength (kPa), W is the vegetation surcharge (kPa), h is the vertical height of the water table (m), γ_{sat} is the unit weight of saturated soil (kN/m^3), and γ_w is the unit weight of water (kN/m^3).

For the area in which dSLAM was initially applied (Oregon Coast Ranges), it was assumed that all unstable elements (FS < 1.0) totally mobilize into debris flows [Ellen and Fleming, 1987]. To predict the deposition of debris flows, Benda and Cundy's [1990] empirical criteria of channel gradient (< 3.5°) and tributary junction angle (angle between the channel or flow tube transporting the failed material with the receiving channel is > 70°) are used. Thus, dSLAM not only calculates and displays spatial distributions of FS and failure probability, but also the runout distance of resultant debris flows.

Application of the Model, Coastal Oregon

The model was tested in a steep forested basin (1.18 km^2) in Cedar Creek drainage of the Oregon Coast Ranges, USA. The basin contains shallow to moderate-depth soils (generally 0.5-1.5 m) overlying sandstone bedrock. This low permeability bedrock acts as a potential failure plane for shallow rapid landslides and promotes the development of a transient shallow groundwater table. The topography is highly dissected with geomorphic depressions that are carved into bedrock representing sites of past landsliding episodes. Soils are quite porous with low cohesion and bulk density. Mean annual precipitation in the area ranges from about 1500 mm along the Pacific coast to as high as 3600 mm at higher inland elevations. Rainfall infiltrates rapidly into the forest soils and drains to streams as subsurface flow. Clearcutting of Douglas-fir forests within the basin has been well documented back to 1953. A series

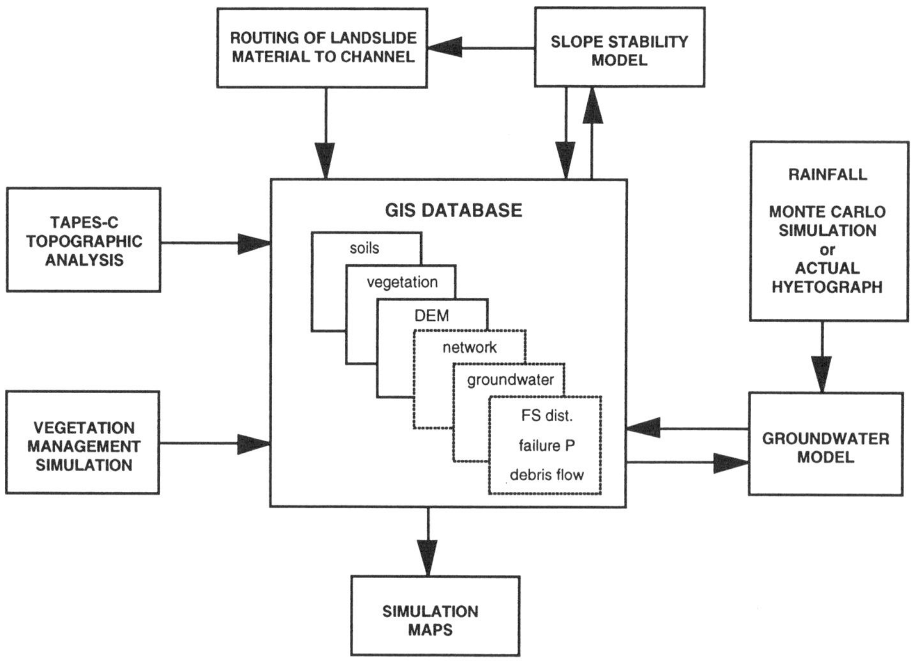

Figure 3. The operational framework of dSlam [from Wu and Sidle, 1995].

of rainstorms in the mid-1970's caused widespread landsliding in the area (especially in clearcuts) which resulted in extensive damage to many streams [Greswell et al., 1979]. A large storm (return interval 17-25 years) on November 29-30, 1975, which triggered many landslides in the region was selected as the rainfall event for various landslide simulations.

The rainfall hyetograph for the November 1975 storm was input to dSLAM to simulate groundwater response and FS for non-channelized elements in the basin. The resolution of the DEM was 30 by 30 m and the average size of discretized element was about 10 m^2. The resolution of the DEM definitely affects the prediction of landslides by dSLAM because of the different distribution pattern of topographic features associated with different DEM resolutions. Total rainfall during the 1975 storm was 178 mm, with intensities ranging from 5-18mm/h during the 21-h duration. Simulated volume (733 m^3) and numbers (4) of landslides in the basin agreed closely with values (734 m^3 and 3, respectively) measured in the field after the 1975 storm. Because the Siuslaw National Forest no longer had the original landslide inventory maps prepared after the 1975 storm, predicted landslide locations could only compared to landslide sites on crude fire maps; however, they were in the general vicinity. All simulated and actual landslides occurred in sites that were clearcut in 1968, seven years prior to the November storm. This timing would coincide with the period of minimum rooting strength after logging at the site. Additionally, all elements in the basin with FS values less than 2.0 at the end of the simulated storm were also clearcut in 1968. The elements that failed in simulations had slope gradients ranging from 37 to 48°. Locations of the simulated failures were in steep areas of groundwater convergence, particularly geomorphic hollows or zero-order basins (Figure 4). Factor of safety declined rapidly in these hollows during the storm

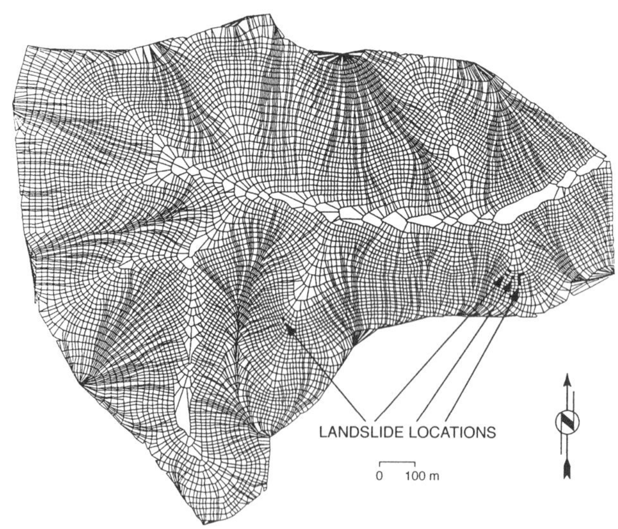

Figure 4. Location of simulated landslides in the November 1975 storm, Cedar Creek basin, Oregon [from Wu and Sidle, 1995].

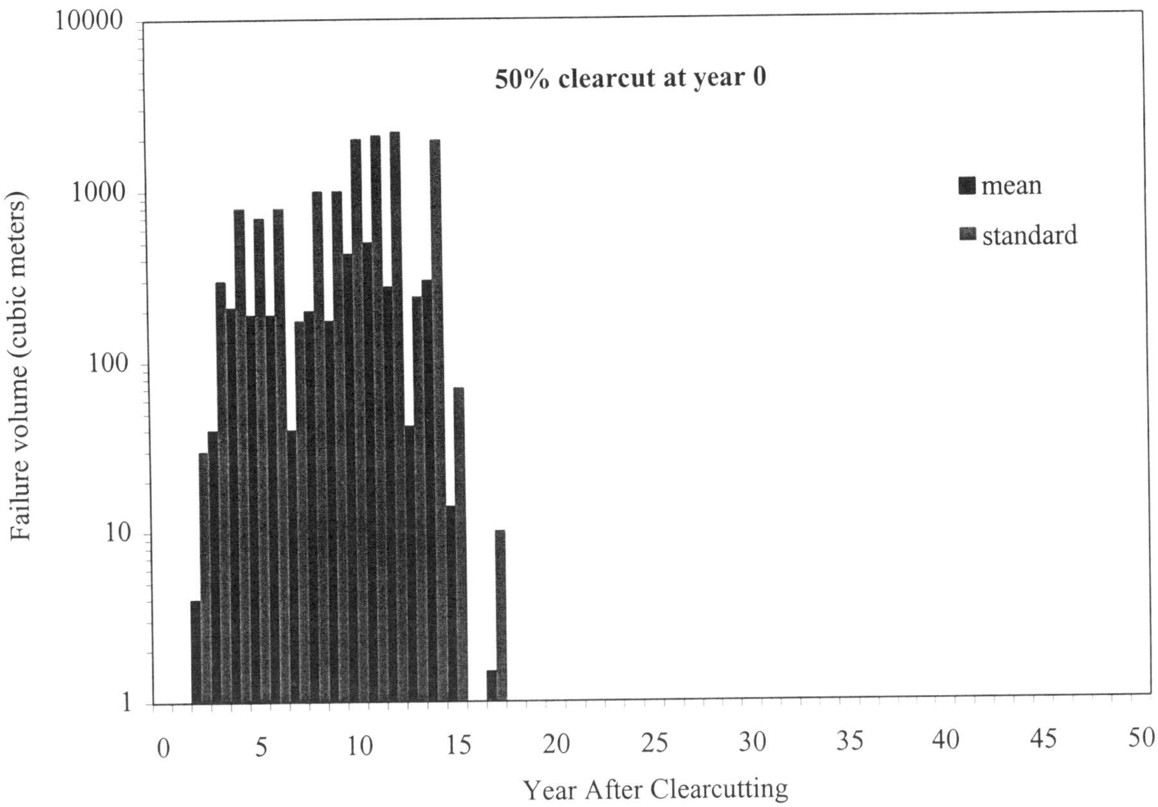

Figure 5. Simulated temporal distribution of landslides during a 50-yr timber harvest cycle; Cedar Creek basin was 50% clearcut (randomly with blocking constraints) in year 0 with no designated leave areas [after Sidle and Wu, 1999].

event and simulations showed that it took more than a day following the storm for FS values to approach pre-storm conditions. Similar accuracy in landslide predictions was achieved in an adjacent drainage basin within the greater Cedar Creek watershed [Wu and Sidle, 1997]. In both basins average values of spatially distributed soil and site parameters were used and no calibration was performed to achieve the accurate and somewhat temporally and spatially explicit results. However, Wu and Sidle [1995, 1997] demonstrated that changes in input parameter values for soil cohesion, root cohesion, and soil depth in the range of ± 20-30% can alter predicted landslide volumes by several fold.

Monte Carlo simulations were also conducted to assess the effects of various timber harvesting scenarios in the coastal Oregon basin [Sidle and Wu, 1999]. Simulation of the temporal distribution of landslide occurrence for a single 50% clearcut is consistent with previous empirical findings: most landslides are distributed in a period of about 3-15 yr after the clearcut (Figure 5). Of the 39,000 simulated rainstorms, 82 triggered landslides. The spatial distribution of failure probability in the Cedar Creek basin was investigated for several cases: the actual pattern of timber harvesting in 1975; a random 50% clearcut with and without leave areas at the beginning of the cycle; and a 100% clearcut at the beginning of the cycle with and without leave areas. Vegetation leave areas, a forest management practice employed to reduce landslide occurrence in unstable portions of proposed clearcuts, was evaluated to be an effective control measure. The simulations indicate that the spatial distribution of failure potential is determined by patterns of timber harvesting and groundwater movement: areas with high failure potential were in sites where vegetation root strength is low and groundwater flow concentrates. Finally, results support the hypothesis that geomorphic hollows are positions that have highest potential of landslide occurrence in steep terrain.

SUMMARY

The potential for shallow, rapid landslides in steep terrain can be analyzed by several methodologies at different levels of detail. Terrain stability classification techniques are generally based on geomorphic information derived either primarily from maps, air photos, and other second-hand sources or from a combination of these data and field investigations. Depending on the detail of the field investigations and the expert knowledge of the terrain mapper, terrain classification can be quite useful in delineating general areas of high landslide potential. However, the impacts of land use practices such as logging and forest roads are quite difficult to interpret from such classification procedures unless detailed historical data on landslides are available together with field data. Stability hazard assessments that employ multiple factor overlays may be developed and applied at wide regional scales where data are sparse or general [as in the case noted in the Lower Himalayas, Gupta and Joshi, 1990] or in areas where very detailed data related to landslide processes (e.g., distributed rainfall, land use, soil characteristics) are readily available [as in the San Francisco Bay area, Nielsen et al., 1979; Keefer et al., 1987]. In the former case only broad generalizations related to landslide potential can be made, while for the most detailed scenario applied to the San Francisco Bay area, real-time landslide predictions are possible [Keefer et al., 1987]. At all levels of detail in such stability hazard assessments, GIS technology is becoming a very useful tool. This GIS technology has greatly enhanced the capabilities for landslide modeling. However, because of these complexities in modeling system components (e.g., shallow groundwater response, rooting strength changes) that contribute to shallow landslide initiation, only a few truly physically-based models have been developed. Additionally, to capture the effects of spatial variability in soil, topography, vegetation, and hydrology, these models must be distributed.

The dSLAM landslide model [e.g., Wu and Sidle, 1995] allows for the simulation of timber harvesting scenarios by incorporating effects of root strength decay and regrowth as well as represents shallow groundwater response by a geomorphically-controlled subsurface flow model. While dSLAM requires a high level of knowledge (or numerous assumptions) related to spatially distributed watershed characteristics, it has the potential to reasonably predict locations, volumes, and timing of shallow landslides as well as debris flow runout distances. As a tool for modeling past or future forest harvesting effects on landslide, dSLAM is quite promising.

REFERENCES

Abe, K., and R.R. Ziemer, Effect of tree roots on a shear zone: Modeling reinforced shear stress, *Can. J. For. Res., 21,* 1012-1019, 1991.

Anderson, L.R., J.R. Keaton, T.F. Saarinen, and W.G. Wells III, The Utah landslides, debris flows and floods of May and June, 1983, in Delineation of Landslide, Flash Flood, and Debris Flow Hazards in Utah, edited by D.S. Bowles, pp. 3-14, *Proc. of Speciality Conf.*, Utah State Univ., Logan, UT, 1985.

Aniya, M., Landslide-susceptibility mapping in the Amahata River basin, Japan, *Ann. Assoc. Am. Geog., 75,* 102-114, 1985.

Ballard, T.M., and R.P. Willington, Slope instability in relation to timber harvesting in the Chilliwack Provincial Forest, *For. Chron., 51,* 59-62, 1975.

Bangxing, T., L. Suqing, and L. Shijjian, Mountain disaster formation in northwest Sichuan, *GeoJournal, 34.1, 41-46,* 1994.

Benda, L. E., and T.W. Cundy, Predicting depositions of debris flows in mountain channels, *Can. Geotech. J., 27,* 409-417, 1990.

Benda, L. and W. Zhang, Accounting for the stochastic occurrence of landslides when predicting sediment yields, *IAHS-AISH Publ. 192*, pp. 115-127, 1990.

Bishop, D.M., and M.E. Stevens, Landslides on logged areas, southeast Alaska, *Res. Rep. NOR-I,* 18 pp., USDA Forest Service, Juneau, Alaska, 1964.

Blong, R.J. and D.L. Dunkerley, Landslides in the Razorback area, New South Wales, Australia, *Geograf. Ann. 58A*, 139-147, 1976.

Burroughs, E.R. and B.R.Thomas, Declining root strength in Douglas-fir after felling as a factor in slope stability, *Research Paper INT-190*, 27 pp., USDA Forest Service, Intermountain Forest and Range Experiment Station, Ogden, UT, 1977.

Cannon, S.H., and S.D. Ellen, Rainfall conditions for abundant debris avalanches, San Francisco Bay region, California, *Calif. Geol., 38*, 267-272, 1985.

Carson, M.A., and D.J. Petley, The existence of threshold hillslopes in the denudation of the landscape, *Trans. Inst. British Geogr., 49*, 71-95, 1970.

Dengler, L. and D.R. Montgomery, Estimating the thickness of colluvial fill in unchanneled valleys from surface topography, *Bull. Assoc. Engr. Geol., 26*, 333-342, 1989.

DeRose, R.C., Relationships between slope morphology, regolith depth, and the incidence of shallow landslides in eastern Taranaki hill country, *Z. Geomorphol. N.F., 105,* 49-60, 1996.

Dietrich, W.E. and T. Dunne, Sediment budget for a small catchment in mountainous terrain, Z. Geomorphol. Suppl. Bd. 29: 191-206, 1978.

Dikau, R., A. Cavallin, and S. Jäger, Databases and GIS for landslide research in Europe, *Geomorphology, 15,* 227-239, 1996.

Dunne, T., Stochastic aspects of the relations between climate, hydrology, and landform evolution, *Trans. Jpn. Geomorphol. Union, 12,* 1-24, 1991.

Ellen, S.D., and R.W. Fleming, Mobilization of debris flows from soil slips, San Francisco Bay region, California, Geological Society of America, *Rev. Eng. Geol., 3,* 31-40, 1987.

Endo, T., and T. Tsuruta, The effect of the tree's roots on the shear strength of soil, *Annual Report*, 1968, pp.167-182, Hokkaido Branch For. Exp. Stn., Sapporo, Japan, 1969.

Finlay, P.J., R. Fell, and P.K. Maguire, The relationship between the probability of landslide occurrence and rainfall, *Can. Geotech. J., 34,*811-824, 1997.

Fujiwara, K., A study on the landslides by aerial photographs, *Res. Bull. Exp., Forest Hokkaido Univ., 27*(2), 297-345, 1970.

Fukuda, F., and H. Ochiai, Landslides caused by the 1993, Hokkaido Nansei-oki earthquake, *Shin Sabo J., 46,* 62-63, 1993.

Garwood, N.C., D.P. Janos, and N. Brokaw, Earthquake induced landslides: A major disturbance to tropical forests, *Science, 205,* 997-999, 1979.

Gillham, R.W., The effect of capillary fringe on water-table response, *J. Hydrol. 67,* 307-324, 1984.

Gomi, T., K. Sassa, and G. Takahashi, The influence of debris movement on the channel forms in the flood plains of a serpentine basin, northern Hokkaido, Japan, *Trans., Jpn.. Geomorph. Union, 19,*1-18, 1998.

Gray, D.H., and W.F. Megahan, Forest vegetation removal and slope stability in the Idaho Batholith, *Res. Pap. INT-271,* 23 pp., USDA Forest Service, Ogden, Utah, 1981.

Gresswell, S., D. Heller, and D. N. Swanston, Mass movement response to forest management in the central Oregon Coast Ranges, *USDA Forest Serv. Resour. Bull. PNW-84,* 26 pp., Portland, Oreg., 1979.

Gupta, R.P., and B.C. Joshi, Landslide hazard zoning using the GIS approach – a case study from the Ramganga Catchment, Himalayas, *Engineering Geol., 28,* 119-131, 1990.

Harp, E.L., and R.W. Jibson, Landslides triggered by the 1994 Northridge, California, earthquake, *Bull. Seismol. Soc. Am., 86,* S319-S332, 1996.

Harvey, A..M.., The influence of sediment supply on channel morphology of upland streams, *Earth Surf. Proc. and Landforms, 16,* 675-684, 1991.

Howes, D.E., and E. Kenk, Terrain classification system for British Columbia (revised edition), *Ministry of Environment Manual 10,* Ministry of Crown Lands, Victoria, B.C., Canada, 1988.

Keefer, D.K., Landslides caused by earthquakes, *Geol. Soc. Am. Bull. 95,* 406-421, 1984.

Keefer, D.K., R.C. Wilson, R.K. Mark, E.E. Brabb, W.M. Brown, S.D. Ellen, E.L. Harp, G.F. Wieczorek, C.S. Alger, and R.S. Zatkin, Real-time landslide warning during heavy rainfall, *Science, 238,* 921-925, 1987.

Keller, E.A., and F.J. Swanson, Effects of large organic material on channel form and fluvial processes, *Earth Surf. Proc. & Landforms, 4,* 361-380, 1979.

Kienholz, H., G. Schneider, M. Bichsel, M. Grunder, and P. Mool, Mapping of mountain hazards and slope stability, *Mountain Res. and Development 4*(3), 247-266, 1984.

Lehre, A.K., Sediment budget of a small Coast Range drainage basin in north-central California, in Sediment Budgets and Routing in Forested Drainage Basins, *Gen. Tech. Rep. PNW-141,* pp. 67-77, Forest Service, U.S. Dep. of Agric., Portland, Oreg., 1982.

Leroi, E., O. Rouzeau, J.-Y. Scanvic, C.C. Weber, and G. Varges C., Remote sensing and GIS technology in landslide hazard mapping in the Columbia Andes, *Episodes, 15,* 32-35, 1992.

Lohnes, R.A., and R.L. Handy, Slope angles in friable loess, *J. Geol., 76*(3), 247-258, 1968.

Maharaj, R.J., Landslide processes and landslide susceptibility analysis from an upland watershed: a case study from St. Andrew, Jamacia, West Indies, *Eng, Geol., 34,* 53-79, 1993.

Marden, M.., and D. Rowen, Protective value of vegetation on Tertiary terrain before and during Cyclone Bola, East Coast, North Island, New Zealand, *N.Z. J. For. Sci., 23,*255-263, 1993.

Mark, A.F., G.A.M. Scott, F.R. Sanderson, and P.W. James, Forest succession on landslides above Lake Thomson, Fjordland, New Zealand, *N.Z. J. Botany, 2,* 60-89, 1964.

McDonnell, J. J. (1990) A rationale for old water discharge through macropores in a steep, humid catchment, *Water Resour. Res., 11,* 2821-2832.

Meehan, W. R., Fish habitat and timber harvest in southeast Alaska, *Naturalist, 25,* 28-31, 1974.

Megahan, W.F., Hydrologic effects of clearcutting and wildlife on steep granitic slopes in Idaho, *Water Resources Research, 19*(3), 811-819, 1983.

Megahan, W.F., N.F.Day, and T.M.Bliss, Landslide occurrence in the western and central northern Rocky Mountain physiographic province in Idaho, in *Proceedings 5th North American Forest Soils Conference,* pp. 116-139, Colo. State Univ., Fort Collins, 1978.

Ministry of Forests, Mapping and assessing terrain stability guidebook, Ministry of Forests, Victoria, B.C., Canada, 1995.

Montgomery, D.R., and W.E. Dietrich, A physically based model for the topographic control on shallow landsliding, *Water Resour. Res., 30*(4), 1153-1171, 1994.

Montgomery, D.R., W.E. Dietrich, R. Torres, S.P. Anderson, J.T. Heffner, and K. Loague, Hydrologic response of a steep, unchanneled valley to natural and applied rainfall, *Water Resour. Res., 33,* 91-109, 1997.

Montgomery, D.R., K. Sullivan, and H.M. Greenberg, Regional test of a model for shallow landsliding, *Hydrol. Processes, 12,* 943-955, 1998

Moore, I.D., E.M. O'Loughlin, and G.J. Burch, A contour based topographic model and its hydrologic and ecological applications, *Earth Surface Processes and Landforms, 13,* 305-320, 1988.

Murphy, W., The geomorphological controls on seismically triggered landslides during the 1908 Straits of Messina earthquake, Southern Italy, *Q. J. Eng. Geol., 28,* 61-74, 1995.

Nielsen, T.H., R.H. Wright, T.C. Vlasic, and W.E. Spangle, Relative slope stability and land-use planning in the San Francisco Bay region, California, *U.S. Geol. Surv. Prof. Pap. 944,* 96pp., 1979.

Okunishi, K., and T. Iida, Evolution of hillslopes including landslides, *Trans. Japanese Geomorph. Union, 2,* 291-300, 1981.

O'Loughlin, C.L., The stability of steepland forest soils in the Coast Mountains, southwest British Columbia. Ph.D. thesis, 147 pp., University of British Columbia, Vancouver, 1972.

O'Loughlin, C.L., The effect of timber removal on the stability of forest soils, *J. Hydrol. N.Z., 13*(2), 121-123, 1974.

O'Loughlin, C.L., and A.J. Pearce, Influence of Cenozoic geology on mass movement and sediment yield response to forest removal, North Westland, New Zealand, *Bull. Int. Assoc. Eng. Geol. 14*, 41-46, 1976.

O'Loughlin, C.L., L.K. Rowe, and A.J. Pearce, Exceptional storm influences on slope erosion and sediment yields in small forest catchments, North Westland, New Zealand, *Natl. Conf. Publ. 82/6*, pp. 84-91, Inst. of Eng., Barton, ACT, Australia, 1982.

O'Loughlin, E.M., Prediction of surface saturation zones in natural catchments by topographic analysis, *Water Resour. Res., 22*, 794-804,1986.

Onda, Y., A. Mori, and S. Shindo, The effects of topographic convergence and location of past landslides on subsurface water movement on granite hillslope, *J. Natural Disaster Sci., 14*, 45-58, 1992.

Pain, C.F., and J.M. Bowler, Denudation following the November 1970 earthquake at Madang, *Z. Geomorphol. Suppl. Bd. 18*, 91-104, 1973.

Pearce, A.J., M.K. Stewart, and M.K. Sklash, Storm runoff generation in humid headwater catchments, 1. Where does the water come from?, *Water Resour. Res. 22*, 1263-1272, 1986.

Pearce, A.J., and A.J. Watson, Medium-term effects of two landsliding episodes on channel storage of sediment, *Earth Surf. Proc. & Landforms, 8*, 29-39, 1983.

Pierson, T.C., Piezometric response to rainstorms in forested hillslope drainage depressions, *J. Hydrol. (New Zealand), 19*, 1-10, 1980.

Pierson, T.C. Soil pipes and slope stability, *Q. J. Eng. Geol. 16, 1-11*, 1983.

Ralph, S.C., G.C. Poole, L.L. Conquest, and R.J. Naiman, Stream channel morphology and woody debris in logged and unlogged basins of western Washington, *Can. J. Fish. Aquat. Sci., 51*,37-51, 1994.

Rice, S., and M. Church, Bed material texture in low order streams on the Queen Charlotte Islands, British Columbia, *Earth Surf. Proc. and Landforms, 21*,1-18, 1996.

Rogers, N.W., and M.J. Selby, Mechanisms of shallow transational landsliding during summer rainstorms: North Island, New Zealand, *Geograf. Ann., 62A*, 11-21, 1980.

Rollerson, T., Relationships between landscape attributes and landslide frequencies after logging – Skidegate Plateau, Queen Charlotte Islands, *Land Management Rep. No. 76*, Ministry of Forests, Victoria, B.C., Canada, 1992.

Rollerson, T.P., B. Thomson, and T.H. Millard, Identification of coastal British Columbia terrain susceptible to debris flows, in *Debris-Flow Hazards Mitigation: Mechanics, prediction and assessment*, Am. Soc. Civ. Engr., San Francisco, Calif., 1997.

Scheroeder, W.L. and J.V. Alto, Soil properties for slope stability analysis; Oregon and Washington coastal mountains, *Forest Science*, 29(4), 823-833, 1983.

Schwab, J.W., Mass wasting: October-November 1978 storm, Rennel Sound, Queen Charlotte Islands, British Columbia, *Publ. 91*, 23 pp. Ministry of Forests, Victoria, B.C., Canada, 1983.

Schweinfurth, U., Uber eine besondere Form der Hangabtragurg im neuseelandischen Fjordland, *Z. Geomorphol., 10*,144-149, 1967.

Shasko, M.J., and C.P. Keller, Assessing large scale slope stability and failure within a geographic information system, in *GIS Applications*, edited by M. Heitand and A. Shartreid, pp. 267-275, GIS World, Inc., 1991.

Shimokawa, E., A natural recovery process of vegetation on landslide scars and landslide periodicity in forested drainage basins, In: Proc. of Symp. on Effects of Forest Land Use on Erosion and Slope Stability, pp. 99-107, East-West Ctr., Honolulu, Hawaii, USA, 1984.

Sidle, R.C., Shallow groundwater fluctuations in unstable hillslopes of coastal Alaska, *Z. für Gletscherkunde und Glazialgeol., 20*, 79-95, 1984a.

Sidle, R.C., Relative importance of factors influencing landsliding in coastal Alaska, *Proc. 21st Annu. Eng. Geol. and Soils Eng. Symp.*, pp. 311-325, Univ. of Idaho, Moscow, 1984b.

Sidle, R.C., A dynamic model of slope stability in zero-order basins, *IASH Publ. 165*,101-110, 1987.

Sidle, R.C., Evaluating long-term stability of previously failed hillslope depressions, in *Proc. 24th symposium on engineering geology & soils engineering*, Idaho Dept. of Transport., Boise, Idaho, 1988.

Sidle, R.C., A conceptual model of changes in root cohesion in response to vegetation management, *J. Environ. Quality, 20*(1), 43-52, 1991.

Sidle, R.C., A theoretical model of the effects of timber harvesting on slope stability, *Water Resources Research, 28*(7), 1897-1910, 1992.

Sidle, R.C., I. Kamil, A. Sharma, and S. Yamashita, Stream response to subsidence from underground coal mining in central Utah, *Environ. Geol., 29,* (in press), 1999.

Sidle, R.C., A.J. Pearce, and C.L. O'Loughlin, *Hillslope stability and land use, Water Resources Monogr.*, Vol. 11, 140 pp., AGU, Washington, D.C., 1985.

Sidle, R.C. and D.N. Swanston, Analysis of a small debris slide in coastal Alaska, *Canadian Geotech. J., 19,* 167-174, 1982.

Sidle, R. C., and Tsuboyama, Y., A comparison of piezometric response in unchanneled hillslope hollows: coastal Alaska and Japan, *J. Japan Soc. Hydrol. & Water Resour., 5,* 3-11, 1992.

Sidle, R.C., Y. Tsuboyama, S. Noguchi, I. Hosoda, M. Fujieda, and T. Shimizu, Seasonal hydrologic response at various spatial scales in a small forested catchment, Hitachi Ohta, Japan, *J. Hydrol. 168*, 227-250, 1995.

Sidle, R.C., Y. Tsuboyama, S. Noguchi, I. Hosoda, M. Fujieda, and T. Shimizu, Progress towards understanding stormflow generation in headwater catchments, in *Environmental Forest Science*, pp. 483-498, Proc. of IUFRO Div. 8 Conf., 19-23 October 1998, Kyoto, Japan, Kluwer Academic Publisher, 1998.

Sidle, , R.C., Y. Tsuboyama, S. Noguchi, I. Hosoda, M. Fujieda, and T. Shimizu, Stormflow generation in steep forested headwaters: a linked hydrogeomorphic paradigm, *Hydrol. Processes, 14,* (in press), 2000.

Sidle, R.C., and W. Wu, Simulating effects of timber harvesting on the temporal and spatial distribution of shallow landslides, *Z. Geomorphol. N.F., 43(2),* 185-201, 1999.

Swanson, F.J., and C.T. Dyrness, Impact of clearcutting and road construction on soil erosion by landslides in the western Cascades, Oregon, *Geology, 3*(7), 393-396, 1975.

Swanson, F.J., S.V. Gregory, J.R. Sedell, and A.G. Campbell, Land-water interactions: the riparian zone, in *Analysis of Coniferous Forest Ecosystems in the Western United States,* edited by R.L. Edmonds, pp. 267-291, Hutchison Ross, Stroudsburg, Pa., 1982a.

Swanson, F.J., R.L. Fredriksen, and F.M. McCorison, Material transfer in a western Oregon forested watershed, in *Analysis of Coniferous Forest Ecosystems in the Western United States,* edited by R.L. Edmonds, pp. 233-266, Hutchison Ross, Stroudsburg, Pa., 1982b.

Swanston, D.N., The forest ecosystem of southeast Alaska. 5. Soil mass movement, *Gen. Tech. Rep. PNW-17,* 22pp., Forest Service, U.S. Dep. of Agric., Portland, Oreg., 1974.

Takasao, T., and M. Shiiba, Incorporation of the effect of concentration of flow into the kinematic wave equations and its applications to runoff system lumping, *J. Hydrol., 102,* 301-322, 1988.

Temple, P.H., and A. Rapp, Landslides in the Mgeta area, western Uluguru Mountains, Tanzania, *Geograf. Ann., 54A,* 157-193, 1972.

Torres, R., W.E. Dietrich, D.R. Montgomery, S.P. Anderson, and K. Loague, Unsaturated zone processes and the hydrologic response of a steep, unchanneled catchment, *Water Resour. Res., 34*(8), 1865-1879, 1998.

Tsuboyama, Y., I. Hosoda, S. Noguchi, and R.C. Sidle, Piezometric response in a zero-order basin, Hitachi Ohta, Japan, in *Proc. Int. Symp. on Forest Hydrology,* edited by T. Ohta, pp. 217-224, Univ. of Tokyo, Japan, 1994a.

Tsuboyama, Y., R.C. Sidle, S. Noguchi, and I. Hosoda, Flow and solute transport through the soil matrix and macropores of a hillslope segment, *Water Resour. Res. 30,* 879-890, 1994b.

Tsukamoto, Y., and T. Ohta, Runoff processes on a steep forested slope, *J. Hydrol. 102,* 165-178, 1988.

Tsukamoto, Y., T. Ohta, and H. Noguchi, Hydrological and geomorphological studies of debris slides on forested hillslopes in Japan, *IAHS-AISH Publ. 137,* pp. 89-98, 1982.

van Westen, C.J., and T.J. Terlien, An approach towards deterministic landslide hazard analysis in GIS: a case study from Manizales (Columbia), *Earth Surf. Proc. and Landforms, 21,* 853-868, 1996.

Ward, T.J., R.Li, and D.B. Simons, Mapping landslide hazards in forest watersheds, *J. Geotech. Eng., ASCE, 108*(GT2), 319-324, 1982.

Wieczorek, G.F., E.W. Lips, and S.D. Ellen, Debris flows and hyperconcentrated floods along the Wasatch Front, Utah, 1983 and 1984, *Bull. Assoc. Eng. Geol., 26,* 191-208, 1989.

Wright, C., and A. Mella, Modifications to the soil pattern of south central Chile resulting from seismic and associated phenomena during the period May to August 1960, *Bull. Seismol. Soc. Am., 53,* 1367-1402, 1963.

Wu, W., Distributed slope stability analysis in steep, forested basins, Ph.D. thesis, 148 pp., Utah State Univ., Logan, 1993.

Wu, T.H., W.P. McKinnel, and D.N. Swanston, Strength of tree roots and landslides on Prince of Wales Island, Alaska, *Can. Geotech. J., 16,* 19-33, 1979.

Wu, W., and R.C. Sidle, A distributed slope stability model for steep forested hillslopes, *Water Resour. Res., 31*(8), 2097-2110, 1995.

Wu, W., and R.C. Sidle, Application of a distributed of a distributed shallow landslide analysis model (dSLAM) to managed forested catchments in coastal Oregon, *IAHS Publ., 245,* 213-221, in, Human Impact on Erosion and Sedimentation, Proc. 5th Sci. Assembly, Wallingford, UK, 1997.

Ziemer, R.R., J. Lewis, R.M. Rice, and T.E. Lisle, Modeling the cumulative watershed effects of forest management strategies, *J. Environ. Qual., 20,* 36-42, 1991.

Ziemer, R.R., and D.N. Swanston, Root strength changes after logging in southeast Alaska, *Res. Note PNW-306,* 10 pp. Forest Service, U.S. Dep. of Agric., Portland, Oreg., 1977.

Validation of the Shallow Landslide Model, SHALSTAB, for Forest Management

William E. Dietrich and Dino Bellugi

Department of Earth and Planetary Science, University of California, Berkeley

Raphael Real de Asua

Stillwater Sciences, Berkeley, California

There are increasing demands at the regional, watershed and local level to reduce the incidence of shallow landsliding resulting from forest management practices. Here we report a validation study of SHALSTAB, a simple mechanistic model for delineating the relative potential for shallow landsliding across the landscape. The cohesionless, infinite-slope model a uses topographically-driven, steady-state shallow subsurface flow theory to estimate the spatial distribution of destabilizing pore pressure. The calculated ratio of the effective precipitation, q, to the soil transmissivity, T at instability is used to map the relative potential for landsliding. We assume that sites with the same q/T value have the same relative potential for instability, and sites with the lowest q/T are least stable. Potential instability (i.e. q/T) is dominated by the drainage area per cell size rather than hillslope gradient for slopes between 20 and 40 degrees. We validated the model (using fixed parameters) in 7 watersheds in Northern California by comparing the frequency of shallow landslides per unit area of q/T classes with that produced using a biased random-placement model of similar sized landslide scars. Over 1100 landslides were mapped and comparisons suggest that for a 10 m grid, a log(q/T) threshold of less than –2.8 covers about 13% of the watershed and predicts on average about 60% of the shallow landslide locations. Hence, the model can discriminate a relatively small part of the landscape that produces the majority of shallow landslides. Model performance depends strongly on the quality of the topographic base map; 10 m grid data derived from USGS 7.5' topographic maps miss important fine-scale topography. This study suggests that SHALSTAB may be useful in regional planning, watershed analyses and timber harvest planning. For site-specific land management practices, field surveys are needed to evaluate the accuracy of the topographic maps and as well other factors affecting local slope stability.

Land Use and Watersheds: Human Influence on Hydrology and Geomorphology in Urban and Forest Areas
Water Science and Application Volume 2, pages 195-227

INTRODUCTION

In the steep, forested coastal mountains of the Pacific Northwest, shallow landslides are a major source of sediment delivered to streams (Figure 1). Individual landslides may mobilize into debris flows, and travel downstream

Figure 1. Debris flow scars resulting from 1996 storms in the Oregon Coast Range. Note that the scars are originating in the hollows lying at the tips of the valley network. Photograph by John Stock.

scouring channels of all sediment and wood, then depositing the acquired mass in a large accumulation where the flow comes to rest [e.g. Dietrich and Dunne, 1978; Hungr et al., 1984; Benda and Dunne, 1997]. Although shallow landsliding and associated debris flows are an integral part of natural geomorphic processes, forest management practices can greatly increase the frequency of their occurrence [e.g. Sidle et al., 1985], which can lead to degradation of stream habitat and loss of habitat features either through scour to bedrock or high sediment loading. Where people live in areas adjacent to lands managed for timber production, there is also an increased chance of forest management actions triggering landslides that pose a risk to human life and property.

As part of landscape-level land use planning efforts mandated by various recent state and federal regulations in the United States, forest managers have been seeking ways to delineate shallow landslide potential on their lands in order to develop management prescriptions that minimize increases in slope instability while sustaining timber harvesting goals. Typically such landslide potential maps must be generated with limited information: geologic, topographic and land use maps and limited aerial photograph coverage. Such things as soil strength, root strength dynamics, and site hydrologic characteristics, all of which strongly influence slope stability are normally not known quantitatively. These properties can vary significantly on spatial scales of only a few meters, making their systematic mapping a daunting task. We are aware of no such spatially registered information in the Pacific Northwest. In response to this absence of data we have explored the use of a simple, mechanistic model for shallow slope stability (typically involving just the soil mantle) [Dietrich et al., 1992, 1993; Montgomery and Dietrich, 1994; Dietrich and Sitar, 1997; Dietrich and Montgomery, 1998, Dietrich et al., 1998]. We suggest that this model, now turned into the computer model SHALSTAB (available at http://socrates.berkeley.edu/~geomorph/), is perhaps the simplest, process-based model that accounts for the topographic control on pore pressure development responsible for shallow slope instability. Federal and state agencies as well as private companies are increasingly using this model. Here we review the basic theory for the model, illustrate how it can be validated and applied in various contexts, and discuss important issues of interpretation and further developments of the model. A central issue here is that SHALSTAB (or any such digital terrain model) is only as good as the available topographic data. In areas with adequate topographic data, we have found that mapped landslides preferentially occur in areas predicted by SHALSTAB to be most prone to slope instability.

PREVIOUS WORK

Montgomery and Dietrich [1994] briefly review the general approaches that have been taken to assessing landslide hazards. These approaches range from field mapping and intuitive extrapolation to multivariate analysis and to mechanistic-based theory. At least four different approaches have been explored in forest management applications.

One approach involves creating a map of observed shallow landslides based on interpretation of aerial photographs and field inspections, using professional judgment and knowledge of local geology and topography to classify the land into landslide hazard categories, and then specifying the types of management that can be conducted in these areas (no road construction, selective harvest, etc.). This is the approach taken in most watershed analyses that follow the Washington State Department of Natural Resources methodology for watershed analysis [WFPB, 1997]. The strength of this approach is that it is based on field investigations, and if the mapper has adequate knowledge of factors influencing landslide processes, reliable interpretations may often be obtained. Its weakness is that landslide maps only reveal where landslides have occurred, not where they are most likely to occur in the future. Hence, the mapmaker must rely on intuition and experience to estimate the full extent of landslide potential existing in a watershed. This resulting lack of objectivity makes the process very dependent on the mapmaker's skills and experience. Furthermore, the mapmaker will be more inclined to create broad categories of land types to avoid the time-consuming and more error-prone process of making detailed interpretations based on inferred local conditions.

Another approach is based on correlating terrain attributes (ranging from hillslope gradient to bedrock type) of polygons mapped in the field with the incidence and character of landsliding [e.g. Rollerson et al., 1997; Fannin et al., 1997; see Carrara, 1983 and Carrara et al., 1995 for discussion of methodology]. This provides a more quantitative empirical basis for likelihood for failure than is conventionally used in the Washington methodology. Because this is an empirical approach, results can not be extrapolated beyond the area of study. Other recent multivariate approaches based on field measurements, such as that reported by Wiczorek et al. [1997], help identify for particular storm events the spatial controls on hillslope instability. There also exists empirical field-based rating systems, such as the Mapleton "Headwall Rating System" that are not based on quantitative testing but rather intuitive assignment of the relative importance of factors [Swanson and Roach, 1987]. While such intuitive systems may tend to identify unstable areas [Martin, 1997], the lack of a mechanistic foundation makes verification problematic.

A third approach that has been explored in the forest management context is the use of a slope stability theory applied to selected area polygons of similar terrain type or to individual sites on a hillslope. In the Pacific Northwest, the programs DLISA, LISA and 3DLISA developed by Prellwitz [1985], Hammond et al. [1992] and Burroughs et al [1985] [see also Cuthbertson,1992] have been used by the US Forest Service. The infinite slope equation is solved with best estimates (DLISA) or with probabilistic assignment of parameters (LISA and 3DLISA), while some values are back-calculated based on the assumption of a factor of safety at failure [e.g. Kohler, 1998]. DLISA and LISA treat the influence of groundwater as an unknown whereas 3DLISA uses an empirical groundwater response model derived from field observations in the Oregon Coast Range [Burroughs et al., 1985].

While more mechanistic than the empirical approaches previously mentioned, these models appear to be of limited value for several reasons. First, it conceptually inconsistent to divide landscapes into terrain polygons with average properties and then use a mechanistic theory that depends strongly on site-specific conditions. In a threshold phenomenon, such as landsliding, small differences in controls, such as local slope, drainage area or site properties can have a large effect on assignment of stability. Furthermore, these parameter-rich models (these programs require values on soil cohesion, root cohesion, soil bulk density, water table level, friction angle, soil depth, hillslope gradient, and more in the case of 3DLISA) are untestable because the spatial distribution of parameters cannot (with the exception of hillslope gradient) be quantified in any realistic time frame. A probabilistic formulation (in LISA and 3DLISA) has been introduced to try to represent this uncertainty, but this effort while logical is perhaps illusory. Parameters may vary systematically spatially (e.g. thick soils in hollows and thin soils on ridges) and their probability density functions are in many cases probably not independent (e.g. thick soils and ground water characteristics may co-vary). Furthermore, such treatment does not make up for the lack of adequate topographic representation in polygon-based models. The fuller mechanistic understanding of controls on slope stability that is the basis of 3DLISA (and of the model proposed by Sidle, 1992) is nonetheless an important contribution, and is most useful when applied to individual sites that are well quantified, or in exploring, theoretically, possible controls on slope stability.

The fourth approach (which includes SHALSTAB) couples digital representation of the topography, an infinite slope stability model, and a shallow subsurface flow model to predict the spatial distribution of relative slope stability. An early such model is that proposed by Ward et al. [1982]. They used a grid-based model with an assumed static groundwater level and randomly distributed soil depth and root strengths. Okimura and his colleagues

[Okimura and Ichikawa, 1985; Okimura and Nakagawa, 1985; Okimura and Kawatani, 1987; Okimura, 1989; Okimura, 1994] appear to have been the first to incorporate a process-based model for shallow subsurface flow into a digital terrain model for slope instability. They applied the infinite slope model to a grid based representation of landscape topography and estimated the spatial pattern of destabilizing pore pressures using a shallow subsurface runoff model for steady rainfall.

Dietrich et al. [1992, 1993] and Montgomery and Dietrich [1994] took a similar, but simpler approach. Their model was based on the digital terrain model TOPOG [O'Loughlin, 1986] which discretizes the landscape by creating elements bounded by two contour lines and two estimated flow lines (a so-called "contour-based" approach). Unlike Okimura and colleagues they ignored cohesion, thereby avoiding the need to estimate soil depth. They also specifically applied the model in the context of assessing effects of forest management by applying the model to lands being managed for timber in Oregon and California [Montgomery and Dietrich, 1994].

Dietrich et al. [1995] subsequently rederived the fundamental equation used by Dietrich et al. [1992] and included soil depth and a vertically varying saturated conductivity (allowing for systematic differences between the soil and underlying bedrock). Runoff was still treated as steady state. To estimate soil depth (which is necessary in an infinite slope model which includes a cohesion term), they used a process-based soil production and transport model to predict the spatial pattern of soil depth (the foundations of this model were later supported with field studies by Heimsath et al., [1997, 1999]). Furthermore, this model was grid-based rather than contour-based. They found that: 1) significantly less rainfall is necessary for instability if the bedrock conductivity is relatively low compared to the overlying soil than if the bedrock is highly conductive, 2) if the bedrock conductivity is high, instability tends to be focused in unchanneled valleys; 3) ridge soils may tend to be thin and stabilized with modest root strength whereas thick soils in unchanneled valleys require large root strength for stabilization; and 4) root strength change (due to fire, disease, climate change, or land use effects) can have a very large effect on the potential pattern of hillslope instability. Although this model is therefore better able to address issues of forest management because of the incorporation of a root strength term (and a spatially varying soil depth), it also requires much more parametrization than can be practically done over large areas. We will return to this issue later in this chapter.

Three Ph.D. dissertation studies were done in the mid-1990's which explored, among other things, modeling dynamic rainfall-runoff response in predicting the pattern of slope instability [Wu, 1993; Hsu, 1994; and Duan, 1996]. Wu [1993] focused on the issue of timber harvesting strategies (and its subsequent effect on root strength distribution and slope stability). He used a contour-based approach [from Moore et al., 1988] and incorporated a kinematic wave driven groundwater model [see also Wu and Sidle, 1995]. Both Hsu [1994] and Duan [1996] use grid-based models and assumed an exponential decline in saturated conductivity with depth. Hsu used the Dietrich et al. [1995] model for predicting soil depth. As part of a sensitivity analysis both Wu [1993] and Duan [1996] identified soil depth, root strength, and saturated conductivity as important parameters affecting model results. Among other conclusions, all of these studies pointed to the difficulty of parameterization of these controlling factors.

Pack and Tarboton [1997] used the Dietrich et al. [1992] and the Montgomery and Dietrich [1994] model in a grid-based analysis of shallow slope stability patterns in British Columbia and found that the majority of the landslides were strongly concentrated in the least stable ground predicted by the model. Building upon that result they reformulated the model to include root strength (as Montgomery et al. [1998b] have done), added uncertainty estimation of parameters and built an ArcView-based model that they made available on the Internet (SINMAP). We suggest here that such models are very useful, but the lack of spatially registered data on soil and root strength and hydrologic properties make them difficult to parameterize.

SHALSTAB THEORY

Empirical studies of shallow landslides find that they occur on steep slopes and commonly in areas of strong planform convergence (hollows) becausethere soils are thick and shallow subsurface flow is concentrated [e.g. Dietrich and Dunne, 1978; Reneau and Dietrich, 1987; Ellen et al., 1988]. This suggests that surface topography is a primary indicator of where shallow landslides are most likely to occur. For practical application in forest management, it is desirable to construct a mechanistic model that can capture the topographic effects described above. As shown below, we have been able to reduce the model, SHALSTAB, to a point where it can be used with fixed parameters over large areas. The value of such a model is that: 1) it can be applied in diverse environments without costly attempts at parameterization - hence it is fully transportable, unlike empirical correlational approaches; 2) results from different sites can be directly compared; 3) it takes little special training to use the model, and 4) it becomes a hypothesis that is rejectable, i.e. the model can fail - rather than just be tuned until it works. The value of a model that can fail is that it can effectively put a spotlight on processes not included in the model that are important. This means the model can help illuminate causality.

SHALSTAB is based on an infinite slope form of the Mohr-Coulomb failure law in which the downslope component of the weight of the soil just at failure, τ, is equal to the strength of resistance caused by cohesion (soil cohesion and/or root strength), C, and by frictional resistance due to the effective normal stress on the failure plane:

$$\tau = C + (\sigma - u)\tan\phi \qquad (1)$$

in which σ is the normal stress, u is the pore pressure opposing the normal load and $\tan\phi$ is the angle of internal friction of the soil mass at the failure plane. This model assumes, therefore, that the resistance to movement along the sides and ends of the landslide is not significant.

A further simplification in SHALSTAB is to set the cohesion to zero. This approximation is formally incorrect in most applications. Although the rocky, sandy soils of colluvial mantled landscapes probably have minor soil cohesion, root strength, which can be treated as an additional cohesive term in (1), plays a major role in slope stability [e.g. , Burroughs and Thomas, 1977; Gray and Megahan, 1981; Sidle, 1992]. We have elected to eliminate root strength in this model for the following reasons. First, root strength varies widely, both spatially (over small scales) and in time. Although field studies show that root strength is quantifiable [i.e. Endo and Tsuruta, 1969; Burroughs and Thomas, 1977; Gray and Megahan, 1981], to do so would involve considerable effort. For watershed scale modeling, field-based parameterization of root strength patterns across the landscape would be very time consuming. It is conceivable that remote sensing of canopy types could be used to estimate possible root strength contributions, but such a method, requiring high spatial resolution information has not, to the best of our knowledge, been developed. Secondly, one effect of forest management is to reduce root strength; by assigning a value of zero to cohesion we effectively model the most extreme case- which is a practical boundary to risk assignment. As discussed below, we have somewhat compensated for the absence of root strength by setting the friction angle to a high, but acceptable value.

This is not to say that there is no value in building models with root strength, and several such models exist which employ digital elevation data have been proposed (as discussed above). We now refer to a subsequent version of SHALSTAB in which the soil depth and cohesion are held spatially constant as SHALSTAB.C [used by Montgomery et al., 1998a,b] (and the program is also available at our website). A version of SHALSTAB in which the soil depth varies spatially, the hydraulic conductivity varies vertically and the cohesion is spatially constant is now referred to as SHALSTAB.V [developed by Dietrich, et al., 1995]. We return to a discussion of cohesion later in this paper and show that models that include cohesion, but have no more information than the topographic setting of the landslide, can not be uniquely parameterized (cohesion and friction angle can not be uniquely determined).

By eliminating cohesion, (1) can be written as

$$\rho_s g z \cos\theta \sin\theta = (\rho_s g z \cos\theta^2 - \rho_w g h \cos^2\theta)\tan\phi \qquad (2)$$

in which θ is the land surface slope, z is soil depth, h is water level above the failure plane, ρ_s and ρ_w are the soil and water bulk density, respectively, and g is gravitational acceleration (see Figure 2). In this form, we have assumed that there is a relatively small difference in saturated bulk density and bulk density of the moist soil above the saturated zone. This assumption is made to simplify the expression by avoiding the need for another parameter: the bulk density of the unsaturated soil. Derivation of the following equations with this term included suggests that neglecting this effect has a small influence on the results. Subsequently we refer to ρ_s as the wet bulk density, suggesting it represents the integral bulk density of a variably saturated soil. Equation (2) can then be solved for h/z, which is the proportion of the soil column that is saturated at instability:

$$\frac{h}{z} = \frac{\rho_s}{\rho_w}\left(1 - \frac{\tan\theta}{\tan\phi}\right) \qquad (3)$$

Equation (3) explicitly states that the soil does not have to be saturated for failure to occur. While saturation is commonly assumed when one analyzes a landslide scar, theoretically it is not necessary. Note that h/z could vary from zero (when the slope is as steep as the friction angle) to ρ_w/ρ_s when the slope is flat ($\tan\theta = 0$). We assume, however, that the failure plane and the shallow subsurface flow is parallel to hillslope, in which case h/z can only be less than or equal to 1.0, and any site requiring h/z greater than 1 is unconditionally stable - no storm can cause it to fail (see Montgomery and Dietrich [1994] for further discussion of terminology of failure potential). Note that $h/z \geq 1.0$ occurs if $\tan\theta \geq \tan\phi(1-(\rho_w/\rho_s))$. We observe in the field that such environments can support saturation overland flow without failing. The value of h/z may locally exceed 1.0 due to exfiltration gradients, such as that observed by Montgomery et al. [1997]. Such processes are not included in the flow model. If the ground slope equals or exceeds the friction angle, then h/z is zero and the site is "unconditionally unstable" or is subject to "chronic"

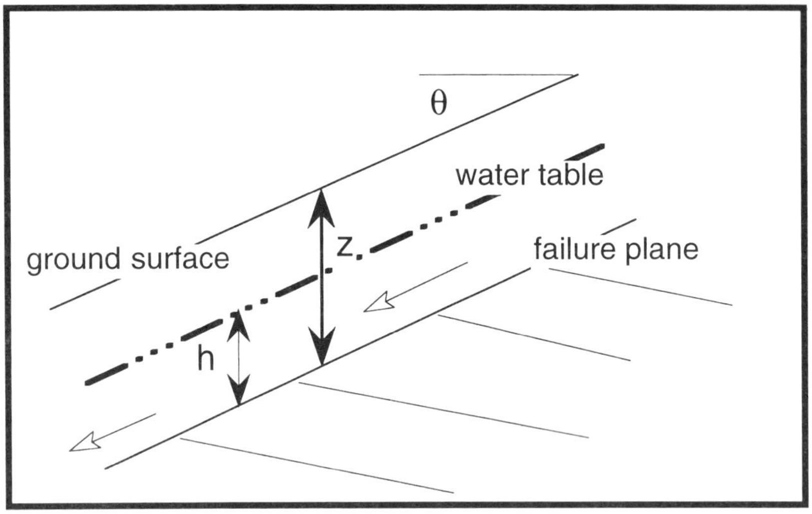

Figure 2. The one-dimensional approximation used in SHALSTAB in which failure plane, water table, and ground surface are assumed parallel. The slope is θ, the height of the water table is h, and the thickness of the colluvium overlying the failure plane is z. Typically the failure plane is at the colluvium-weathered bedrock or saprolite boundary. Open arrows show that flow is assumed parallel to the ground surface.

potential failure; this commonly corresponds to sites of bedrock outcrop.

SHALSTAB links the slope stability model of (3) with a hydrologic model that predicts the topographic controls on h/z. It uses a steady state shallow subsurface flow model to estimate h/z and assumes that the flow is driven by a head gradient equal to the topographic slope. Flow is therefore parallel to the slope (Figure 2). The model actually calculates a water table that is less steep than the ground surface as the h/z increases downslope, but this effect on head gradient on steep slopes is small. Conservation of mass still applies. A key assumption here is that the steady state runoff model effectively mimics what the relative spatial pattern of wetness (h/z) would be during a landslide-producing storm that is not in steady state. If precipitation events are sufficiently intense and of short duration such that thin soils on non-convergent sites can quickly reach destabilizing values of h/z before shallow subsurface flow can converge on unchanneled valleys, then the model will be incorrect. Such a case may have occurred in the 1995 Madison County, Virginia rainstorm, of 770 mm in 16 hours, in which according to Wieczorek et al. [1997] landslides were most common on thin soils mantling planar slopes close to the ridge divide.

In the steady state shallow subsurface flow model, the effective precipitation, q (precipitation minus evapotranspiration and deep drainage), times the drainage area, a, must equal the flow in the conductive soil layer across a cell of width, b, given by Darcy's law, i.e.

$$qa = k_s h \cos\theta \sin\theta b \qquad (4)$$

in which sinθ is the head gradient, k_S is the saturated conductivity, and hcosθ is the saturated thickness measured normal to the ground surface. At saturation the shallow subsurface flow will equal the transmissivity, T (the vertical integral of the saturated conductivity), times the head gradient, sinθ, and the width of the outflow boundary, b. We can approximate this as follows:

$$Tb\sin\theta = k_s z \cos\theta \sin\theta b \qquad (5)$$

Combining (4) and (5) leads to:

$$\frac{h}{z} = \frac{q}{T}\frac{a}{b\sin\theta} \qquad (6)$$

Here we see that the pattern of h/z for a given storm is governed by two ratios: one hydrologic ratio and the other topographic. The hydrologic ratio, q/T is the magnitude of the precipitation event, represented by q, relative to the subsurface ability to convey the water downslope for a given head gradient, i.e. the transmissivity. All else being equal, the larger the q relative to T the higher the water

TABLE 1. Conversion values for the hydrologic ratio

T/q (m)	q/T (1/m)	*log (q/T) (1/m)
2512	0.00040	-3.4
1259	0.00079	-3.1
631	0.00158	-2.8
316	0.00316	-2.5
158	0.00633	-2.2
79	0.01266	-1.9

**If the transmissivity is about 65 m^2/day [Montgomery and Dietrich, 1994], a value of log(q/T) of -3.4 means that the steady state rainfall was 26 mm/day, where as a value of -1.9 was 818 mm/day. Each interval of log(q/T) is about a 2 times change in precipitation for a given transmissivity.

table in the soil, and consequently the greater the number of sites on a hillslope that will become unstable (i.e., the h/z specified by equation (6) exceeds that given by equation (3)). The topographic ratio, a/bsinθ, captures the essential effects of topography on runoff and is composed of two terms: a/b is the topographic convergence that concentrates subsurface flow and elevates the pore pressures, while sinθ is the ground slope for which the steeper the ground, the faster the subsurface flow and consequently the lower the relative wetness defined by h/z. The topographic ratio is nearly identical to that identified by TOPMODEL [Beven and Kirkby, 1979; Beven, 1997] and is very widely used in local and regional hydrologic modeling. The important difference is that TOPMODEL uses tanθ rather than sinθ (as pointed out by Dietrich et al. [1995]). Physically, tanθ is incorrect and while this has no impact on low gradient systems, the error on hillslopes is significant if tanθ is used instead of sinθ. The report by Dietrich and Montgomery [1998] illustrates the behavior of the hydrologic model (and is available at http://socrates.berkeley.edu/~geomorph/).

Setting (6) equal to (3) and solving for the hydrologic ratio gives

$$\frac{q}{T} = \frac{\rho_s}{\rho_w}\left(1 - \frac{\tan\theta}{\tan\phi}\right)\frac{b}{a}\sin\theta \tag{7a}$$

while solving for the drainage area per outflow boundary length (grid size) for instability gives

$$\frac{a}{b} = \frac{\rho_s}{\rho_w}\left(1 - \frac{\tan\theta}{\tan\phi}\right)\frac{T}{q}\sin\theta \tag{7b}$$

Equation 7 is the coupled hydrologic-slope stability equation solved by SHALSTAB. The model has three topographic terms that are defined by the numerical surface used in the digital terrain model: drainage area, a, outflow boundary length, b, and hillslope angle, θ. There are potentially four parameters that need to be assigned to apply this model: the soil bulk density, ρ_s the angle of internal friction, ϕ the soil transmissivity, T, and the effective precipitation, q. As we will discuss below, we have found it useful to assign bulk density and friction angle values to be the same everywhere, and compare q/T values, making (7) a fixed-parameter model easily applied across large areas. Of course, if data on soil properties are available, then locally appropriate values could be used.

The ratio of q/T is equal to length per time over length squared per time, i.e. it has the dimensions of 1/length. Throughout this report we will use the metric system, and the unit of q/T will be 1/meters or for T/q it is meters. Likewise, the dimension of a/b is meters. Because q/T is always a small number, we normally use the logarithm of the value in plots (Table 1). The upper and lower bounds of the model are the "unconditionally unstable" or chronic condition and the "unconditionally stable" condition, respectively.

Figure 3 illustrates the relationships between the hydrologic ratio (q/T) and the topographic ratio (broken into a/b and sinθ), for a bulk density ratio of 1.6 and a friction angle of 45 degrees. For a given a/b, the value of log(q/T) at instability for a given slope is shown. In heavy lines are the log(q/T) intervals given in Table 1. The model predicts that hillslopes gentler than 20.6 degrees are unconditionally stable and slopes steeper than 45 degrees are unconditional unstable. Between these two end members, the value of log(q/T) for instability is dominated by the convergence term, a/b, and virtually independent of slope up to hillslope gradients of nearly 35 degrees. Between 30 and 40 degrees the decrease in log(q/T) is only slightly greater than the -0.3 class interval we use in plotting our results on maps. Hence, up to nearly 40 degrees, the SHALSTAB log(q/T) values that would be determined across a landscape are dominated by the convergence term, a/b. Where the landscape becomes very steep (greater than 40 degrees) the gradient term dominates the log(q/T) value for instability.

In order to use SHALSTAB with a digital elevation model, the topographic terms, θ and a/b must be calculated for each grid or element. There is no unique procedure, but there are ones that are less prone to create artifacts. We

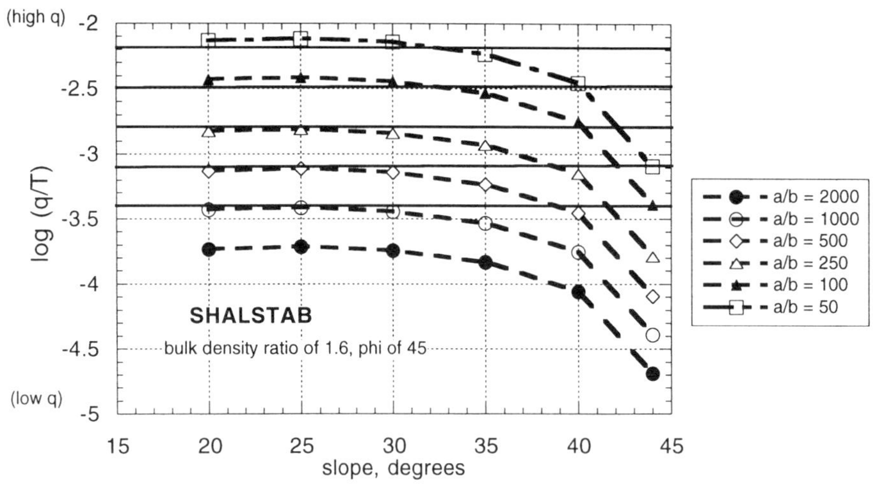

Figure 3. Relationships among log(q/T), surface slope, and a/b for a bulk density ratio of 1.6 and a friction angle of 45 degrees for the model SHALSTAB. The heavy horizontal lines correspond to the log (q/T) values used as boundaries to class intervals used in map production and data analysis.

currently use a grid-based model. Local slope (θ in equation (7)) is estimated as the geometric mean of the slopes in the two cardinal and two diagonal directions across the grid cell, hence all eight surrounding cells are used. This procedure produces results that differ little from the ARC/INFO SLOPE function in the GRID module. Because it uses the eight surrounding cells it will tend to smooth out local variations (associated with just a few cells) in slope, and in some cases will therefore underestimate the local slope. This approach, however, reduces grid artifacts associated with grid noise and with orientation of topography relative to the grid.

We also use a multiple-direction algorithm rather than maximum fall method of distributing area [see discussion in i.e. Quinn et al., 1991; Costa-Cabral and Burges, 1994; and Tarboton, 1997]. The method is similar to that proposed by Quinn et al., in that the proportion of the drainage area each cell distributes to a lower cell, f_i, is equal to the local slope to that cell, S_i, divided by the sum of slopes to all lower cells, i.e. $f_I = S_i/\Sigma S_j$. Extensive testing has shown that this approach gives results that are weakly dependent on the orientation of the topography relative to the grid, as long as cells are not close to the boundaries of the data field (a concern raised by Costa-Cabral and Burges [1994]). As Costa-Cabral and Burges [1994] and Tarboton [1997] point out, this procedure will tend to be "dispersive", i.e. it tends to spread the calculated flow across the slope more than Tarboton's procedure which only permits flow into at most two adjacent cells. Given the diffusive nature of the flow, the broad heterogeneity of the subsurface conductivity field of the subsurface, and the large error in local topography inherent to most digital elevation data, we feel this more dispersive approach which minimizes grid artifact is acceptable.

In SHALSTAB, the proportion of slope in each direction is first calculated. Then starting at a low point in the topography, the contributing line is followed to the divide and then the area to the point at the bottom is calculated. This process is repeated for all cells. The specific catchment area, a/b, in our model is the total drainage area for each cell divided by the cell width.

SHALSTAB VALIDATION

The fundamental premise of SHALSTAB is that sites with the lowest q/T values for instability (least amount of precipitation for instability) should be the least stable and consequently the incidence of shallow landsliding should be highest in these sites. According to SHALSTAB (equation 7), the least stable sites (lowest q/T) have the largest drainage area per unit contour width and steepest slopes. SHALSTAB can be validated once the angle of internal friction and bulk density are fixed, as this leaves only topographic terms on the right hand side of equation (7a).

There are four parameters that could be evaluated in the use of equation (7): tanϕ, ρ_s, T and q. The first three are soil properties and the last is effective steady state precipitation. Each of these parameters also varies spatially, but so far in most applications of SHALSTAB a single value has been assigned to the entire landscape [see Tang and Montgomery, 1995, for an exception]. The three soil properties on average also vary between different landscapes.

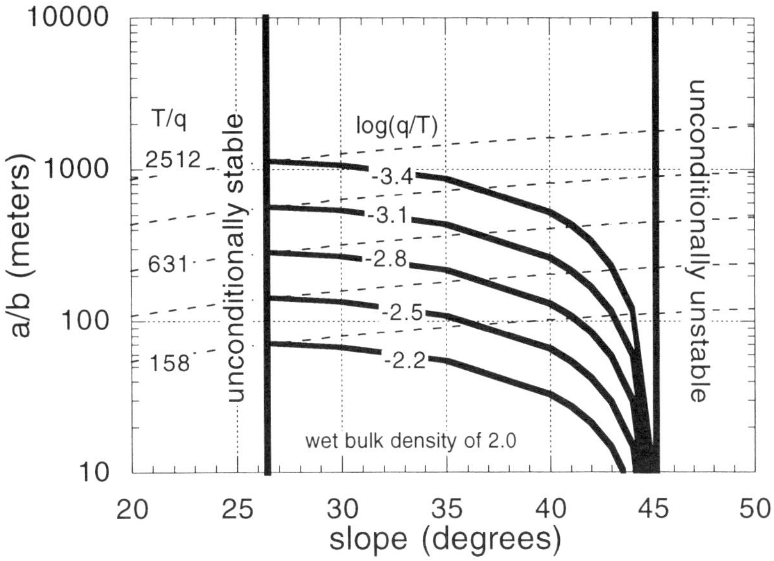

Figure 4. SHALSTAB stability field for bulk density ratio of 2.0 and friction angle of 45 degrees. Vertical heavy lines delineate slopes below which failure will not occur even when the ground is saturated (unconditionally stable) or above which no rainfall is needed for instability to occur (unconditionally unstable). The threshold line above which instability is predicted is shown for a range of log(q/T) values. Note that with increasing q/T the threshold line lowers and a smaller a/b is needed for instability. The dashed lines delineate the threshold of saturation (for each labeled log(q/T) line that intersects it) and is labeled by the corresponding T/q value as expressed in equation (8b). Above the dashed line for the given T/q the land is predicted to be saturated.

In the temperate rainforests of Coastal Oregon, for example, wet soil bulk density is about 1600 kg/m^3 [Torres, et al., 1998] and the friction angle could be as low as the mid 30's [Schroeder and Alto, 1983], whereas in parts of the California coast, the wet bulk density is about 2000 kg/m^3 and the friction angle is in the 40's [Reneau et al., 1984]. Root strength is an important contributor to overall strength, but as mentioned earlier it was eliminated in order to simplify parameterization of the model.

If cohesion is not considered, we have found it useful to set the friction angle equal to 45 degrees, and not let it vary between landscapes. This accomplishes three things: 1) it increases the threshold slope to be more similar to that which would be obtained by a soil with lower friction angle but with some strength contribution from cohesion (see later), 2) by holding it constant, it no longer needs to be parameterized, and the model can be run with only digital elevation data as necessary input, and 3) relative potential hillslope instability can be compared across different landscapes.

The low range of wet soil bulk density that is likely to be encountered in the field has a relatively small effect on the predicted pattern of slope instability when cohesion is neglected. The lower the bulk density, however, the gentler the slope at which instability is predicted to occur. In the absence of field data, it is recommended that a value of 2000 kg/m^3 be used. When comparing results of SHALSTAB run by different groups it is, nonetheless important to note what values of bulk density and angle of internal friction have been used.

Model predictions and field observations on landslide locations can be compared in two ways: comparing field data with model results using graphical representation of equation (7b) or by using equation (7a) to create a map of q/T values. The graphical representation method is described by Dietrich et al. [1992, 1993], and Montgomery and Dietrich [1994]. Note that on a graph of a/b against slope, equation (7b) would appear as a curved line, above which all points would be unstable and below which they would be stable. Figure 4 shows such a plot. The two vertical lines represent the upper and lower bounds to the SHALSTAB model, corresponding to unconditionally unstable or "chronic" and unconditionally stable fields, respectively. The dashed sloping lines which show a/b increasing with slope delineate the threshold for saturation for selected values of T/q, i.e. at saturation

$$qa = Tb \sin\theta \tag{8a}$$

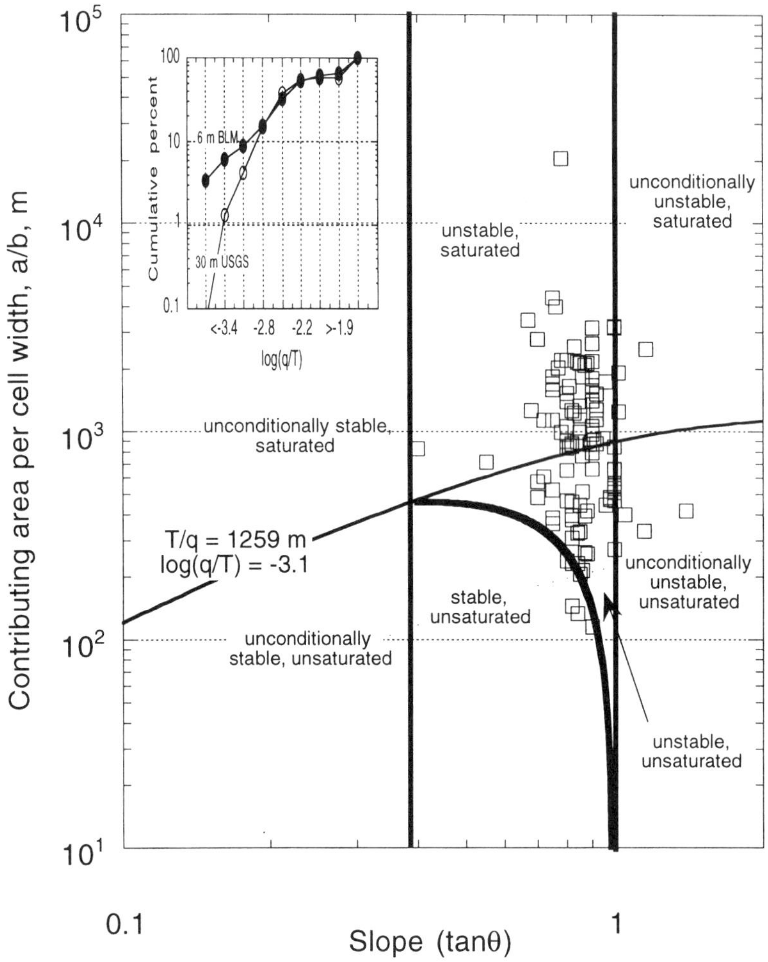

Figure 5. Field observations on a/b and slope for shallow landslides in the Oregon Coast Range plotted against the threshold lines for instability and soil saturation (data collected by the Bureau of Land Management (BLM)). Nearly all data points fall above the threshold curve in one of the four stability fields labeled on the plot. The insert shows the cumulative percent of land area in log(q/T) classes for the study area for 30 m USGS grid data and for data provided from the BLM that could be gridded at 6 m.

which can be written as

$$\frac{a}{b} = \sin\theta \frac{T}{q} \tag{8b}$$

Sites that fall above the dashed lines are predicted to be saturated by the steady state hydrologic model for the corresponding T/q value. The strongly curved solid lines, which terminate at the corresponding saturation lines, define the threshold between stable and unstable topography for selected q/T value, as defined by equation 7b. Each of these lines is labeled with a log(q/T) value which correspond to the T/q values on the corresponding lines of saturation. Figure 4 can easily be used in the field by documenting the drainage area above the scar (a) the width of the scar (b) and the local slope of the slide. If the landslides are concentrated in area of least stability (lowest q/T) then the data should tend to cluster in the upper right hand corner of the plot (above say the -3.1 log(q/T) curve).

Figure 5 shows an example of data obtained from analysis of field data sheets collected by crews working for the Bureau of Land Management in the forested lands of the central Oregon Coast Range. This shows that nearly all of the 93 landslides fall above the curved threshold line for a log(q/T) of -3.1. Each of the seven fields of instability defined by whether the a/b and slope values for a site would cause it to be above or below the thresholds of saturation and slope stability are labeled on the figure. The data are separated by the curve that defines the threshold for saturation (equation 8b). The inset graph shows the cumulative percent of the landscape in successively higher q/T categories for an area of the Oregon Coast range typical of the area where the field data were collected. The 30 m United States Geological Survey (USGS) data are compared with 6 m grid data generated from topographic maps created from aerial photography for the Bureau of Land Management (BLM). The higher resolution topographic map more accurately reproduces the a/b and $\sin\theta$ than that obtained from field observations, although some fine scale features such as small hollows are still missed by this map. The threshold log(q/T) value of < -3.1 on the 6 m map covers about 9% of the landscape. With a more accurate map, this percent would probably be larger. If one used this field determined threshold value of -3.1 on the 30 m map, the area of potential high instability would be greatly underestimated. If only 30 m data were available, it would be necessary to determine the corresponding log(q/T) value for each landslide site from the digital map and use that to define a threshold log(q/T). We discuss this method below.

Data from aerial photographs can also be plotted on such graphs, but the drainage area and local slope will then be determined from the digital terrain model (or by hand from topographic maps) rather than from field observations. As discussed below, this procedure can introduce relatively large errors because of uncertainty of landslide placement and errors in the digital topography.

An alternative method for model testing is to overlay the location of mapped landslides (either from field or aerial photography) onto a grid map of q/T values. Figure 6a shows an example from the Oregon Coast Range in which landslides were sufficiently small that they were treated as point data and assigned to individual cell locations (The location symbols are much larger than the grid cells). The value of such maps is that one can get a visual impression of model performance, which may have large spatial variation. In this data set, because of survey errors and inaccuracies in the topographic map (see below), there were significant location uncertainties in some of the data points. Figure 6a shows the best estimate of location of the slides. To portray the performance of the model, in Figure 6b we have computed the number of landslides counted in each log(q/T) class (given in Table 1) divided by total area occupied by that log(q/T) class in the study area and plotted that ratio against the corresponding log(q/T) class (using the upper value). If the landslides were randomly distributed with respect to log(q/T) classes the ratio would be the same everywhere (here it would be about 7 slides/km^2). Clearly the landslide occurrence is much more numerous in the log(q/T) classes strongly supporting the model. The two curves in Figure 6b represent the two efforts to locate the slides on the map. In either case, nearly 50% of all mapped landslides fell in log(q/T) values less than -2.8 which represents only 13% of the area of the study.

Two important issues emerge in validation studies. First, the performance of the model is strongly influenced by the quality of the topographic data. Figure 7 shows a shaded relief view of a study site near Coos Bay, Oregon [e.g. Montgomery et al., 1997; Anderson et al., 1997, 1998; Torres et al., 1998] for four different grid scales: 30 m USGS data, 10 m USGS data (obtained from digitized 7.5' quadrangles), 10 m data generated from aerial photography for the Oregon Department of Forestry, and 2 m data generated from an airborne laser altimetry survey [see also Roering et al. 1999; Montgomery et al., 2000]. This illustrates just how crude even 10 m data are compared to the actual topography (which the laser altimetry approximates: note that the roads along the ridge can be seen in this map). Most noticeable is the loss of the fine scale ridge and valley topography (which strongly dictates shallow landslide location at this site) with scale coarsening. This is perhaps best illustrated in Figure 8 where the contour lines of the 7.5' quadrangle and that of the data obtained from the airborne laser altimetry are shown at the same scale. Note the much greater planform curvature in the high-resolution topography (hence greater a/b) and the greater number of valleys. Airborne laser altimetry is just becoming commercially available and techniques for aerial surveying and data analysis are still being developed in forested landscapes. Nonetheless, as figure 7 and 8 clearly illustrate, the technology has arrived and should be exploited for improved landslide mapping.

Figure 9 shows the field of log(q/T) for the four cases shown in Figure 7 on a topographic map where the mapped channel network and landslide scars are shown. This illustrates how the pattern of potential slope instability is strongly affected by the resolution of the topographic data. Note that in the 30 m data, many of the landslides do not occur in the least stable log(q/T) class of < -2.8 (because the convergent areas are not correctly defined), whereas in the laser altimetry data, all the slides fall in the least stable class. The inaccuracy of available topographic maps degrades validation studies, hazard rating assignment, and field studies. All digital terrain based landslide models will suffer from these limitations.

The second issue that arises in validation studies is the problem that landslide size is often larger than a single grid cell, or at least may be mapped as lying in more than one grid cell. This then raises the question of what is the appropriate q/T value for the instability. One could argue, for example, that the average log(q/T) of all the cells lying under a landslide scar polygon should be used. We have used as standard practice that the lowest q/T value in contact with the landslide polygon is the q/T that controls in-

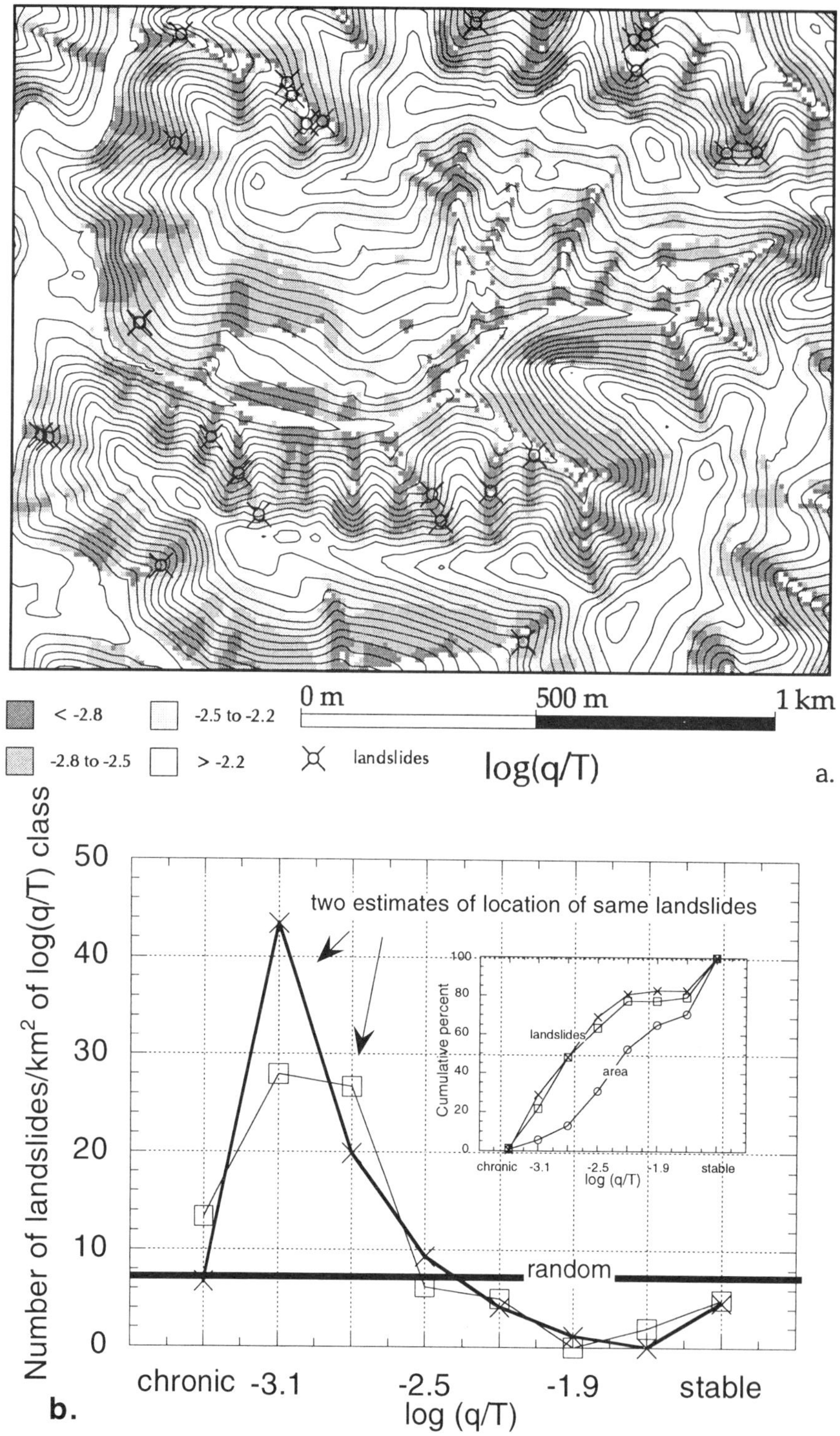

Figure 6. Shallow landslides in the Elk Creek area of the Oregon Coast Range (data provided by the Oregon Department of Forestry). (a) shows the approximate location of the landslides relative to the calculated pattern of log(q/T). (b) shows the landslide density as function of log(q/T). The two curves represent the effect of different estimates regarding the precise location of the landslides mapped in the field. The inset graph shows the cumulative percent of the area in each log(q/T) class (labeled as 'area') and the cumulative percent of number of landslides in each log(q/T) class for each of the two estimates of landslide location. Topographic map is derived from digital data obtained from aerial photographs and then gridded at 10 m.

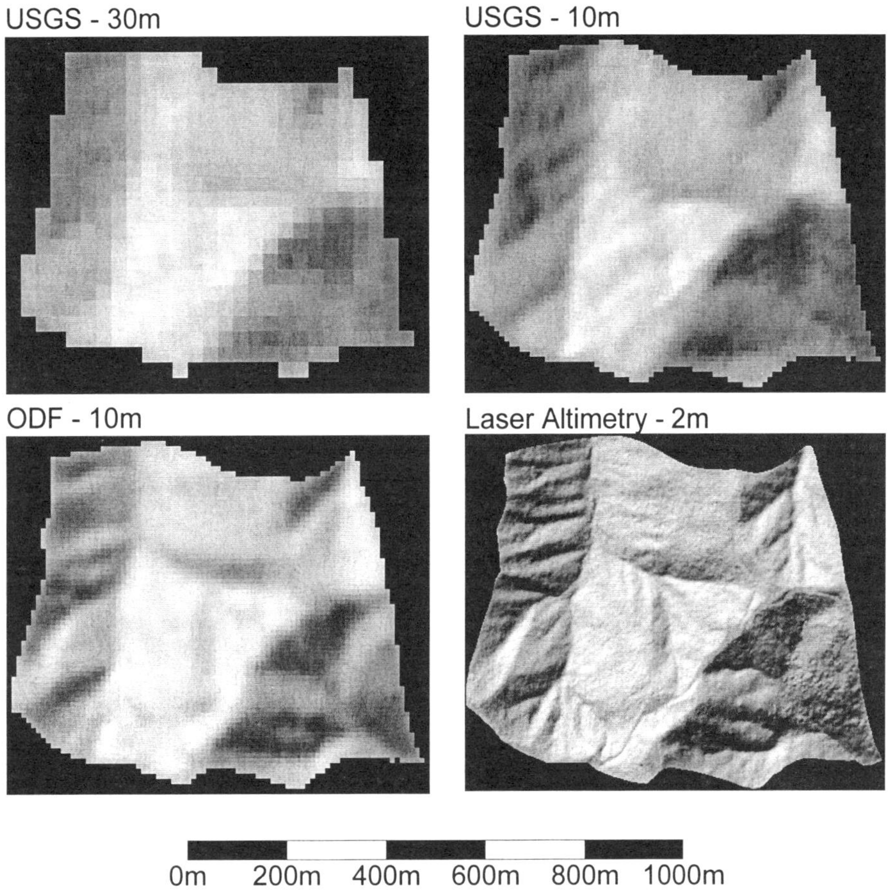

Figure 7. Shaded relief maps of the Coos Bay study area in the Oregon Coast Range based 30 m data from the United States Geological Survey (USGS 30m), digitized contours from the 7.5' USGS quadrangle which was then gridded at 10 m (USGS 10m), 10 m grid data derived from digital data provided by the Oregon Department of Forestry (ODF) derived from aerial photographs (ODF 10m), and 2m girded data obtained from an airborne laser altimetry survey of the site (laser altimetry 2m). Road tread and cuts are visible along the ridge tops on the right side of the laser altimetry image.

stability. There are several reasons for this. Digital elevation generated topographic maps are generally less steep and less convergent than real topography. Hence, the lowest q/T value is perhaps most likely to represent the actual local topography where failure occurs. Including all the other values in a landslide polygon by, say, taking some average, median or some other statistical measure, would tend to systematically include poorly mapped topography, leading to an elevated threshold q/T value and likely lead to a significant overestimation of the extent of potential slope instability. Equally important, most landslide polygons are crudely located. This is because the base maps are inaccurate and because mapmakers commonly don't have the time or resources to carefully locate each landslide on topographic maps. Few mapmakers distinguish between the slide scar (to which the model applies) and the runout track (and this is difficult to do with poor aerial photographs). Scars are commonly mapped as bigger than they actually are- leading to inclusion of topography that is more stable than that associated with failure. Also landslide scars may progressively expand upslope and the landslide mass may run some distance downslope and stop – both of these processes tend to spread the interpreted size of the scar into areas not associated with the initial failure. We propose that picking the lowest q/T value in a polygon may give the most accurate estimate of the local failure condition and avoid problems associated with map resolution and inaccurate mapping of landslide scars.

Our procedure, however, produces a bias toward low q/T. Even randomly located landslides would tend to be

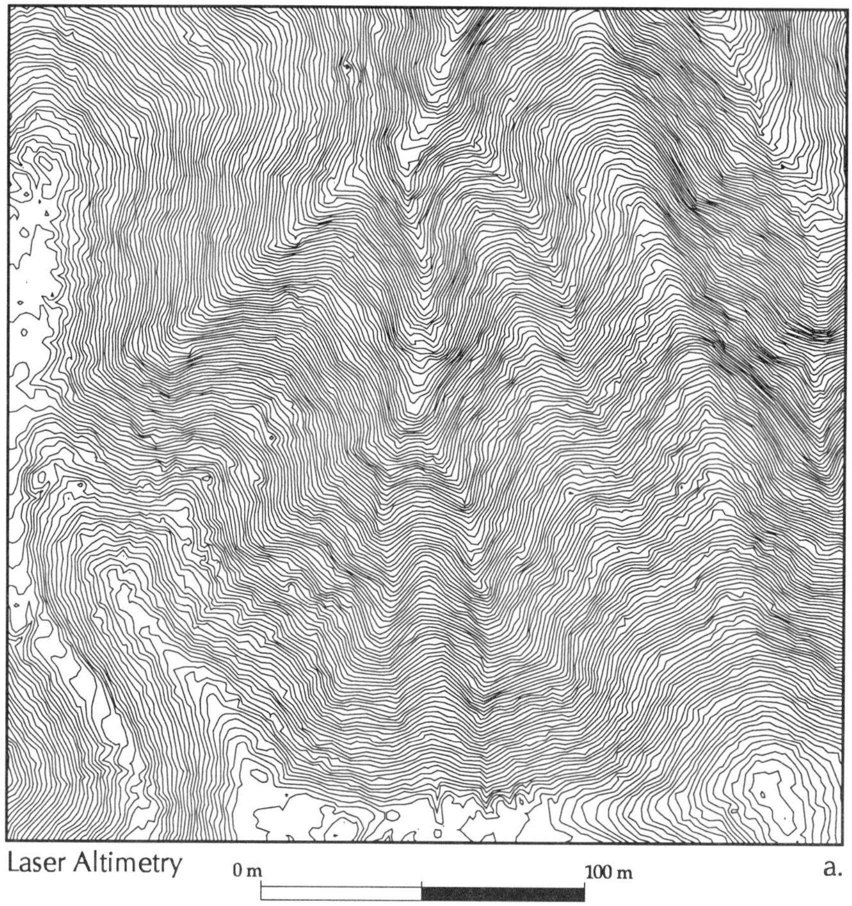

Figure 8. Local comparison of best commonly available data (USGS 10m data) with that obtained from laser altimetry. Contour interval in the USGS data is 12 m while that in the laser altimetry is 1 m.

concentrated in areas having the lowest q/T values because for each randomly located landslide the lowest q/T value would be chosen to represent relative instability category.

Therefore, as a second phase of data analysis, we needed to see if the model would perform significantly better than a similar biased-random model. To answer this question we developed a biased-random landslide generation model to compare with the statistics of the actual mapped landslides. Groups of grid cells of approximately the same size as the median landslide size in a given watershed are randomly placed throughout the watershed until the number of landslides equals the number that had been observed. As is done for the observed landslides, the minimum q/T value within each landslide polygon is selected to represent the value for each landslide generated by the random model, hence we retain this bias. This process is repeated an average of 10 times for each site and the median and standard deviation of the number of landslides found in each log(q/T) category listed in Table 1 are determined. A comparison is then made between observed and randomly generated landslide scars to ensure that any apparent success of the model would not be due solely to bias created by the selection of minimum q/T values. If this bias is large and the model does not perform significantly better than the random model, there would be no observable difference between q/T values for the populations of observed landslides versus the randomly generated landslides. If landslides are preassigned to specific grid cells (often landslides are sufficiently small that they are treated as point data) then a random model would produce a constant landslide density for all log(q/T) classes, as shown in Figure 6b. The random model is a stringent test of the SHALSTAB's results which ensures that we do not draw spurious conclusions about model performance; it is included with our program on our website. We have not attempted to formulate other empirical models against which to test our model performance. While we could find correlations with topography that may perform better than our model, they

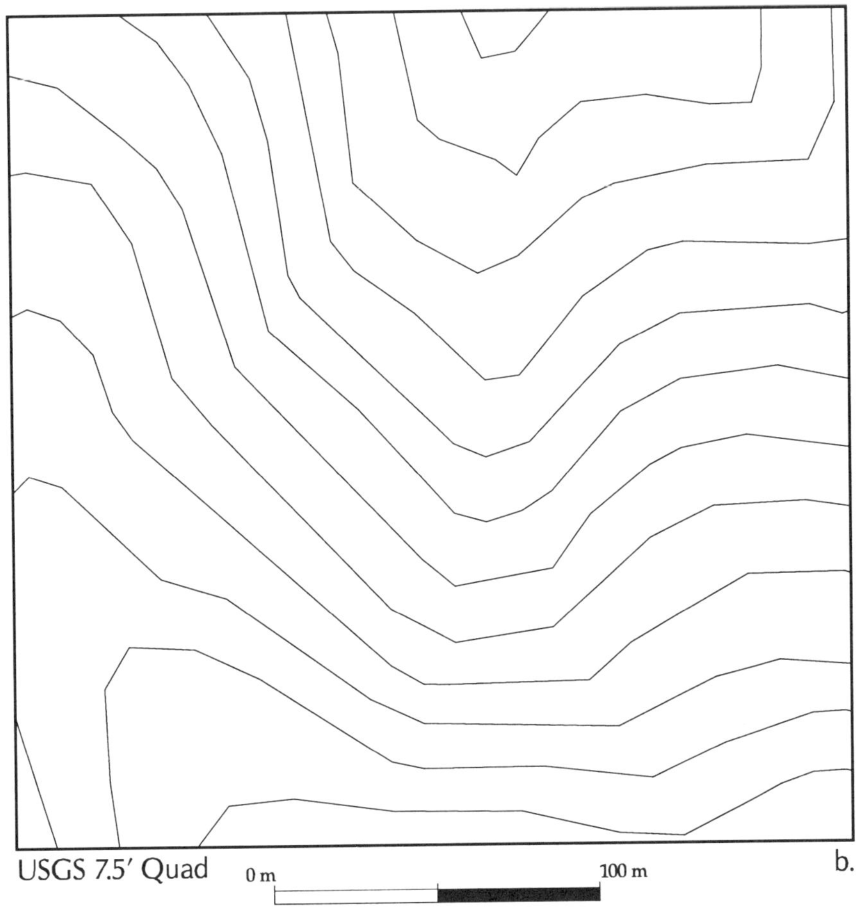

Figure 8 (continued)

would lack the generality of the mechanistic-based approach. Furthermore, the biasing in the random model makes it a difficult model to beat. If model landslides are sufficiently large, then randomly dropped polygons of the same size will have a large probability of intersecting low log(q/T) values. In this case, even if SHALSTAB is accurate, it may not give results significantly different from the biased random model.

The decision to use a single value of log(q/T) to characterize a landslide polygon introduces another concern. The model may be judged as successful because a large percentage of the mapped landslides have minimum log(q/T) values below some low threshold value, such as –2.8, which also differs significantly from the biased-random model. But significant portions of each landslide polygon may overlie higher log(q/T) values. If a hazard map is made using the minimum log(q/T) as a threshold, this map may tend to underestimate the extent of observed landslide area. Hence, by picking the least stable cell to account for a landslide we will tend to underestimate the actual extent of the landslide prone areas [Mark Reid, pers. com., 2000]. This becomes particularly important when using a log(q/T) threshold to delineate high risk areas that may in turn require restrictive land management practices (such as no timber harvest). Without better topographic base maps that show actual topography and permit accurate mapping of landslide location (and avoid the problems with inferior maps described above), we feel that the approach proposed here may be a reasonable compromise between estimating correctly where landslides are likely to occur and not overestimating the extent of high risk area.

Validation Study in Northern California

As part of an assessment of the validity of using SHALSTAB as a tool to guide forest management prescriptions, landslide maps were made from aerial photographs of 7 watersheds in the Northern California Coast Range (Table 2) and compared with q/T predictions using digitized USGS 7.5' quadrangle maps gridded at 10 m [Dietrich et al., 1998]. Aerial photographs taken in 1978 and 1996 of the watersheds, which ranged from 4.8 km^2 to

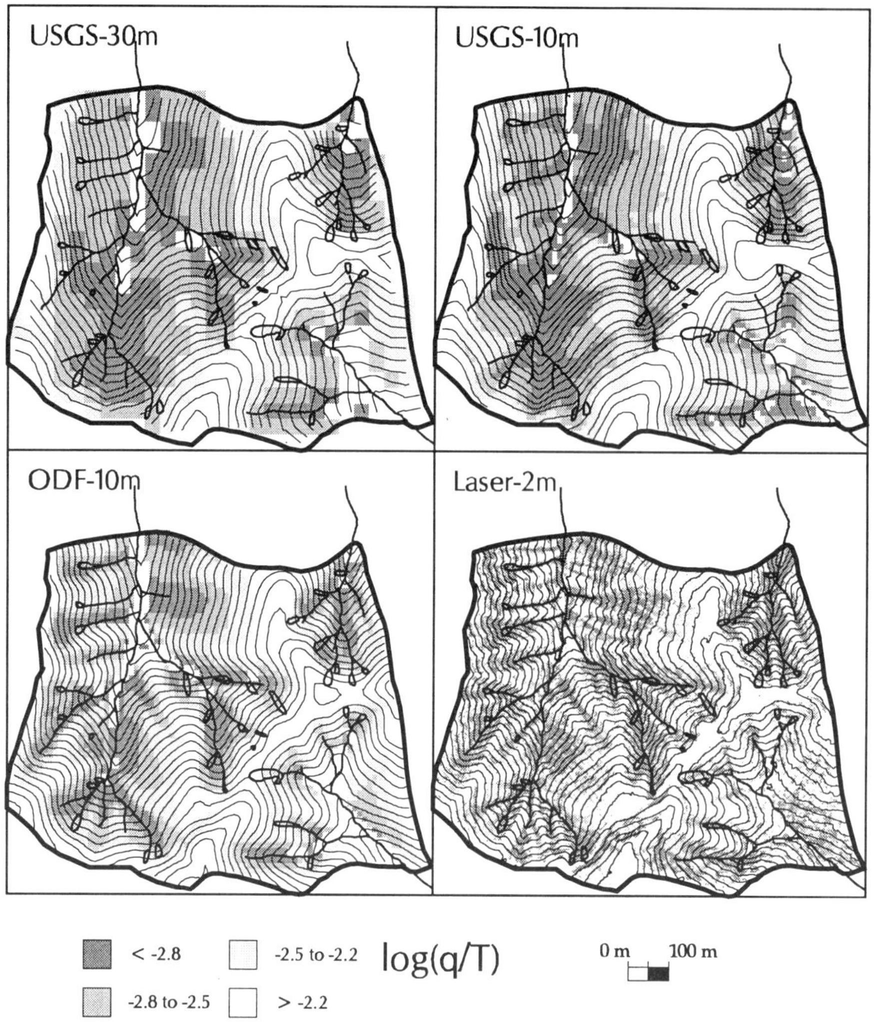

Figure 9. Comparison of SHALSTAB predictions of log(q/T) for the four cases shown in Figure 7 with the location of landslide scars and the channel network (field mapping done by David Montgomery). Contour interval is 5m and bulk density ratio is 1.6.

143 km^2 in drainage area, were used. A total of 844 in-unit failures (i.e. landslides occurring within timber harvest units that were not associated with roads) and 354 road-related failures were mapped in the total study area of 281 km^2. Details of the mapping procedures and all resulting field maps are reported in [Dietrich et al., 1998]. All landslides not associated with roads were classed as in-unit failures because there are no uncut forests in the study sites. Landslides ranged in size from 36 to 17,045 m^2, with a median size of about 500 m^2. Figure 10 shows the results from the largest watershed (the Noyo basin) for the 1996 landslides to illustrate the analysis performed on each watershed. For each site, the number of cells in each log(q/T) category was determined and the resulting cumulative frequency (or percent area) of the total watershed area falling into each successive category is shown in Figure 10a (see curve labeled "area"). This curve shows the predicted potential slope instability across the entire watershed. For the Noyo basin, only about 55 percent of the watershed area is predicted to be unstable. The remaining lands are characterized by gradients too low to fail even when saturated (i.e., classified as "stable"). A classification of "chronic" denotes that the cell is sufficiently steep to be potentially unstable even without the addition of precipitation or runoff (i.e. equivalent to "unconditionally unstable" as defined by Montgomery and Dietrich [1994]). The curve generated by random placement of landslides differs from the total watershed area curve because of the bias that results from selecting only the minimum q/T value in each cluster of cells randomly placed on the landscape. This difference is

TABLE 2: Validation sites in Northern California

WATERSHED	DRAINAGE AREA (Km^2)	NUMBER OF IN UNIT LANDSLIDES	NUMBER OF ROAD RELATED LANDSLIDES	INNER GORGE INCLUDED
Caspar (Spittler) (Field checked)	4.8	29	14	yes
Caspar (Coyle) (1978 &1996)*	21.7	103	none mapped (115 total count)	yes
James (1978 and 1996)	18	72	15 mapped (117 total count)	yes
Noyo (1978)	143	207	42	no
Noyo (1996)	143	222	56	no
Rockport (Juan & Howard) (1978 and 1996)	34	148	214	no
Maple (1978 and 1996)	49.9	43	not counted	no
McDonald (1978 and 1996)	14.6	18	not counted	no

* The dates for each site refer to the date of the aerial photograph used. In each case, the 1978 photographs were black and white and the 1996 were color

large: 26 percent of the total watershed area has an assigned instability value at or smaller than log(q/T) of -2.5, whereas about 51 percent of the randomly placed landslides were assigned log(q/T) values of -2.5 or smaller. The curve for the minimum q/T value for each observed landslide is labeled as "landslides" in Figure 10a and is distinctly different from both the total watershed area curve and the random model curve. The difference largely results from the much greater incidence of observed landslides assigned to the chronic and -3.1 categories. By log(q/T) of -3.1, 58 percent of the observed landslides have been counted, whereas only 21 percent of the random slides and 5.4 percent of the total watershed area has smaller q/T values. This is clear evidence that SHALSTAB has successfully predicted areas with greater probability of failure.

Figure 10b and 10c show landslide density as a function of slope instability category for in-unit and the road-related failures using the Noyo 1996 landslide data. Landslide density is the number of landslides found in a given log(q/T) interval divided by the total area (km^2) included in that category. The density is plotted as a function of the larger bound of that category (e.g., density for the category -3.1 to -2.8 is plotted as a function of -2.8). If the model is not successful at identifying unstable areas (and if there was no bias due to selection of minimum q/T value for each slide), then landslide density should be the same for all instability categories. Because of the bias resulting from using the minimum q/T values, the random landslide density shows a progressively greater concentration of landslides in areas of the highest instability ratings. The curve for observed landslide densities, however, is much different. For log(q/T) values of -2.8 and smaller, the incidence of landsliding was much higher than that estimated from the random placement model. For areas mapped as "chronic" or those falling into the category of log(q/T) < -3.1, the incidence of landsliding is high, equivalent to 9 and 7 landslides per km^2, respectively, for the period recorded by the Noyo 1996 aerial photographs. The large difference in landslide density at low q/T values between the mapped and biased-random placement model demonstrates that SHALSTAB is successful at identifying unstable areas of the landscape and that this finding holds true for both in-unit and road-related shallow landslides. We did not expect the relatively successful performance on road-related landslides because of the large local effects roads have on hydrology, soil strength and topography. This suggests that, at least in our study area, road failures are more likely in steep areas with high a/b – like those found associated with in-unit failures.

Figures 11 through 17 and Table 3 summarize SHALSTAB model performance in this validation study. In each watershed, modeled landslide density was greatest in the most unstable categories and differed substantially from that determined by the random placement model. Figure 11 shows landslide density for observed landslides for each watershed. Landslide density was very high (up to nearly 150 landslides per km^2 of the log(q/T) category) for areas assigned to categories of highest instability. Observed landslide density was greater than that for randomly

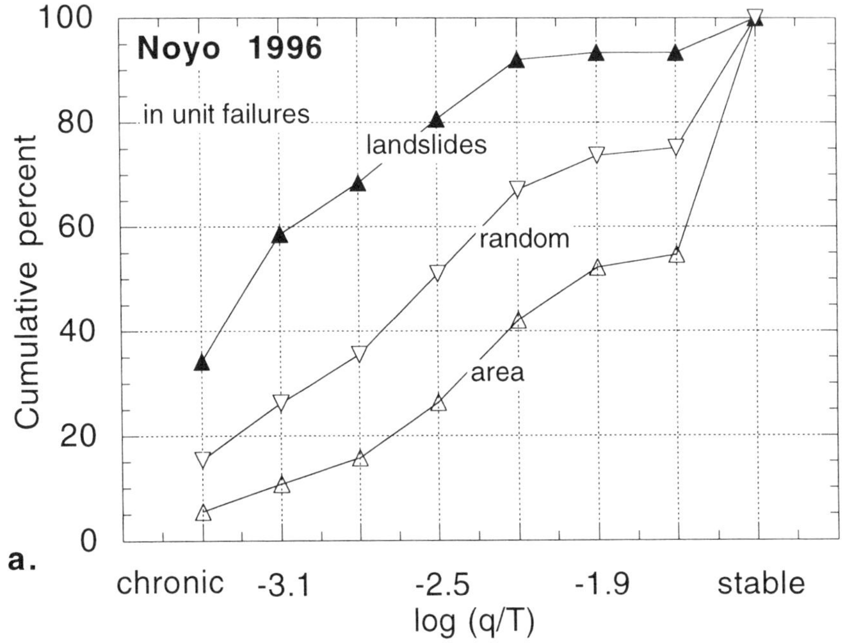

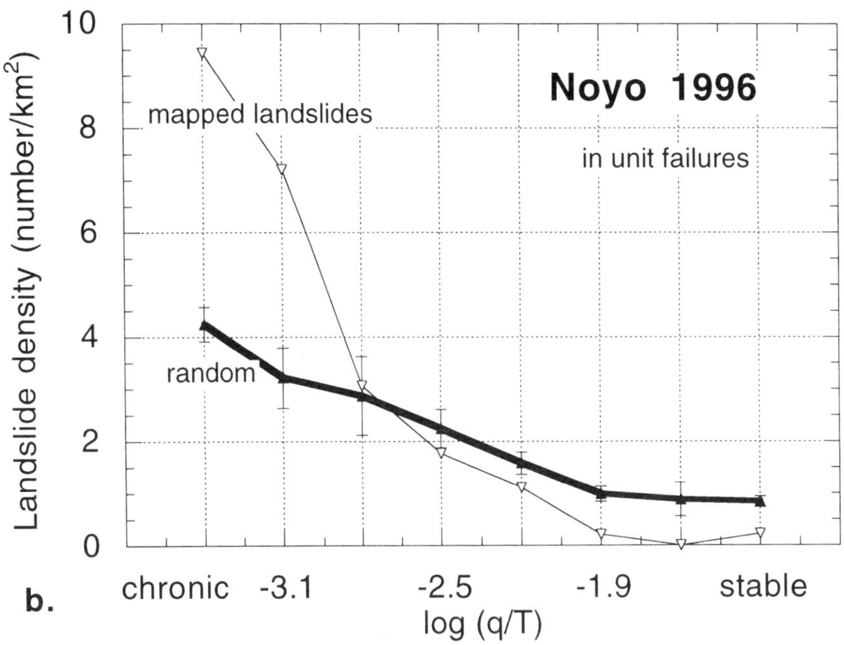

Figure 10. Log (q/T) values for mapped and randomly placed shallow landslides in the Noyo watershed of Northern California. Landslides were mapped from 1996 aerial photographs. Chronic and stable refer to unconditionally unstable and unconditionally stable conditions, respectively. (a) Cumulative percent of watershed area, of random landslides, and of mapped landslides for in unit failures (no influence of roads) as a function of log(q/T) class. (b) Landslide density (i.e. the number of landslides per area of log (q/T) class) for in-unit failures for mapped and randomly placed slides (standard deviation is shown on the random density function). (c) Landslide density for mapped and randomly placed road landslides.

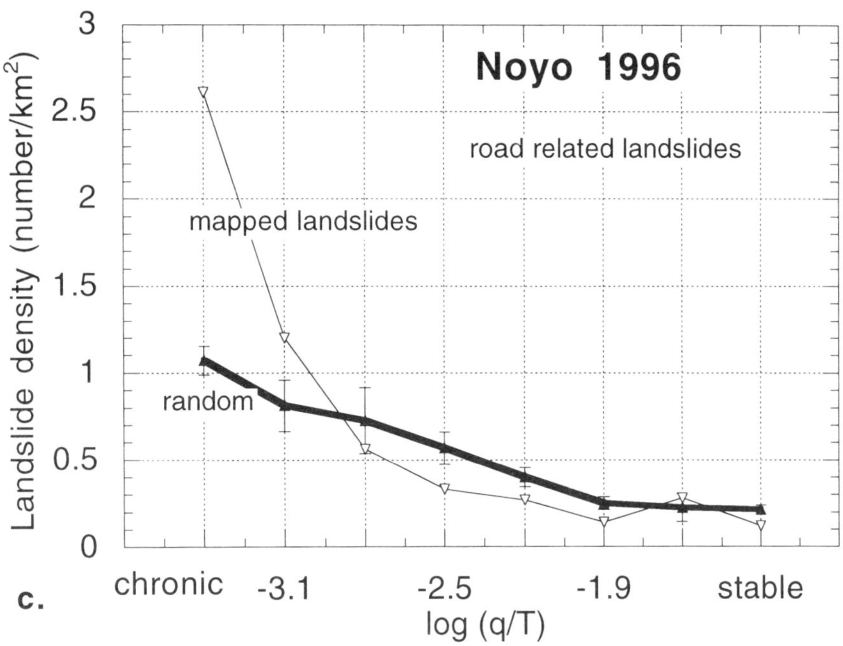

Figure 10 (continued)

placed landslides for log(q/T) values less than or equal to -2.5 for Caspar (Coyle and Spittler). Observed landslide density was also greater than random for values less than or equal to -3.1 for Noyo (1978 and 1996), McDonald (1978 and 1996 combined), Rockport (1978 and 1996 combined) and James. In the Maple watershed, only the landslide density in areas within the chronic category differed from random; however, 26 percent of all landslides occurred in lands of this category.

Figure 12 shows cumulative percentages of landslides found in each log(q/T) category for in-unit landslides in the watersheds. Rockport had the greatest proportion (85%) of landslides in the lowest log(q/T) category whereas Maple had over 40 percent of the landslides falling in the stable category. The high number of landslides mapped in the stable class in the Maple watershed appears to be partly due to the poor quality of the topographic map (and perhaps also the influence of deep-seated landsliding). For all watersheds, the average cumulative percentage of mapped in-unit landslides for the chronic, -3.1, -2.8, and -2.5 categories is 23, 46, 58, and 73 percent, respectively (with a standard deviation of about 19 percent for each category). These numbers are similar for road-related landslides.

Figure 13 shows the cumulative percentage of total area in each landslide instability category for each watershed and the corresponding landslide density based on 1978 aerial photographs. This suggests that the potential for shallow landslide instability is greatest in the Rockport watershed and least in the Maple watershed. In the 1978 aerial photograph coverage landslide density did vary in a manner consistent with watershed potential with Rockport landslide density equal to 4.2 landslides per km^2 and in Maple landslide density of only 0.3. This pattern differed greatly in the 1996 photographs, however, because very little management activity had occurred in the Rockport area since the 1970's and no landslides were detected. Hence, SHALSTAB maybe used as a tool for regional or landscape-scale classification of watersheds for potential landslide hazard, but site specific conditions which dictate when landslides occurs will depend on the management and storm history.

Figure 14 summarizes the validation results and illustrates one measure of risk, by showing the relationship between log(q/T) threshold and watershed area affected. For each instability category, the cumulative percentage of landslides found in that category and cumulative percentage of watershed area were calculated. These two attributes were then plotted against each other to reveal how much of the watershed area would have to be categorized as unstable in order to account for a certain percentage of the mapped landslides. For example, to account for 40 percent of all mapped landslides, about 3 to 8 percent of the total watershed area would have to be categorized as unstable. For 60 percent of the landslides to be accounted for, about 7 to 20 percent of the watershed would be have to be cate-

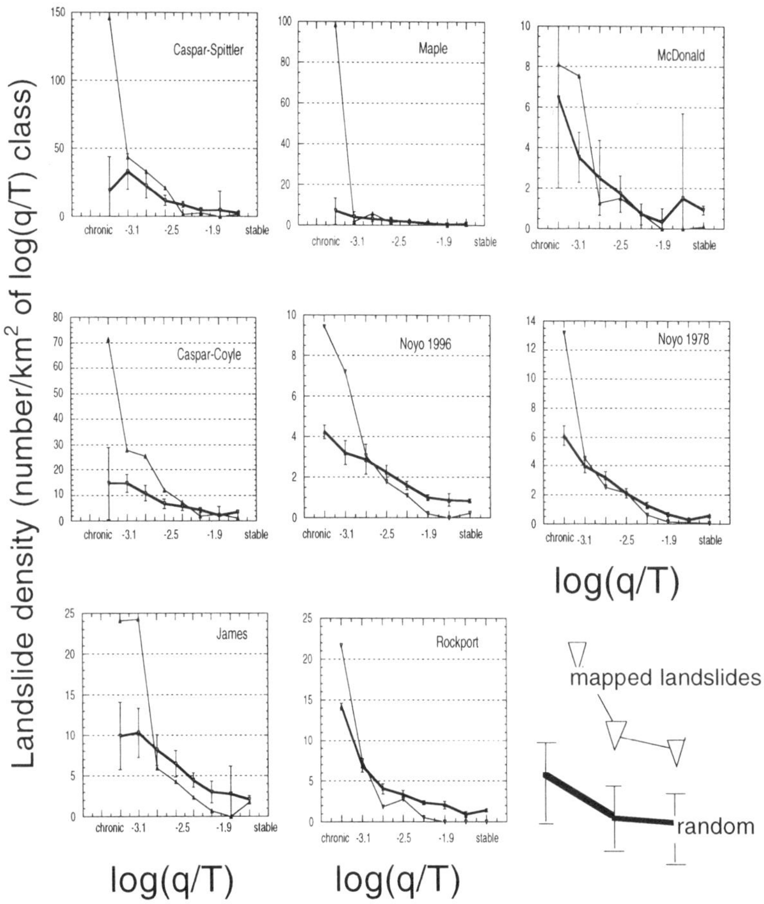

Figure 11. Mapped and randomly-placed landslide density for all the basins in the Northern California study area. Name in each graph refers to watersheds listed in Table 2. The lower, heavy line with the standard deviation bars is the random density. Note that the vertical axis scale varies considerably between watersheds.

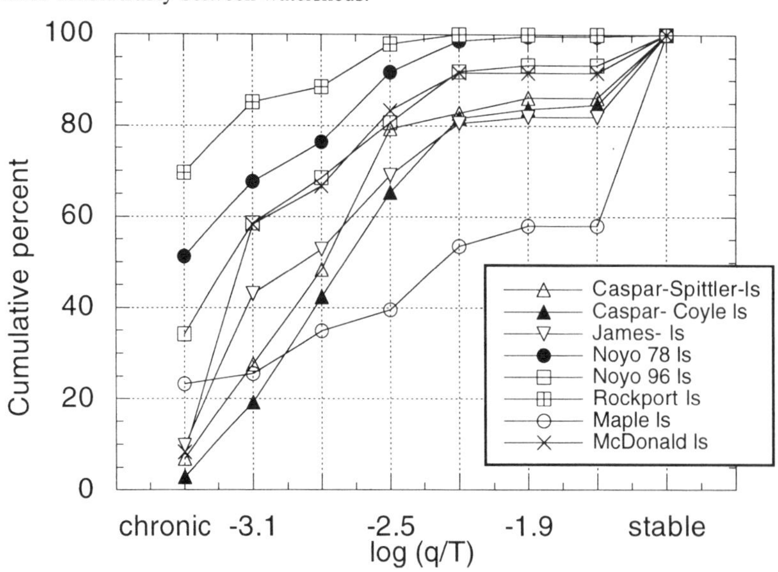

Figure 12. Cumulative percent number of landslides as a function of corresponding log(q/T) for each watershed.

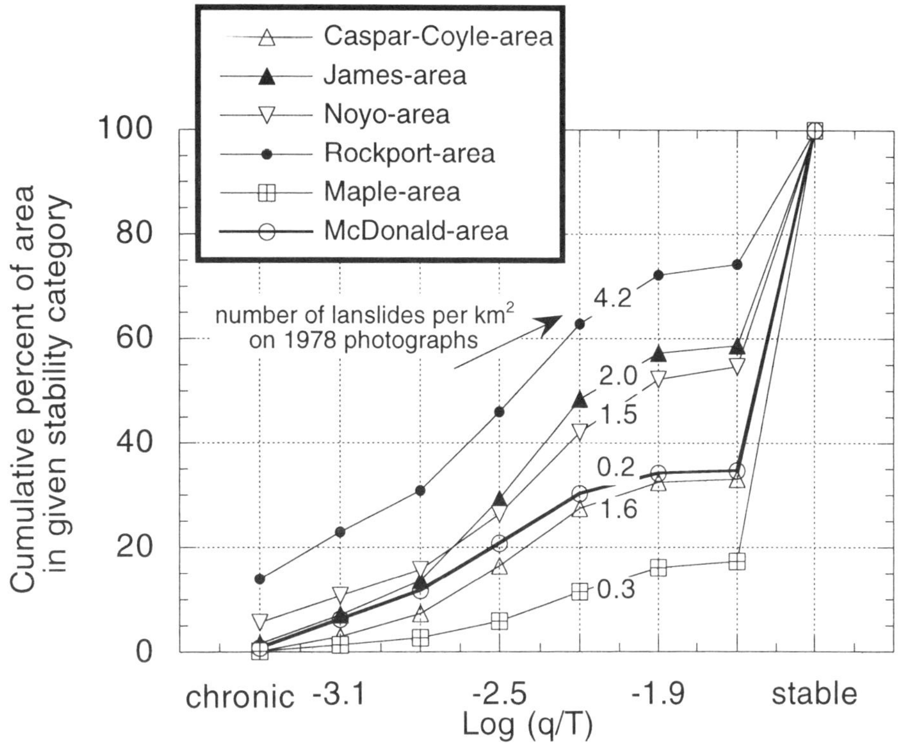

Figure 13. Cumulative percent of the watershed area for a corresponding log(q/T) category and the number of landslides per unit area of the watershed based on 1978 aerial photographs.

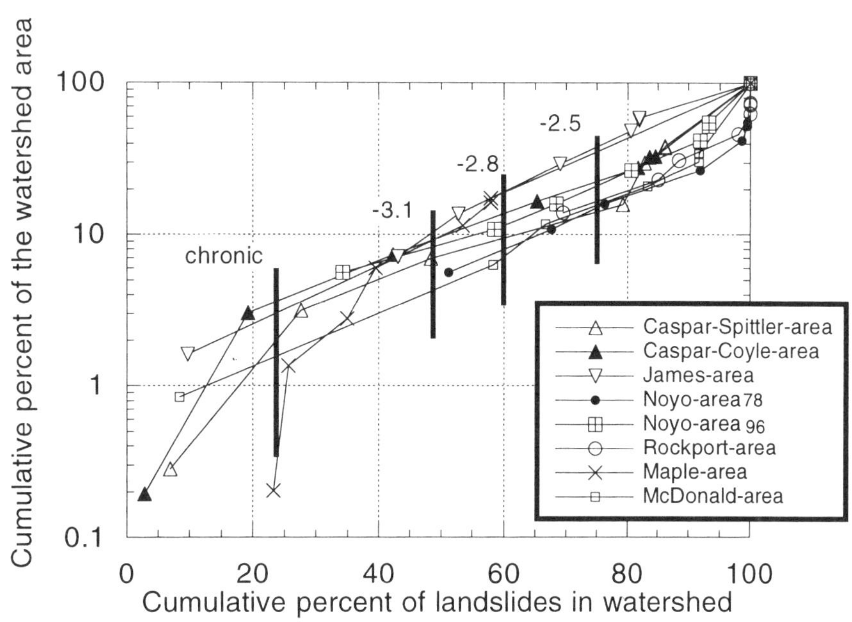

Figure 14. Cumulative percent of the watershed area as a function of the cumulative percent of the number of landslides for progressively larger log(q/T) classes. The vertical lines show the mean cumulative percent of landslides for each log(q/T) class. Each curve starts with the chronic class and each successive symbol along an individual curve records the next larger log(q/T) value (see Table 1 for log(q/T) classes).

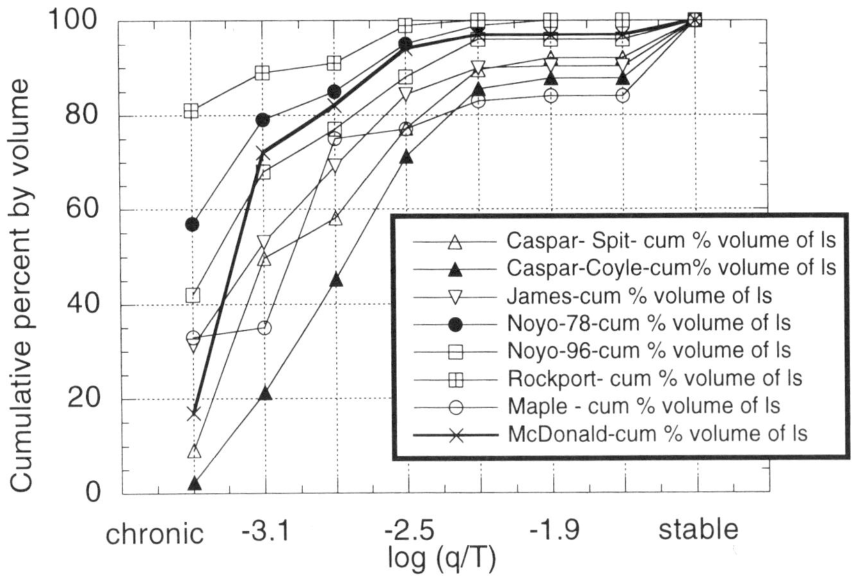

Figure 15. Cumulative percent volume of landslides as a function of corresponding log(q/T) for each watershed.

gorized as unstable. For a given log(q/T) class, however, the range of percent of landslides and corresponding affected area is large. For example, In the Noyo basin about 60% of the landslides fall in the less than -3.1 class, which covers about 11% of the watershed, while in the Rockport watersheds 85% of the landslides fall in the less than -3.1 class, which covers about 23% of the watershed. The average percent of landslides that fall into a given log(q/T) class for the 8 curves is shown as bold vertical lines. An expression fit to all the data shown in Table 3a (ignoring the inner gorge cases) gives $L = 20\ A^{0.43}$ ($R^2 = .87$, $n = 21$), in which L is the cumulative percentage (by number) of landslides a given cumulative drainage area for any given threshold log(q/T). Hence, this correlation, treating all watersheds as one data set, gives the cumulative number of landslides "hit" as a function of the percentage of the watershed areas classified as high risk. While individual watersheds have different relationships, taken together these data suggest that putting 13% of the area as high risk would include 60% of the landslides, while 20% of the area includes 73%.

The same analysis can be done with the more relevant measure of cumulative landslide volume instead of landslide number. These numbers differ because in four of the watersheds there is a well defined relationship of decreasing landslide size with increasing log(q/T). Such a relationship would be expected if the larger landslides are associated with instability in unchanneled valleys which are typically the least stable elements of the landscape. Landslide area was converted to volume by multiplying the measured plan area by the approximate colluvium depth of 1.0 m (the average of 28 measurements made in the field). Figure 15 shows the cumulative percent of landslide volume and Figure 16 is like Figure 14 except for landslide volume is used instead of landslide number. In general the model performance is improved, i.e. for the < -3.1 class the percent volume is on average 58 percent (as compared to 46 percent by cumulative number), for < -2.8 it is 73 percent (as compared to 58 percent) and for < -2.5 it is 86 percent (as compared to 73 percent). A correlation of cumulativelandslide volume (V) against cumulative watershed area gives $V = 33\ A^{0.3}$ ($R^2 = 0.67$, $n = 21$). Hence, the plots shown in Figures 14 and 16, and the related correlations can be used to define the tradeoff between maximizing the number of landslides predicted to occur in a high hazard zone and the minimizing the amount of area that must be classified as high hazard.

Table 3 summarizes model performance by log(q/T) class for each watershed by number (Table 3a) and by volume (Table 3b). For comparison among the watersheds, the cumulative percent watershed area for which the corresponding cumulative log(q/T) class accounts for approximately two-thirds of the number or volume of mapped landslides is labeled. This area ranged from as little as 3% to as high as 23% of the total watershed area. In Northern California, slopes immediately adjacent to channels are often quite steep, and noticeably steeper than upslope areas. These inner gorges are rarely accurately portrayed on

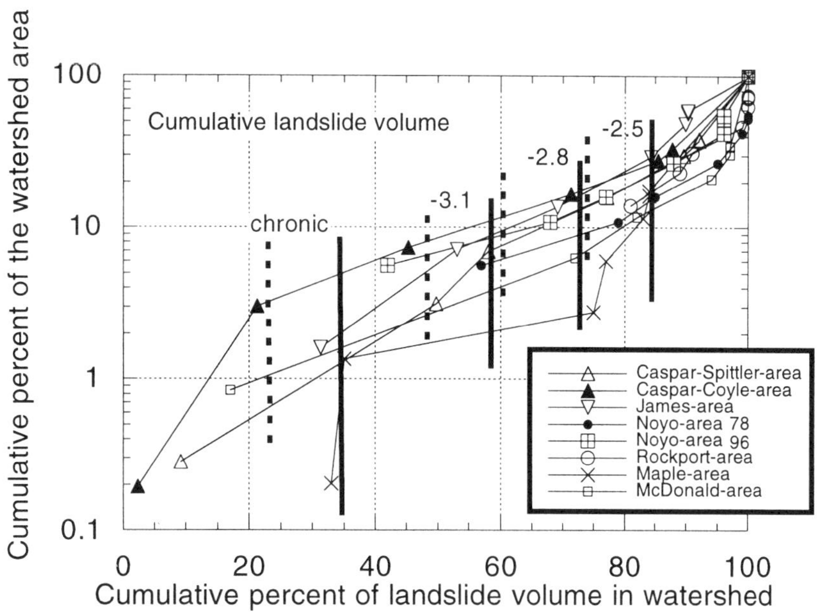

Figure 16. Cumulative percent of the watershed area as a function of the cumulative percent of the volume of landslides for progressively larger log(q/T) classes. Heavy vertical lines record the average cumulative percent volume of landslides for each log(q/T) class; the dashed lines show the same thing for cumulative percent by number (as in Figure 14).

standard 7.5' USGS topographic maps but are common places for shallow landsliding. Hence, further delineation of high-risk areas may be obtained by mapping inner gorges and adding them to the high-risk category. For two of the basins (Caspar and James), the inner gorge was mapped from aerial photographs by the California Division of Mines and Geology (CDMG) and reported in quadrangle-based landslide maps. Although a large percentage of the landslides (35 to 69 percent) fell within inner gorge areas, so did a large percentage of total drainage area (12 to 20 percent). Consequently, a greater percentage of the landslides in each watershed (i.e., Caspar-Spittler, Caspar-Coyle, and James) could be associated with a smaller percentage of the landscape area if SHALSTAB alone (without including the inner gorge) was used without separate delineation or classification of inner gorge areas using the available CDMG maps. This outcome probably depends on the quality of the topographic map and how accurately inner gorge areas are delineated on the CDMG maps. The best solution is to obtain higher resolution topographic maps (from laser altimetry for example) so that the inner gorge is well represented in the topographic base.

The cumulative percent area versus percent landslides for a given log(q/T) for the Northern California study area is similar to that found for similar quality topographic maps in the Oregon Coast Range (Elk Creek-Figure 6a and Coos Bay-Figure 9) (Figure 17). The use of higher quality (laser altimetry) data creates a very different relationship. For the laser altimetry-based analysis, all 36 of the landslides fell in the < -3.1 log(q/T) class.

APPLICATION OF SHALSTAB IN FOREST MANAGEMENT

In all applications a decision must be made as to what constitutes "high", "medium" and "low" hazards, i.e., the log(q/T) class cutoff that delineates the sites of greatest concern. Typically these different classifications will receive different land use limitations. For example, high risk areas may be classified as no timber harvest areas or they may be simply used to flag areas that should be reviewed by geologists. One approach would be to: 1) obtain the best possible topographic base; 2) use field observations and aerial photographs to create a map of landslide scars (noting which slides are due to land use activities) and locate accurately these scars on the topographic data base; 3) use output from SHALSTAB to determine a log(q/T) value for each scar; and 4) use the number of landslides associated with different log(q/T) values and some measure of risk (both environmental and economic) to guide in the decision as to what threshold values to assign. Additionally, to further evaluate model performance it can be

TABLE 3a. Summary of validation results based on number of landslides in each category

Watershed	cumulative percentage (by number) of mapped landslides (in unit) in modeled slope stability category (log q/T)			
	Inner Gorge	-3.1**	-2.8	-2.5
Caspar (Spittler)	69 *(20%)**	76 *(21%)*	86 *(23%)*	97 (30%)
	---	---	---	---
	n.a.	28 *(3%)*	48 *(7%)*	79 *(16%)*#
Caspar (Coyle)	42 *(20%)*	45 *(22%)*	59 *(24%)*	75 *(30%)*
	---	---	---	---
	n.a.	19 *(3%)*	42 *(7%)*	65 *(17%)*#
James	35 *(12%)*	63 *(17%)*#	68 *(23%)*	79 *(36%)*
	---	---	---	---
	n.a.	43 *(7%)*	53 *(14%)*	69 *(29%)*
Noyo (1978)	n.a.	68 *(11%)*#	76 *(16%)*	92 *(26%)*
Noyo (1996)	n.a.	59 *(11%)*	68 *(16%)*#	81 *(26%)*
Rockport	n.a.	85 *(23%)*#	89 *(31%)*	98 (46%)
Maple	n.a.	26 *(1%)*	35 *(3%)*	40*(6%)*
McDonald	n.a.	58 *(6%)*	67 *(12%)*#	83 *(21%)*

TABLE 3b. Summary of model results (for volume)

Watershed	cumulative percentage of volume of mapped landslides (in unit) in modeled slope stability category (log q/T)		
	-3.1**	-2.8	-2.5
Caspar (Spittler)	50 *(3%)**	58 *(7%)*#	77*(16%)*
Caspar (Coyle)	21 *(3%)*	45 *(7%)*	71 *(17%)*#
James	53 *(7%)*	69 *(14%)*#	84 *(29%)*
Noyo 1978	79 *(11%)*#	85 *(16%)*	96 *(26%)*
Noyo 1996	68 *(11%)*#	77 *(16%)*	88 *(26%)*
Rockport	89 *(23%)*#	91 *(31%)*	99 (46%)
Maple	35 *(1%)*	75 *(3%)*#	77*(6%)*
McDonald	72 *(6%)*#	82 *(12%)*	94 *(21%)*

*percentages (in itallics) refer to cumulative percent of area in this slope stability category
**cumulative percent includes the chronic category
#log(q/T) category which accounts for about two-thirds of the landslides

useful to compare the results from mapped scars with those produced by other means, such as the biased-random model proposed here, or to compare the performance of SHALSTAB with other models. Both maps and plots (showing where the landslides occur on the a/b versus slope graph (i.e. Figure 5)) are useful. This analysis can be done in a fixed parameter mode, or optimization can be attempted by allowing parameters to vary (producing location specific results).

Experience with SHALSTAB [Montgomery and Dietrich, 1994; Dietrich et al., 1998a,b] suggests that a threshold value for which a large proportion (about 60% to 80%) of shallow landslide scars occur depends on the quality of the base map, but in general, a value of log(q/T) -2.5 will capture the vast majority of the scars (up to 100%). In the three small test sites reported by Montgomery and Dietrich [1994] between 83 and 100% of all scars fell below the -2.5 threshold (using 5 m contour interval data). A study in the upper Chehalis watershed in Washington in which 629 landslides were mapped (including 470 that were road-related) found 86% of all the scars in values below a log(q/T) of -2.5 using 30 m grid data [K. Sullivan, pers. com., 1994]. A modified version of SHALSTAB (using a spatially constant cohesion and soil depth) was applied to 3224 landslides in 14 watersheds in Oregon and Washington and about 66 percent of the landslides occurred in less than the log(q/T) of -2.5 using mostly 30 m grid data [Montgomery et al., 1998b]. Montgomery et al., [2000] also show that the cohesion based SHALSTAB.C was better than the biased-random model below log(q/T) values of –2.2. In the Oakland hills, just south of the University of California, 84% of the 78 scars were found below log(q/T) of -2.5 using 10 m grids (unpublished data). In the Northern California validation study described above 71 to 99% of landslide volume occurred in cells with log(q/T) less than -2.5 using 10 m grid data.

For forest management decisions, the general goal would be to reduce shallow landsliding to an appropriate level by restricting management on as small a fraction of the landscape as possible. Based on our analysis in the Pacific Northwest, in order to capture more than two-thirds of the landslides, for 30 m grid data, a threshold of -2.5 appears to be needed, for 10 m data (from digitized 7.5' quadrangles) a threshold of -2.8 may be adequate, and for still higher resolution data this threshold may be pushed to -3.1. Comparison with the biased-random model indicates

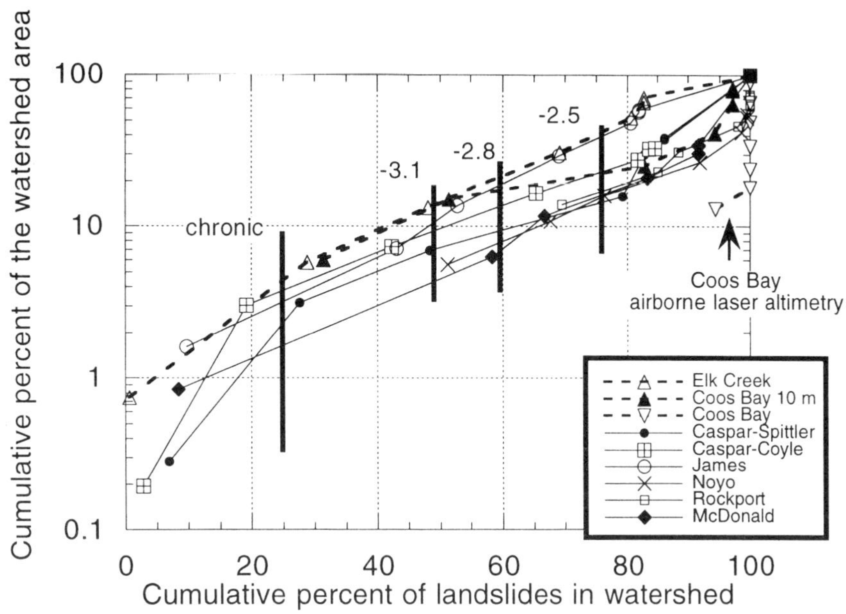

Figure 17. Cumulative percent of the watershed area as a function of the cumulative percent of the number of landslides for progressively larger log(q/T) classes for Northern California and Oregon Coast Range watersheds. The Coos Bay 10m data are for the USGS 10 m data shown in Figure 9, where as the data labeled 'Coos Bay' are derived from the laser altimetry, also shown in Figure 9. Elk Creek refers to the data shown in Figure 6.

that for the 10 m case and with relatively large mapped landslides, SHALSTAB can't distinguish observed from randomly placed sites for a log(q/T) > -2.8. Lacking additional data, we recommend the above threshold values of log(q/T). Specific site studies may find otherwise. For example, Pack and Tarboton [1997] using their own computer code based on SHALSTAB accounted for 91% of their mapped landslides using a threshold log(q/T) of -3.3 for a 20 m grid of the 730 km^2 Trout Lake Basin in British Columbia.

Ultimately, the choice of the high hazard cutoff and consequent land use prescription will implicitly or explicitly reflect the user's perspective on risk. The higher the threshold, the greater the likelihood that all shallow landslides will be accounted for, but also the greater the watershed area that is classed as high hazard. Perception of risk will probably differ depending on the value of the resource that shallow landsliding may threatened (e.g. endangered fish, roads, or houses).

Three kinds of applications of SHALSTAB to forest management issues have been explored or considered. One application has been its use by public agencies (e.g. Bureau of Land Management) or by private companies in watershed analyses that lead to site-specific prescriptive measures. For example, Mendocino Redwood Company (MRC), which was created from lands purchased from Louisiana-Pacific in Northern California, is using maps produced from SHALSTAB in a variety of ways. The maps have been combined with a channel rating system to delineate watersheds that have the greatest potential for high value aquatic resources and high potential shallow landsliding (either from in-unit failures or road failures). These watersheds are considered most at risk and are to be given priority for watershed analysis [Olsen and Orr, 1999]. The maps have been used as part of MRC's watershed analysis process. SHALSTAB maps in combination with aerial photograph analysis and field mapping are used to develop mass wasting map units which are assigned watershed specific land management prescriptions [Chris Surfleet, pers. com., 2000]. SHALSTAB is also used as a tool, along with other information to determine risk of instability and assign local prescriptions. Specifically the MRC policy is: "no harvest activity will occur, with the exception of cable or helicopter harvesting that retains over 50% of the pre-harvest basal area, or any construction of roads and landings in areas defined in the field as having a significant likelihood of sediment delivery from mass wasting unless a site-specific assessment is conducted and operations approved by a registered geologist" [Chris Surfleet, pers. com., 2000].

Another application is in regional planning analyses performed by government agencies. For example, we have

Table 4. Cumulative percent of landscape in each landslide hazard category

log(q/T)	***Allegany	Golden Falls	Greenleaf	Mapleton	Heceta Head	Cedar Butte
*chronic	<1	<1	<1	<1	<1	1
<-3.4	1	2	1	1	2	7
-3.4 to-3.1	3	5	4	4	6	14
-3.1 to -2.8	11	13	15	14	17	29
-2.8 to -2.5	30	28	38	33	35	53
-2.5 to -2.2	49	40	55	47	47	71
-2.2 to -1.9	54	42	58	49	49	77
> -1.9	54	42	58	49	49	77
**stable	100	100	100	100	100	100

* chronic = slope equal to or greater than 45 degrees
**stable = slope less than 20.6 degrees
*** 7.5' quadrangle in Oregon Coast Range (30 m data)

run SHALSTAB on the entire Oregon Coast Range using 30m USGS data to provide basic data on relative potential slope instability to the National Marine Fisheries Service. This was done to help inform the debate about regional plans for protecting and restoring coho salmon in the Oregon Coast Range. Table 4 shows examples of results from selected quadrangles, showing that in the southern highly dissected part of the range, the proportion of area in each log(q/T) class is similar, whereas in the very steep topography in the northern part of the range, a much greater proportion of the area is in the low log(q/T) classes. If the log(q/T) threshold for high hazard were -2.5, then, in the highly dissected portions of the range, upwards of 30% of the landscape would require field review by a specialist because it would be classed as high hazard. This value is halved if the threshold is lowered to -2.8. These tables and accompanying maps provide a useful quantitative assessment for regional planning purposes.

A third application is in hazard mapping with regard to structures and people. With increasing development in areas adjacent to lands managed for timber production, the risk that forest practices will lead to destruction of houses and loss of life is increasing. SHALSTAB only maps the spatial pattern of relative potential for shallow landslides: it does not delineate debris flow runout path or distance. The simplest approach to include the runout is to define a threshold channel slope below which debris flows typically stop and then map separately all channels above and below this threshold [i.e. Montgomery and Dietrich, 1994; Dietrich and Sitar, 1997]. Other procedures exist, most notably that of Benda and Dunne [1997] which includes the effects of tributary junction angle on runout distance. Steep valleys fed from drainage areas with areas of low q/T present the greatest risk. In order to include debris flows that run down unchanneled valleys, the threshold for channel initiation used in a digital terrain prediction of the channel network needs to be set fairly low.

Ideally, the forest practices decision (or prescription) for a site that is rated as high potential for instability could be conditioned by the risk downslope. If structures or people are at risk, then clearly the most stringent restrictions should apply. Presumably high restrictions would apply for river systems in which aquatic resources have been greatly diminished. If the sediment that could be released from a shallow landslide cannot be delivered to the river system (for example if the sediment were to be deposited on a debris fan on a terrace that does not drain via a channel to the river system) there is reduced value in restricting forest practices at the potentially unstable site. If sediment arrives in a steep, non fish-bearing stream that is connected to the river network, however, that sediment has been "delivered" and should not be discounted, as it still can contribute to sediment loading downstream.

Presently, there seems to be little knowledge about what might be an acceptable frequency of timber harvest related increase in landsliding and resultant river sedimentation to a river ecosystem. In fact, some have argued that debris flows can be beneficial to aquatic habitat [Reeves et al., 1995]. While the concept that debris flow delivery of coarse sediment and wood to sediment deficient and wood free streams can create important aquatic habitat is useful, its practical application requires considerable knowledge of the river system. Specifically one might argue that it is desirable to cut and destabilize an area because of the perceived need for coarse sediment downstream (in, say, bedrock dominated channels). There are several problems with this idea. The notion that further disturbance of potential failure sites is beneficial ignores the fact that there are no extensive areas under industrial forestry in the Pacific Northwest that have not already experienced some level of timber harvesting or land use related fires, and consequently are in a reduced state of stability. It would take extensive, detailed field studies of all potential sites to demonstrate that without further timber harvesting the rate of debris flow generation would be too low. Cutting and removal of wood from potential landslide source areas also eliminates what might be a primary source of large woody debris replenishment. Furthermore, most river systems have been altered through some combination of such things as large woody debris removal, splash damming, gravel mining, stream side timber harvesting, and dams such that considerable care must be taken to infer simple causal relationships between debris flows and channel habitat.

DISCUSSION

The model SHALSTAB is an attempt to create a simple mechanistic model that can be widely used to delineate relative potential for shallow landsliding in the absence of spatially registered soil strength and hydrologic properties. Because of its simplicity, it can be validated with field studies. At present we know of no application in which the model demonstrably failed, but we can think of cases in which it should. In areas in which shallow landsliding is associated with the active toes of large-deep seated landsliding the model may fail because of the low gradients and strong lithologic control on location of deep-seated landsliding. The mechanisms of mass wasting are different in deep-seated slides. If short, intense rains dominate shallow landsliding, then convergence effects of surface topography on shallow subsurface flow may not operate, and, consequently, the shallow landsliding may not depend strongly on a/b (as is suggested by Wieczorek et al.[1997] for a storm in Virginia). There are circumstances in which the topography is favorable for the generation of shallow landsliding, but the incidence is very rare due to lack of soil mantle, high soil strength, or insufficient rainfall. One such area is the glaciated, bedrock-dominated landscape of the Sierra and Cascade Mountains.

Several points need to be stressed about the interpretation and limits of SHALSTAB. Even if all mapped shallow landslides in a region occur in the lowest log(q/T) classes, the vast majority of the low log(q/T) sites in the study area will show no evidence of shallow landsliding. The things that have been eliminated from the slope stability model dictate the timing and size of failure at individual sites. Spatial and temporal variability in such properties as soil cohesion, root strength, soil depth, hydraulic conductivity fields of the soil mantle and underlying bedrock, antecedent moisture and storm history, and land use alterations of hydrology (e.g. changes in evaportranspiration, diversion of road runoff to unstable sites) are unknowable at all potential sites across a landscape, making the prediction of the exact time and location of shallow landsliding impossible. Hence, the SHALSTAB map is solely a portrayal of the effect that surface topography has on the relative potential for instability. All these other factors bear on whether the site actually fails in some interval of time. If most shallow landslides occur in low log(q/T) areas, then the SHALSTAB map is simply showing those places on the landscape that are topographically similar to those that failed, and therefore, without further information about individual sites, should be considered equally likely to fail.

It is often thought that the effective precipitation term, q, in the model can be interpreted in terms of precipitation characteristics in an area, i.e. the rainfall amount can be related to a particular storm frequency and magnitude. The hydrologic model in SHALSTAB is based on steady state precipitation and runoff, hence q specifically would be a precipitation event of sufficient duration that steady state runoff is obtained, a condition rarely if ever experienced under natural storms. Nonetheless, it can be useful to estimate an average transmissivity (T) and compare the calculated precipitation (q) for a given log(q/T) category to rainfall characteristics in an area. For example, based on detailed fieldwork at an experimental site in coastal Oregon [e.g. Montgomery et al., 1997; Anderson et al., 1997, 1998; Torres et al., 1998], Montgomery and Dietrich [1994] estimated the transmissivity to be about 65 m^2/day. This gives an effective q of 0.05 m/d for a log(q/T) of -3.1, a reasonable multi-day precipitation rate, but gives a value of 0.2 m/d for a log(q/T) of -2.5, which simply does not occur over the several day period that would be needed to reach equilibrium. With sufficiently detailed topographic data (such as shown in Figure 7), this kind of calculation may place some limit on reasonable values of log(q/T) for instability to occur.

SHALSTAB does not predict either the size of individual scars nor the rate of landsliding. The model is based on the infinite slope approximation, and therefore, it can not specify a specific size, other than the artificial size of the grid chosen for the analysis. Okimura [1994] discusses a method using a three-dimensional analysis in a digital terrain model to estimate size. Furthermore, the rate of landsliding remains problematic, as this will depend on such things as storm history, potential site evolution (soil accumulation) and vegetation dynamics.

Because of its application in forest management, it would be desirable to have a term that explicitly reflects possible changes in root strength contribution to slope stability. Building upon earlier development of the SHALSTAB theory, Dietrich et al. [1995] proposed an infinite slope model that includes the effects of vertical root strength and of vertical varying saturated conductivity. They couple a steady state hydrologic model with a infinite slope model in which soil cohesion, C_{sw} and apparent cohesion, C_r, due to root strength are included to derive a slope stability model that accounts for soil depth variation and effects of decreasing saturated conductivity with depth below the surface:

$$\frac{q}{k_1} = \frac{b \sin\theta}{a n_1}(e^{-n_1 \beta z \cos\theta} - e^{-n_1 z_0 \cos\theta} + \frac{k_2 n_1}{n_2 k_1} e^{-n_2 z_0 \cos\theta}) \quad (9)$$

where

$$\beta = 1 - \frac{\rho_s}{\rho_w}(1 - \frac{1}{\tan\phi}(\tan\theta - \frac{C_r + C_{sw}}{z \rho_s g \cos^2\theta})) \quad (10)$$

The saturated conductivity field is assumed to be described by an exponential decline in which the slope (n_1,n_2) and intercept (k_1,k_2) may change at a depth, z_0, due to changes in soil bulk density or transition to the underlying bedrock. All terms are the same as used in previous equations with the addition of C_{sw} and C_r, which are the soil cohesion and the apparent cohesion due to roots, respectively. With the use of an exponential decline in saturated conductivity, the hydrologic model is very similar to TOPMODEL [Beven and Kirkby, 1979]. Equation (9) is dimensionless as k_1 is the saturated conductivity at the ground surface and has the same units as the effective precipitation. Clearly, equation (10) requires much more information about site conditions that does SHALSTAB.

To model the soil depth influence on slope stability, Dietrich et al. [1995] also proposed a process-based theory for predicting the spatial pattern of soil depth. It is based on the assumptions that the rate of soil production varies inversely with soil thickness and that soil is redistributed across hillslopes by a linear diffusive transport model. Subsequently, the soil production assumption has been verified through cosmogenic radionuclide analysis [Heimsath et al., 1997], while Roering et al. [1999] have argued that the linear diffusion assumption may only apply on modestly inclined hillslopes. This modeling approach to including soil depth in slope stability analysis overcomes the challenging problem of mapping the soil depth, but it undoubtedly oversimplifies the actual pattern of depth variability across the landscape [Schmidt, 1999].

Most digital terrain based models use the assumption that soil depth is spatially constant in order to perform calculations [e.g. Okimura and Kawatani, 1987; Duan, 1996; Wu and Sidle, 1995; Pack et al., 1998] but this assumption greatly reduces the distinctive role of cohesion on slope stability. A modified version of SHALSTAB can illustrate this point. If all the same assumptions are made that led to SHALSTAB but in addition it is assumed that vertical root cohesion contributes to strength, then the slope stability equation can be written as:

$$\frac{q}{T} = \frac{b}{a \sin\theta}\left[\frac{\rho_s}{\rho_w}(1 - (1 - \frac{C_r}{\rho_s g z \sin\theta \cos\theta})\frac{\tan\theta}{\tan\phi})\right] \quad (11)$$

This form of the slope stability equations has been referred to as SHALSTAB.C [Dietrich and Montgomery, 1998)] and has been used in various analyses by Montgomery et al. [1998a,b; 2000] in which field estimates of cohesion and friction angle were used. If the ratio $(C_r/\rho_s gz) = C^*$, (11) becomes

$$\frac{q}{T} = \frac{b}{a \sin\theta}\left[\frac{\rho_s}{\rho_w}(1 - (1 - \frac{C^*}{\sin\theta \cos\theta})\frac{\tan\theta}{\tan\phi})\right] \quad (12)$$

If C^* is treated as a spatial constant, (12) differs little from SHALSTAB which does not have a cohesion term. To illustrate this point, we assume (12) and (7a) produce the same q/T for failure, then by setting (12) equal to (7a) and solving for friction angle in SHALSTAB yields

$$\tan\phi = \frac{\tan\phi_c}{(1 - \frac{C^*}{\sin\theta \cos\theta})} \quad (13)$$

in which $\tan\phi_c$ is the friction angle for the case with cohesion in (12) (whereas $\tan\phi$ is for the cohesionless case). The product $\sin\theta\cos\theta$ only varies from about 0.3 to 0.5 for 20 to 45 degrees, respectively. Therefore, if the ratio of cohesion to the soil depth-bulk density product is a spatial constant then there is a friction angle in the cohesionless case (SHASTAB) that will produce similar results. Hence, adding the cohesion term provides relatively little additional information about the system and there is no unique solution of cohesion and friction angle to the slope stability problem. Such a model may be, nonetheless, useful in illuminating how change in root strength associated with land management effects relative slope stability [Wu, 1993].

Pack et al., [1998] have released a digital terrain program based on equation (12) on the World Wide Web (http://www.engineering.usu.edu/cee/faculty/dtarb/sinmap.htm). They call the program SINMAP and have provided a well designed ArcView interface for handling data, making maps, and performing calibration of (12). Pack et al. let q/T, cohesion (divided by soil depth and bulk density, i.e. C^*) and friction angle be variables. They have the user display their landslide data on a plot of a/b versus slope (like that shown in Figure 5) and then interactively pick the combination of C^*, q/T and friction angle that visually best fits the data (i.e. for the majority of mapped landslides plot in the a/b and slope field that is predicted to be unstable). Each landslide is assigned a unique value of a/b and slope: the problem of landslides being bigger than the grid size and therefore having variable values of these topographic factors is not addressed. SINMAP asks the user to identify

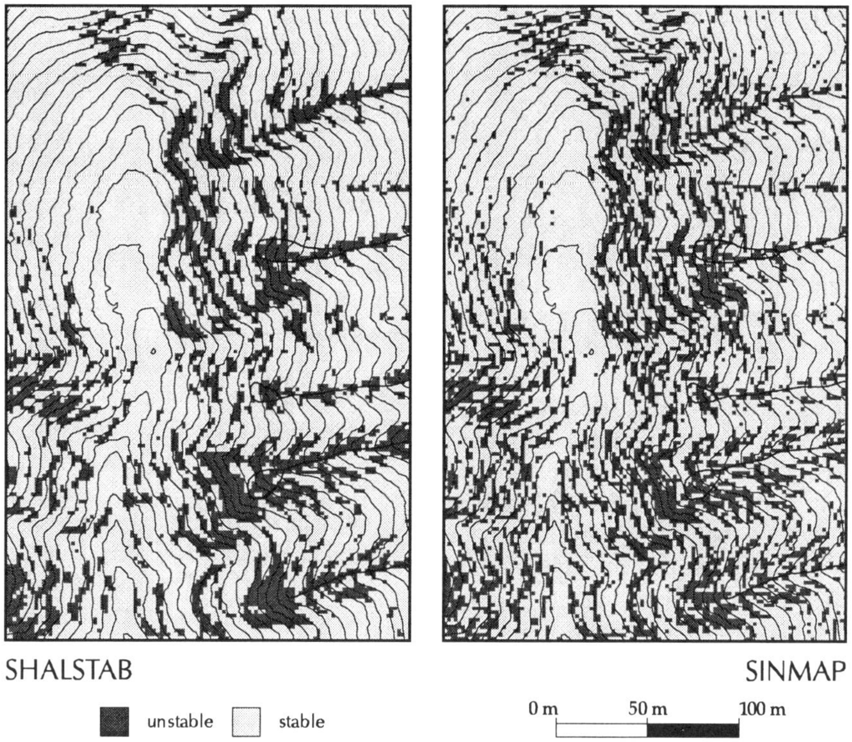

Figure 18. Comparison of predicted sites of instability using identical parameters (log(q/T) of -3.1, friction angle of 45 degrees, bulk density ratio of 1.6 and cohesion equal to zero) for SHALSTAB and for SINMAP. Site is a small portion of the Coos Bay study area. Contour interval is 5m and the thin lines in the valleys show the mapped channel location and the location of landslide scars.

upper and lower values of each of these three variables, and then assuming the values are normally distributed between these ranges and that the distribution functions are independent, the program calculates probability of failure. This probability is then used to assign a stability index to each site. The final product is a map of relative slope stability hazard based on assigned ranges of variables. While SINMAP is quite useful, it is not possible to uniquely determine cohesion (divided by bulk density and soil depth) and friction angle from data just on the location of landslides in a landscape unless the spatial variation in soil depth is also known. Even with that information, uniqueness is difficult to obtain [Dietrich et al., 1995]. It is also not clear how reliable the probability assignment is given the large uncertainty in parameter values and their covariance.

Differences between SHALSTAB, SINMAP, and other such digital terrain models may arise even if they use the same equations and parameters because the models may use difference procedures for estimating slope and drainage area. Tarboton [1997], for example, reports a detailed comparison between the performance of the algorithm for calculating area used in SINMAP and other methods including one called the multiple flow path procedure which is similar to that used in this paper. He argues in favor of his method because it is numerically efficient, and strikes a balance between highly directional and more dispersive flow paths. As mentioned earlier, we have chosen the multiple path procedure to minimize grid artifacts, i.e. we wanted to minimize results being dependent on the orientation of the topography relative to the grid. The SINMAP procedure is more sensitive to relative orientation than is the procedure we have used in SHALSTAB. In land use applications, it seems desirable to minimize orientation artifacts. We also use different procedures for estimating local slope. Figure 18 shows a comparison of the two programs for the same place with identical parameters. Three landslide scars are shown, and both models predict unstable cells where the scars are located. The more clustered appearance of the SHALSTAB model results from the dispersive area calculation procedure and the large area used to estimate local slope. This illustrates that details of model predictions will depend on the specific methods used to calculate topography and users should be aware that such differences could occur.

CONCLUSION

The model SHALSTAB is a mechanistic, yet simple tool for delineating the relative potential for shallow landsliding using digital terrain models. This combination of being mechanistic and simple permits broad application in the absence of landslide maps or data on soil strength and hydrologic properties. Useful models that include root strength and dynamic runoff (non-steady flow) have been developed and provide insight about the land use and climatic controls on slope stability, but they are difficult to parameterize accurately across a landscape. We suggest further that because surface topography (local drainage area and slope) can have such a large effect on local slope stability that the most valuable information to improve model performance is increased topographic resolution. Even the best currently available 7.5' USGS quadrangles miss the fine scale topography that dictates local shallow subsurface flow paths. While validation studies support its use, they also show that the currently available topographic maps in the Pacific Northwest (7.5' USGS quadrangles) do not provide satisfactory topographic resolution to permit the application of site specific land use measures without field inspection. Hence, the digital terrain model can serve as a planning tool at the regional, watershed timber harvest level and as a guide for fieldwork when specific land use measures are being applied. Higher resolution topography, such as that obtained from laser altimetry, would greatly improve modeling and would have great utility in other land management activities (e.g. road design) and watershed analysis (e.g. mapping of channels, modeling wood recruitment to channels). These other applications could make it cost effective to obtain such higher quality topographic data.

Because this model can be used in a fixed parameter mode, a standard value of log(q/T) can be used to define high risk landslide areas. There remains the policy decision about what percentage of the observed landslides should be placed in the high risk class (i.e., what value of log(q/T) should be used to define the bound of high risk) and what should be the appropriate land use prescription for the relative stability classes. Plots of cumulative percentage of watershed area as a function of cumulative percentage number (or volume) of landslides associated with a log(q/T) threshold can be used to examine potential costs of choosing different threshold values. Restrictions should be greater where landslides could threaten structures and people. If we use the criteria that two-thirds of the mapped landslides (by volume) should fall in the high risk category, then for 10 m grid maps (used in this validation study), log(q/T) should be < -2.8 (placing on average 13% of the watershed area of our validation sites into high hazard). To obtain a similar level of performance we suggest a threshold of log(q/T) < -2.5 for 30 m grid data and -3.1 for 5 m or less grid data. Digital maps can be gridded to any scale, so these grid scales refer to the original topographic data density from which maps are made. These log(q/T) thresholds are based on model performance, however, not on perceived environmental or economic risk. It is our view that uniform guidelines could be established in the absence of data using the model we have validated here, but that flexibility should remain in the field application in order to account for inaccuracies and to include the issue of risks to downstream resources as criteria for specific forest management decisions. Such combined use of digital terrain hazard mapping and field investigation should lead to the successful reduction in forest management related shallow landsliding even as active timber harvesting continues.

Acknowledgments. The development and testing of SHALSTAB reported here has been supported by Weyerhaeuser, Louisiana-Pacific, Mendocino Redwood Company, NCASI, Bureau of Land Management, Stillwater Sciences, and the Water Resources Center of the State of California. Landslide mapping of the Northern California watersheds was done by John Coyle. Martin Trso helped with data analysis. The Oregon Department of Forestry provided topographic data and maps of landslide locations and we specifically thank Jim Paul and Keith Mills. Conversations with many individuals including James McKean, Kathleen O. Sullivan, Thomas Spittler, Barry Williams, Bruce Orr, Mark Reid, and Frank Ligon were helpful. Reviews by John Stock, Josh Roering, Fred Booker, Douglas Allen, Mauro Casadei, Frank Ligon, and two anonymous readers improved the manuscript. Discussions with David Montgomery were, as always, valuable.

REFERENCES

Anderson, S. P., W. E. Dietrich, D. R. Montgomery, R. Torres, M. E. Conrad, and K. Loague, Subsurface flow paths in a steep unchanneled catchment, Water Resour. Res., 33, 2637-2653, 1997.

Anderson, S. A., W. E. Dietrich, R. Torres, D. R. Montgomery and K. M. Loague, Concentration-discharge relationships in runoff from a steep, unchanneled catchment, Water Resour. Res., v 33, 211-225,1997,

Benda, L, and T. Dunne, Stochastic forcing of sediment supply to channel networks from landsliding and debris flow, Water Resources Research, v. 33, p. 2849-2863, 1997.

Beven, K. and M. J. Kirkby, A physically based, variable contributing area model of basin hydrology, Hydrol. Sci. Bull., 24, 43-69, 1979.

Burroughs, E. R., Jr., and B. R. Thomas, Declining root strength in Douglas-Fir after felling as a factor in slope stability, U.S. Dept. of Agriculture, Forest Service Research Paper INT-190, 27p. 1977.

Burroughs, E. R., Jr., C. J. Hammond, G. D. Booth, Relative stability estimation for potential debris avalanche sites using field data, Internat. Symposium on erosion, debris flow and disaster prevention, 335-339, Tsukuba, Japan, 1985.

Carrara, A., Multivariate models for landslide hazard evaluation, Math. Geol., 15, 403-426, 1983.

Carrara, A., M. Cardinalli, F. Guzzeti and P. Reichenbach, GIS technology in mapping landslide hazard, in Geographical Information Systems in Assessing Natural Hazards, edited by A. Carrara and F. Guzzetti, pp. 135-175, Kluwer Academic Publishers, Netherlands, 1995.

Costa-Cabral, M. C., and S. J. Burges, Digital elevation model networks (DEMON): A model of flow over hillslopes for computation of contributing and dispersal areas, Water Resources Research, 30, 1681-1692, 1994.

Cutherbertson, J.G., Geotechnical evaluation of the slope stability computer program 3-D LISA, M.S. thesis, University of Idaho, Boise, 271p., 1992

Dietrich, W. E., and T. Dunne, Sediment budget for a small catchment in mountainous terrain, Zeit. für Geomorph., Suppl. 29, 191-206, 1978.

Dietrich, W. E., Dunne, T., Humphrey, N. F., and Reid, L. M., Construction of sediment budgets for drainage basins, in Sediment Budgets and Routing in Forested Drainage Basins, edited by F. J. Swanson, R. J. Janda, T. Dunne, and D. N. Swanston, General Technical Report PNW-141, Forest Service, U.S. Dept. of Agriculture, p. 5 - 23, 1982.

Dietrich, W. E., Wilson, C. J., Montgomery, D. R., McKean, J., and Bauer, R., Channelization thresholds and land surface morphology, Geology, v. 20, 675-679, 1992.

Dietrich, W. E., Wilson, C. J., Montgomery, D. R., McKean, J., Analysis of erosion thresholds, channel networks and landscape morphology using a digital terrain model, Journal of Geology, v. 101, 259-278, 1993.

Dietrich, W. E., R. Reiss, M.-L Hsu, and D. R. Montgomery, A process-based model for colluvial soil depth and shallow landsliding using digital elevation data, Hydrological Processes, v. 9, 383-400, 1995.

Dietrich, W. E., D. R. Montgomery, SHALSTAB: a digital terrain model for mapping shallow landslide potential, National Council of the Paper Industry for Air and Stream Improvement, Technical Report, 26p., 1998.

Dietrich, W. E., R. Real de Asua, J. Coyle, B. Orr, M. Trso, A validation study of the shallow slope stability model, SHALSTAB, in forested lands of Northern California, project report to Louisiana-Pacific Corporation, June 1998, 16p.

Dietrich, W. E. and N.Sitar, Geoscience and geotechnical engineering aspects of debris-flow hazard assessment, in Debris flow hazard mitigation: mechanics, prediction, and assessment, in C. Chen (edt), American Society of Civil Engineers, p. 656-676, 1997.

Duan, J., A coupled hydrologic-geomorphic model for evaluating effects of vegetation change on watersheds, unpublished Ph.D. dissertation, Oregon State University, 133p., 1996.

Dunne, T., 1991, Stochastic aspects of the relations between climate, hydrology and landform evolution, Trans., Japanese Geomorphological Union 12, 1-24.

Ellen, S. D., S. H. Cannon, S. Reneau, Distribution of debris flows in Marin County, in, Landslides, floods, and marine effects of the storm of January 3-5, 1982, in the San Francisco Bay Region, California, Ellen, S. D. and G. F. Wieczorek, 113-131, 1988

Endo, T., and T. Tsuruta, The effects of tree roots on the shearing strength of soil, Annual Report, Forest Experiment Station, Hokkaido, 167-182, 1969.

Fannin, R. J., M. P. Wise, J. M. T. Wilkinson, B. Thomson, E. D. Hetherington, Debris flow assessment in British Columbia, in Debris flow hazard mitigation: mechanics, prediction, and assessment, in C. Chen (edt), American Society of Civil Engineers, 197-206, 1997

Gray, D. H., and W. F. Megahan, Forest vegetation removal and slope stability in the Idaho Batholith, Res. Pap. INT-271, Forest Service, U.S. Dept. of Agriculture, Ogden, UT, 23pp., 1981.

Hammond, C., D. E. Hall, S. MIller, and P. Swetik, Level I stability analysis (LISA) documentation for version 2.0, U.S. Dept. of Agric. Forest Service, Intermount. Res. Sta., Gen. Tehcn. Rep. INT-285, Ogden, UT, 190p.

Heimsath, A. M., W. E. Dietrich, K. Nishiizumi, and R. C. Finkel, The soil production function and landscape equilibrium, Nature 388, 358-361, 1997.

Heimsath, A. M., W. E. Dietrich, K Nishiizumi, R. C. Finkel, Cosmogenic nuclides, topography, and the spatial variation of soil depth, Geomorphology, 27, 151-172, 1999..

Hsu, M., A grid-based model for predicting soil-depth and shallow landslides, unpublished Ph.D. dissertation, University of California, Berkeley, 253p., 1994.

Koler, T. E., Evaluating slope stability in forest uplands with deterministic and probabilistic models, Environ. and Eng. Geoscience, IV, 185-194., 1998.

Johnson, K. A. and N. Sitar, Hydrologic conditions leading to debris flow initiation, Can. Geotechn. J., 27(6), p. 789-801, 1990.

Martin, K., Forest management on landslide prone sites: the effectiveness of headwall leave areas and evaluation of two headwall risk rating methods, M.S. thesis, Oregon State University, Corvallis, 93p., 1997.

Montgomery, D. R., W. E. Dietrich, R. Torres, S. P. Anderson and J. T. Heffner, Hydrologic response of a steep unchanneled valley to natural and applied rainfall, Water Resources Research, 33, 91-109, 1997.

Montgomery, D. R., and W. E Dietrich, .A physically-based model for the topographic control on shallow landsliding, Water Resources Research, v. 30, 1153-1171, 1994.

Montgomery, D. R., W. E. Dietrich, and K. Sullivan, The Role of

GIS in Watershed Analysis, in Landform Monitoring, Modelling and Analysis, edited by S. Lane, K. Richard, and J. Chandler, John Wiley & Sons, pp. 241-261, 1998a.

Montgomery, D. R., K. Sullivan, and H. Greenberg, Regional test of a model for shallow landsliding, Hydrological Processes special issue on GIS in Hydrology, 12, 943-955, 1998b.

Montgomery, D. R., K.M. Schmidt, H. M. Greenberg and W. E. Dietrich, Forest clearing and regional landsliding, Geology, 28, 311-314, 2000.

Moore, I.D., E. M. O'Loughlin, and G. J. Burch, A contour-based topographic model for hydrological and ecological applications, Earth Surf. Process. Landforms, 13, 305-320. 1988.

Okimura, T., and R. Ichikawa, A prediction method for surface failures by movements of infiltrated water in a surface soil layer, Natural Disaster Science, 7, 41-51, 1985.

Okimura, T. and T. Kawatani, Mapping of the potential surface-failure sites on granite mountain slopes, in International Geomorphology 1986, Part I, V. Gardiner (edt.), J. Wiley and Sons, 121-138, 1987.

Okimura, T. Prediction of slope failure using the estimated depth of the potential failure layer, Jour. Natural Disaster Sci., 11, 67-79, 1989.

Okimura, T., and M. Nakagawa, A method for predicting surface mountain slope failure with a digital landform model, Shin-Sabo, v. 41, 48-56, 1988.

Okimura, T. Prediction of the shape of a shallow failure on a mountain slope: the three-dimensional multi-planar sliding surface method, Geomorphology, 9 , 223- 233, 1994.

O'Loughlin, E. M., Prediction of surface saturation zones in natural catchments by topographic analysis: Water Resour. Res., 22, 794-804, 1986.

Olson, C.M. and B.K. Orr., Combining tree growth, fish and wildlife habitat, mass wasting, sedimentation, and hydrologic models in decision analysis and long-term forest land planning. Forest Ecology and Management, 114, 339-348, 1999.

Pack, R. T. and D. G. Tarboton, New developments in terrain stability mapping in B.C., Proc. of the 11th Vancouver Geotechnical Soc. Symp- Forestry Geotechnique and Resource Engineer., 12p., 1997.

Pack, R.T., D. G. Tarboton, and C. N. Goodwin, SINMAP: a stability index approach to terrain stability hazard mapping, available at http://www.engineering.usu.edu/cee/faculty/dtarb/sinmap.htm , 1998.

Pierson, T. C., Factors controlling debris-flow initiation on forested hillslopes in the Oregon Coast Range, unpublished Ph.D., University of Washington, 166p., 1977.

Prellwitz, R. W., A complete three-level approach for analyzing landslides on forest lands, in Proceedings of a workshop on slope stability: problems and solutions in forest management, U.S. Dept. of Agric. For. Serv. Pac. Northwest, Res. Sta., Gen. Tech. Report PNW-180, 94-98, 1985.

Quinn, P., K. Beven, P. Chevalier and O. Planchon, The prediction of hillslope flow paths for distributed hydrological modelling using digital terrain models. Hydrol. Proc., 5, 59-79, 1991.

Reeves, G.H., L.E. Benda, K.M. Burneet, P. A. Bisson,and J. R. Sedell, A disturbance-based ecosystem approach to maintaining and restoring freshwater habitats of evolutionary significant units of anadromous salmonids in the Pacific Northwest, Symp. Evolution and the Aquatic System: Defining Unique Units in Population Conservation, Am. Fish. Soc. 17, 334-349, 1995.

Reneau, S. L., W. E. Dietrich, C. J. Wilson, and J. D. Rogers, Colluvial deposits and associated landslides in the northern San Francisco Bay area, California, USA: *in* Proceedings of IVth International Symposium on Landslides, International Society for Soil Mechanics and Foundation Engineering, Toronto, Canada, p. 425-430, 1984.

Reneau, S. L., and W. E. Dietrich, Size and location of colluvial landslides in a steep forested landscape, in Erosion and Sedimentation in the Pacific Rim, edited by R. L. Beschta, T. Blinn, G. E. Grant, F. J. Swanson, and G. G. Ice, Int. Assoc. Hyd. Sci. Pub. 165, 39-49, 1987.

Reneau, S. L., and W. E. Dietrich, Depositional history of hollows on steep hillslopes, coastal Oregon and Washington, National Geogr. Res., 6, 220-230, 1990.

Roering, J J., J. W. Kirchner, and W. E. Dietrich, Evidence for non-linear, diffusive sediment transport on hillslopes and implications for landscape morphology, Water Resour Res, 35, 853-870, 1999.

Rollerson, T. P., B. Thomson, T.H. Millard, Identification of Coastal British Columbia terrain susceptible to debris flows in Debris flow hazard mitigation: mechanics, prediction, and assessment, in C. Chen (edt), American Society of Civil Engineers, 484-495, 1997

Schroeder, W. L., and J. V. Alto, Soil properties for slope stability analysis; Oregon and Washington coastal mountains, Forest Sci., 29, 823-833, 1983.

Sidle, R. C., A. J. Pearce, C. L. O'Loughlin, Hillslope stability and land use, Amer. Geophy. Union, Wat. Res. Mon. 11, 140p., 1984.

Sidle, R. C., A theoretical model of the effects of timber harvesting on slope stability, Water Resour. Res., 28, 1897-1910, 1992.

Swanson, F. J. and C. J. Roach, Administrative report, Mapleton leave area study, USDA Forest Service, Pacific Northwest Res. Sta., Corvalis, Oregon, 139p., 1987.

Tang, S. M., and D. R. Montgomery, Riparian buffers and potentially unstable ground, Environmental Management, 19, 741-749, 1995.

Tarboton, D. G., A new method for the determination of flow directions and upslope areas in grid digital elevation models, Water Resources Research, 33, 309-319, 1997.

Torres, R., W. E. Dietrich, D. R. Montgomery, K. Loague , and S. P. Anderson,, Unsaturated zone processes and the hydrologic response of a steep, unchanneled catchment, Water Resour.Res., v. 34, p.-1865-1879, 1998.

Ward, T.H., R. Li, and D. B. Simons, Mapping landslide hazards in forest watersheds, J. Geotech. Engin., Am. Soc. Civil Engin. 108, 319-324, 1982.

Wieczorek, G.F., G. Mandrone, L. Decola, The influence of hill-slope shape on debris-flow initiation, in Debris flow hazard mitigation: mechanics, prediction, and assessment, in C. Chen (ed.), American Society of Civil Engineers, 21-31, 1997.

WFPB (Washington Forest Practices Board), Board manual: Standard methodology for conducting watershed analysis under Chapter 222-22 of the Washington Administrative Code (WAC), Versioni 4.0, Washington Department of Natural Resources, 1997.

Wu, W., Distributed slope stability analysis in steep, forested basins, Ph.D. thesis, Utah State University, 134p., 1993

Wu, W., and R. C., Sidle, A distributed slope stability model for steep forested basins, Water Resour. Res., 31, 2097-2110, 1995.

Zhang, W., and D. R. Montgomery, Digital elevation model grid size, landscape representation, and hydrologic simulations, Water Resources Research, 30, 1019-1028, 1994.